Organofluorine Compounds in Biology and Medicine

Organofluorine Compounds in Biology and Medicine

V. Prakash Reddy
Department of Chemistry, 341 Schrenk Hall
Missouri University of Science and Technology, Rolla, MO 65409, USA

ELSEVIER

AMSTERDAM • BOSTON • HEIDELBERG • LONDON
NEW YORK • OXFORD • PARIS • SAN DIEGO
SAN FRANCISCO • SINGAPORE • SYDNEY • TOKYO

Elsevier
Radarweg 29, PO Box 211, 1000 AE Amsterdam, Netherlands
The Boulevard, Langford Lane, Kidlington, Oxford OX5 1GB, UK
225 Wyman Street, Waltham, MA 02451, USA

ISBN: 978-0-444-53748-5

British Library Cataloguing-in-Publication Data
A catalogue record for this book is available from the British Library

Library of Congress Cataloging-in-Publication Data
A catalog record for this book is available from the Library of Congress

For information on all Elsevier publications
visit our web site at http://store.elsevier.com/

Typeset by TNQ Books and Journals
www.tnq.co.in.

To My Parents and Teachers

Contents

5. Organofluorine Pharmaceuticals

6. Organofluorine Anesthetics

9. Organofluorine Compounds as Anticancer Agents

Short Biography

Professor Prakash Reddy is currently active in research and teaching at the Missouri University of Science and Technology. He received his Ph.D. in organic chemistry from Case Western Reserve University, Cleveland, Ohio and M.Phil. from the University of Hyderabad, India. He was appointed as adjunct and visiting professor at internationally reputed universities prior to joining Missouri University of Science and Technology in 2000. He received the Golden Key National Honor Society Award for teaching excellence at Case Western Reserve University and was a NASA Faculty Fellow at the Jet Propulsion Laboratory, California Institute of Technology, Pasadena, CA. His research interests are in synthetic, mechanistic, and biological organic chemistry, and he has significantly contributed to the area of carbocations and electrophilic reactions, superacids, and organofluorine chemistry.

Preface

Organofluorine compounds have unique physicochemical and biochemical properties, often distinct from their corresponding nonfluorinated analogues, and are the widely used components of agrochemicals, pharmaceuticals, and advanced materials. Strategic incorporation of fluorine or fluoroalkyl groups into drug candidates enhances their binding affinity at the active sites of enzymes. ^{19}F NMR spectroscopy and the X-ray structures of enzyme–substrate complexes are increasingly used to monitor the biochemical reaction mechanisms as well as to screen the fluorinated drug candidates for their improved potency, towards designing more effective therapeutics.

Ever since the first organofluorine inhalation anesthetic halothane was introduced in 1955, fluorinated inhalation anesthetics have dramatically influenced the modern clinical practice. In recent years, there have been growing applications of protein engineering in the design of peptides and proteins with enhanced thermal and proteolytic stabilities and with unique biological properties. The noncanonical peptides and proteins, such as fluorinated collagen-like peptides and fluorinated versions of the glucagon-like peptide GLP-1, have potential therapeutic applications. These potential applications are rapidly developing thanks to the advances in the solid phase peptide synthesis techniques and the recombinant DNA technology.

The blockbuster pharmaceuticals such as the cholesterol-lowering drug atorvastatin (Lipitor), the COX-2-specific nonsteroidal antiinflammatory drug celecoxib (Celebrex), the antimalarial drug ciprofloxacin (Cipro), the antidepressant drug fluoxetine (Prozac), and the antidiabetic drug sitagliptin (Januvia) are all fluorinated compounds. Interestingly, about 20% of the currently marketed drugs are organofluorine compounds, and the number of fluorinated drugs is rapidly growing each year. Incorporation of fluorine at or in the vicinity of the metabolically vulnerable sites enhances the metabolic stability of the drugs, resulting in their increased bioavailability. Although the effect of fluorine on the biochemical properties of organic compounds is not completely predictable, it is possible to design organofluorine compounds that are tailored to specific applications. For example, the antibacterial flurithromycin differs from the erythromycin by a single fluorine substitution two carbons away from the tertiary hydroxyl group. This substitution provides greater stability to the flurithromycin in the acidic medium and therefore is preferred over erythromycin for the treatment of gastritis caused by *Heliobacter pylori*. The cholesterol-lowering drug ezetimibe owes its effectiveness to the optimal metabolic stability afforded by the aryl fluorines. Strategic incorporation of aryl-fluorine also increases the drug-potency, as in the case of the FDA-approved antibacterial drug, linezolid, used to treat vancomycin-resistant bacteria. Many recently FDA-approved anti-cancer agents such as ponatinib, clofarabine, afatinib, sorafenib, regorafenib, trametinib, dabrafenib, vemurafenib, and ponatinib owe their high potency to the strategic incorporation of fluorine.

Due to the longest half-life of the ^{18}F isotope among all other positron emitting isotopes, the ^{18}F–radiolabeled pharmaceuticals or their analogues are ideally suitable for positron tomography imaging (PET imaging) for medical diagnostics. Although ^{18}F-PET tracers are lagging behind the ^{11}C-labeled PET tracers in the medical practice, the recent FDA approval of the three ^{18}F tracers for monitoring the amyloid plaques in the brains of Alzheimer's disease, Parkinson disease, and epilepsy opens the door for introduction of many other task-specific ^{18}F-PET tracers, not only for their use in personalized medicine in various pathologies, but also in the drug discovery, especially for neurodegenerative diseases, where the biochemical mechanisms of the disease-onset are vastly unknown.

This monograph book, partly based on my teaching of a graduate level course in organofluorine chemistry at the Missouri University of Science and Technology is intended to be used as a text or reference book for graduate level courses in organofluorine chemistry, synthetic organic chemistry, advanced organic chemistry, and medicinal chemistry. It is also intended as a reference book for researchers interested in the biomedical applications of organofluorine compounds.

I am grateful to the editors and editorial staff, particularly Drs. Kostas Marinakis, Anita Koch, Jessica Vaughan, and Christine McElvenny for their continued encouragement and help throughout the process of the manuscript preparation, and to Dr. M. Rajendran, Manager of the production department, for patiently reviewing the manuscript and carefully incorporating many corrections. I thank Professor G. K. Surya Prakash (University of Southern California), Professor Ekk Sinn (Western Michigan University), Professor Narayan Hosmane (Northern Illinois University), and Dr. Shobha Kantamneni (US Patent Office) for reading parts of the manuscript and for their helpful suggestions. I also thank all my colleagues in the department of chemistry at Missouri S&T, Dr. Samsher Prakash (Missouri S&T), and current and former members of my research group for their encouragement. Finally and most importantly, the main credit goes to numerous researchers that contributed to the organofluorine chemistry. Although many of them are listed in the references provided, the important contributions of many other researchers may have been inadvertently omitted, and I welcome any suggestions and corrections from the readers.

Importance of this Book

Nearly one-fifth of all the pharmaceuticals, as well as many of the blockbuster drugs and inhalation general anesthetics, consist of organofluorine compounds. Accordingly, this book highlights the importance of organofluorine compounds in biomedical research, focusing on rational drug design, synthetic methods, ^{18}F-positron emission tomography tracers for clinical diagnosis, protein engineering targeted to potential therapeutics, and anti-cancer agents, and offers mechanistic insight into the therapeutic effect of various fluorinated pharmaceuticals. This book offers a comprehensive coverage of up-to-date literature on the biologically and medicinally important organofluorine compounds. Thus, it is intended for use as a reference text not only for researchers interested in the applications of organofluorine compounds in the biomedical areas and drug discovery, but also for graduate students taking courses in synthetic organic chemistry, advanced organic chemistry, and medicinal and biochemistry.

Chapter 1

General Aspects of Organofluorine Compounds

Chapter Outline

1. INTRODUCTION

Organofluorine compounds, often due to their unique and favorable physicochemical and chemical reactivity characteristics, and due to their favorable membrane permeability and enhanced bioavailability, are widely used as components of materials, agrochemicals, and a large number of pharmaceuticals.[1–4] The carbon–fluorine (C–F) bond is relatively highly stable toward chemical or metabolic transformations due to its high bond strength and high oxidative stability, that is, the C–F bond strength is significantly greater than that of the C–H or other carbon–heteroatom (and C–X) bonds. Further, due to its highest electronegativity and thereby strong electron-withdrawing inductive effect, the oxidative stability of the neighboring C–H bonds also is dramatically increased. Many drug candidates with improved oxidative and metabolic stability are designed based on this criterion (vide infra). High stability of the C–F bond, altered lipophilicities, enhanced binding affinity to the targeted enzymes, and inertness to enzymatic defluorination are the hallmark features of the organofluorine compounds that are used for medicinal applications. Further, the pathophysiological effects of organofluorine compounds could be monitored,

Organofluorine Compounds in Biology and Medicine. http://dx.doi.org/10.1016/B978-0-444-53748-5.00001-0

in vivo, using positron emission tomography (PET) and positron-emitting ^{18}F-isotope-labeled compounds.[5,6]

Due to the high electronegativity of the fluorine, the hydrogen bonding efficiency of fluorine is significantly lower than that of the oxygen or nitrogen atoms, and in many cases the hydrophobic interactions of the C–F bonds, rather than the C–F···H hydrogen bonding interactions, play a dominant role in the stabilization of the enzyme–substrate complexes.[7–10]

The average C–F bond length of ~1.39 Å is relatively much smaller than that of the C–C bond, ~1.54 Å, and is comparable with that of the average C–O bond length of ~1.43 Å. Based on this observation, the C–F bond has been used as an isosteric and isopolar replacement for the hydroxy groups in biologically relevant compounds. Similarly, the difluoromethylene (CF_2) moiety has been used as isosteric and isoelectronic replacement for carbonyl groups. The CF_2 moiety is considered to be bioisosteric and isopolar with respect to the carbonyl, ether, and hydroxyl groups, but unlike the latter functional groups the difluoromethylenated compounds exhibit enhanced hydrophobicity and lipophilicity, and therefore these compounds have unique physicochemical and biological properties as compared to their hydrocarbon derivatives.[11]

The unique biological properties exhibited by the organofluorine compounds, such as increased hydrophobicity, lipophilicity, and thermal and metabolic stability play a key role in chemical biology and medicinal chemistry. The increased lipophilicity of amines when fluorine substituents are in proximity to the amino group is due to the lowered basicity of the amines so that the amines exist mostly in the neutral form at the biological pH.[12] The decreased hydrogen-binding ability of the fluorinated amines and aromatics, in general, results in their increased lipophilicities. However, in the case of aliphatic hydrocarbons or alcohols, substitution of the fluorine far from the functional groups results in the decreased lipophilicity, as measured by their log *P* values (partition coefficients of the compounds between *n*-octanol and water); the log *P* value for $CH_3(CH_2)_5OH$ is 2.03, while it is 1.14 for $CF_3(CH_2)_5OH$, showing that the latter compound is relatively less lipophilic.[3]

The substrate-mimetic effect of the monofluorinated compounds (usually the aryl-F), however, does not apply to compounds that have *gem*-difluoromethylene (CF_2)- or trifluoromethyl (CF_3) moieties, because these groups have significantly higher van der Waals radii as compared to the CH_2 or CH_3 groups. On the other hand, the CF_2 group is considered to be isosteric and isopolar with the carbonyl group, and the CF_3 group has similar van der Waals volume as that of the isopropyl group. The similarity of the steric crowding of the isopropyl and CF_3 groups was inferred from the similarity of the observed rotational barriers in the corresponding 1,1'-disubstituted biphenyls (21.9 and 22.2 kcal/mol, respectively for the trifluoromethyl and isopropyl derivatives).[13–15]

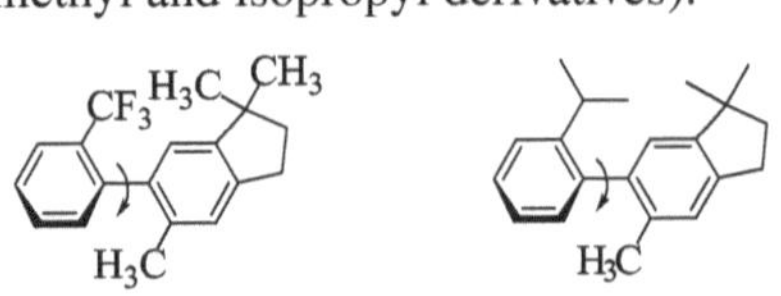

Rotational barriers: 21.9 kcal/mol 22.2 kcal/mol

The trifluoromethyl group also significantly increases the rotational barriers for the interconversion of the syn- and anti-conformers of the 9-trifluoromethylepiquinidine favoring the *syn* conformer as the major species, the isomeric ratio being dependent on the solvent used.[16] The solvent-dependent conformational interconversion could be conveniently monitored by ^{19}F nuclear magnetic resonance (NMR) spectroscopy and substantiated by the empirical Kamlet–Taft solvation model.[16]

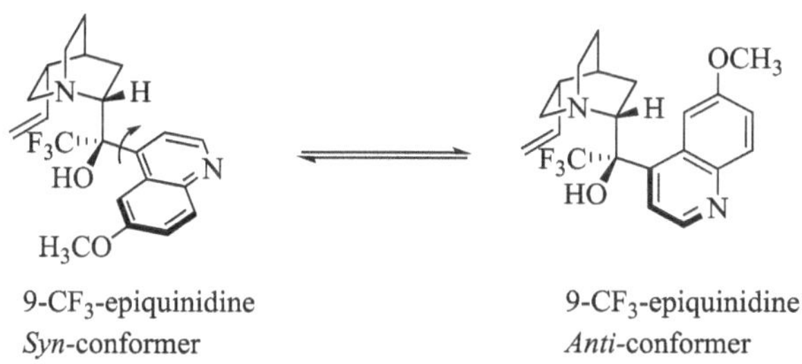

The *gem*-difluoromethylene groups also serve as isosteric and isopolar replacement for the oxygen in the C–O–P bond of the phosphate groups in nucleoside monophosphates and triphosphates. Through substitution of the P_α–O–P_β phosphate oxygen by the CF_2 group in the deoxynucleoside triphosphates (dNTPs), the dissociation of the pyrophosphate-leaving groups is prevented.[17–18] The X-ray crystal structures of these compounds showed substantially similar dNTP-like conformation (Figure 1). Further, the CF_2-triphosphate group exerts minimal steric perturbations and has an overall polarity similar to that of the nonfluorinated triphosphate moiety, and is, therefore, a useful probe to study the deoxyribonucleic acid polymerase–nucleoside interactions, as demonstrated by Olah, Prakash, and coworkers.[19]

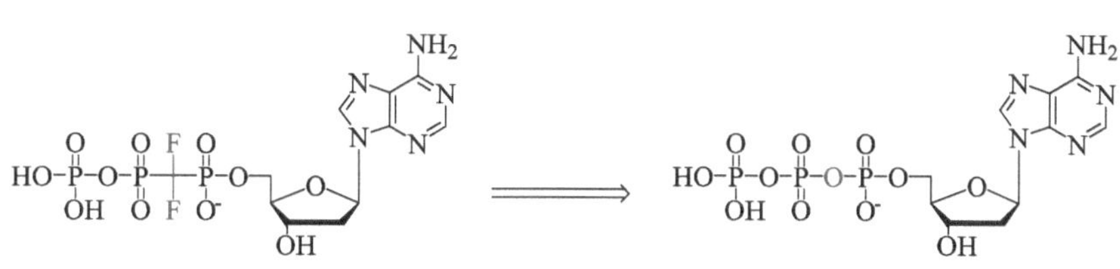

FIGURE 1 *gem*-Difluoromethylene (CF_2)-triphosphate isostere of the 2′-deoxadenosine-5′-triphosphate.

In polyfluorinated organic compounds, as in the case of the perfluorinated crown ethers, and perfluoroaromatics, polarity is reversed inside the ring, as compared to the nonfluorinated analogs. Crown ethers, for example, 18-crown-6, selectively bind cationic species, whereas the corresponding perfluorinated crown ethers strongly bind to anions such as fluoride or superoxide anions (Figure 2).[20] Perfluoro-15-crown-5, has a substantially weak binding to Na^+ and K^+ cations (complexation constants of 5.5 and 1.7 M^{-1}, respectively, in

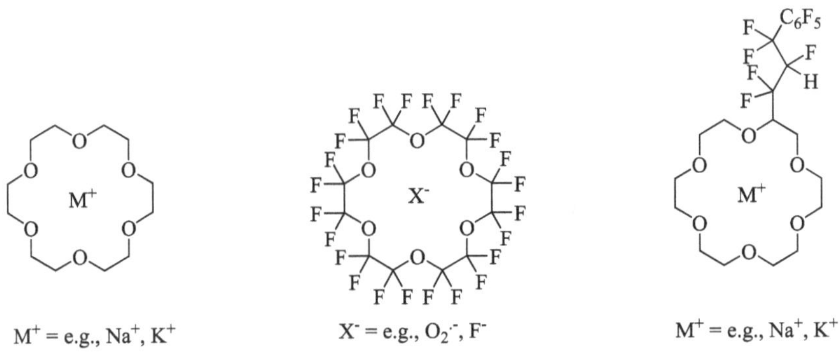

FIGURE 2 Structures of the ion complexes of 18-crown-6 and its fluorous analogs; fluorous crown ethers have reversal of polarity at the center and bind preferentially to anionic species.

contrast to the expected values of ~$10^5\,M^{-1}$ for the corresponding nonfluorinated 15-Crown-5).[21,22] Use of a pendant polyfluoroalkyl (or polyfluoroarylalkyl) side chain, also called as ponytail, would retain the cation-binding affinity of the crown ethers. These types of hydrocarbons with fluorous side chains can be used as fluorous biphasic catalysts. They are useful not only as heterogeneous catalysts in various organic functional group transformations but also for increasing the stability of proteins toward chemical denaturation or proteolytic degradation.

As evident from the above discussion, due to the high C–F bond polarity, and due to its being isosteric with C–O bonds, fluorinated organic compounds have unusual and favorable physicochemical and biochemical properties, as compared to other hydrocarbons of similar molecular mass; in general, fluorinated analogs of organic compounds have relatively higher lipophilicities, reduced surface tensions and refractive indices, higher acidities (lower basicities), and lower boiling points as compared to their hydrocarbon analogs. These unique physicochemical characteristics are utilized advantageously in the design of numerous agrochemicals, pharmaceuticals, refrigerants, and specialty chemicals on the industrial scale, and the number of organofluorine compounds continues to grow at a rapid rate each year.

Many reviews and monographs address the detailed physical properties of organofluorine compounds, and the interested reader is referred to the literature.[4,23-28] In this chapter, we will present physicochemical properties of fluorinated compounds as related to their significance in biochemical interactions, and an overview of the applications of the ^{19}F NMR in drug design and in the study of biochemical mechanisms.

2. C–F BOND STRENGTHS

Fluorine, being the most electronegative element, imparts relatively stronger bond dipole moments to the C–F bonds. Due to the strong electrostatic

TABLE 1 Bond Dissociation Energies for Selected C–X Bonds[27]

C–X Bond	Bond Dissociation Energy (kcal/mol)	C–X Bond	Bond Dissociation Energy (kcal/mol)
C–F	105	C–C	83
C–H	99	C–Cl	79
C–O	84	C–N	70

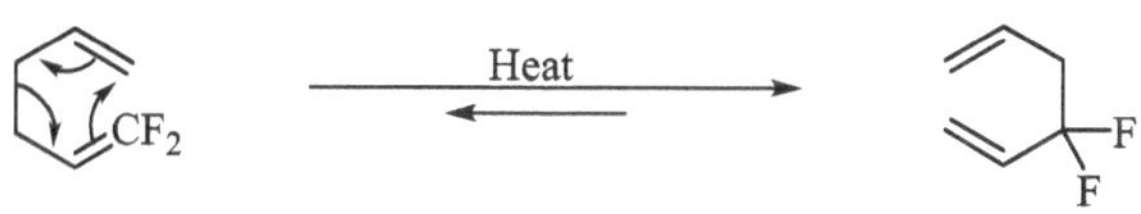

FIGURE 3 Cope rearrangement (a 3,3-sigmatropic rearrangement) of 1,1-difluoro-1,5-hexadiene; at equilibrium the sp^3 C–F bonds are favored over the sp^2 C–F bonds, which shows that the sp^3 C–F bonds are relatively stronger than the sp^2 C–F bonds.

attractions between these bond dipoles the C–F bond has the highest bond strength as compared to that of any other C–X (X = any atom including H) bond (Table 1).[27] The C–F bond strength in monofluorinated compounds is relatively greater than that of the C–H bond by 6 kcal/mol, and the C–F bond is stronger than the other carbon–halogen or carbon–nitrogen bonds by over 25 kcal/mol. It is also important to note that the C–F bonds in the *gem*-difluoro- or trifluoromethyl groups are much stronger than the isolated C–F bonds; that is, the C–F bond strength increases as the number of geminal fluorines increases. For example, the C–F bond energies for CH_3F, CH_2F_2, CHF_3, and CF_4 are 107, 110, 115, and 116 kcal/mol, respectively. The higher the *p*-character (i.e., the higher the hybridization index) of the C–F bonds, the greater is the polarization of the C–F bonds. In other words, the hybridization index of the C–H bond in CH_3F is relatively smaller than that of the C–F bond and thus the percentage of *p*-character of the C–H bonds decreases as the number of the geminal fluorines increases, while the percentage of *p*-character of the C–F bonds increases across the series, CH_3F, CH_2F_2, CHF_3 and CF_4, with the resultant gradual increase in their C–F bond strength in this order. This is also in agreement with the observed HCH bond angle of 114 Å in CH_2F_2. (that is, the C–H bond hybridization in the latter compound is close to sp^2).

The sp^3-C–F bonds are relatively stronger than the sp^2-C–F bonds, as shown by the thermodynamically favored Cope rearrangement, a 3,3-sigmatropic rearrangement, of 1,1-difluoro-1,5-hexadiene to 3,3-difluoro-1,5-hexadiene (Figure 3).[29] Diels–Alder reactions also conform to the same trend.[30] This trend is in opposite direction as that for C–H bonds where sp^2-C–H bonds are stronger than the sp^3-C–H bonds, further supporting the increased bond polarization in case of C–F bonds. The strong thermodynamic driving force for the conversion of the sp^2 CF_2 into the sp^3 CF_2 is also translated into its relatively lower enthalpy of activation of 29.3 kcal/mol, which is about 3.5 kcal/mol lower than that for the Cope rearrangement of 1,1-dideuterio-1,5-hexadiene.[29]

3. EFFECT OF THE NEIGHBORING C–F BONDS ON THE C–H BOND DISSOCIATION ENERGIES

It is instructive to compare the effect of the α- and β-fluorine atoms on the corresponding C–H bond dissociation energies; the higher the stability of the derived radicals, the lower are the C–H bond dissociation energies for the corresponding hydrocarbons, and therefore these bond dissociation energies reflect the relative stabilizing or destabilizing effects of the neighboring fluorine atoms. α-Fluorine atom generally stabilizes the corresponding free radical by electron-releasing resonance effect, while a β-fluorine destabilizes the corresponding free radical through its electron-withdrawing inductive effect. In other words, in case of the α-monofluorinated radicals the resonance stabilization of the α-fluorine overrides its inherent electron-withdrawing inductive destabilization, but there is no such resonance stabilization effect in the case of the β-fluorinated radicals and, therefore, β-fluorination has a destabilizing effect on the corresponding radicals. High-level ab initio calculations predict the C–H bond dissociation energies with a reasonable accuracy and these values are, in most cases, comparable with the experimental values (Table 2).[31]

Using the following isodesmic reaction at MP2/6-311+G(3df,2p)//MP2/6-31G(d) theory (Eqn (1)), the C–H bond dissociation energies are calculated for a series of hydrocarbons, and the selected C–H bond dissociation energies obtained through these calculations are listed in Table 2.[31]

$$\text{R-H} + \dot{C}H_3 \rightleftharpoons \dot{R} + CH_4 + \Delta E \ (298\ K) \quad (1)$$

It is evident that a single fluorine or a double fluorine substitution on the α-carbon decreases the bond dissociation energy by 3 kcal/mol, whereas the presence of three fluorine atoms on the α-carbon results in an increase of its C–H bond dissociation energy by 2 kcal/mol. In case of the CF_3 radical the electron withdrawing inductive effect (–I) of the fluorine atom slightly overwhelms its electron releasing resonance effect (+R) so that the C–H bond dissociation in CHF_3 is relatively less favorable as compared to that of CH_4, that is, the resonance stabilization of the radicals is nearly saturated with the mono-α-fluorine, and additional fluorine atoms, on the other hand, contribute to the destabilization of the radicals through their –I effects. The fluorines on the β-carbons destabilize the corresponding radicals and therefore the C–H bonds in $(CF_3)_3C$–H and $(CF_3)_2CH$–H are about 10 kcal/mol stronger than those of the corresponding nonfluorinated hydrocarbons.

Figure 4 shows the plots of the bond dissociation energies of the tertiary C–H bonds of $(CH_3)_3C$–H, $(CH_3)_2(CF_3)C$–H, $(CH_3)(CF_3)_2C$–H, and $(CF_3)_3C$–H as a

TABLE 2 Effect of Fluorine on the C–H Bond Energies (kcal/mol)

Compound	Theory[31]	Experiment[32]
CH_4	103.9[a]	104.9
CH_3F	99.9	102.0
CH_2F_2	101	102.3
CHF_3	105.8	107.1
CH_3CH_2–H	100.4	102
CH_2FCH_2–H	103.5	104
CHF_2CH_2–H	106.0	105.9
CF_3CH_2–H	106.7	107.1
CF_3CF_2–H	103.1	104.6
$(CH_3)_2CH$–H	98.6	100
$CH_3(CF_3)CH$–H	103.6	–
$(CF_3)_2CH$–H	107.7	107.6
$CH_3(CH_3)_2C$–H	98.8	96.7
$CF_3(CH_3)_2C$–H	101.4	–
$(CF_3)_2CH_3C$–H	105	–
$(CF_3)_3C$–H	107.6	107.6

[a]*At CBS/QB3 level (coupled cluster calculations with triple excitations for the calculations at the highest level).*[33]

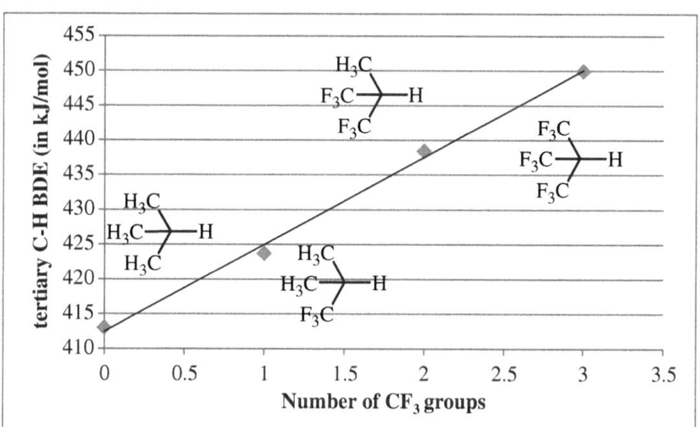

FIGURE 4 Effect of the substitution of α-CH_3 by α-CF_3 groups on the *tert*-C–H bond dissociation energies. *Based on the values in Table 2.*[31]

function of the number of the CF_3 groups (based on theory values in Table 1).[31] It is evident from Figure 4 that there is a linear increase in the tertiary C–H bond dissociation energies with progressive substitution of the CH_3 groups in 2-methylpropane by the CF_3 moieties, in accord with expectation based on the strong electron-withdrawing inductive effect of the CF_3 groups. A similar trend is also observed in case of the primary C–H bonds, that is, the α-C–H bond strengths increase with successive substitution of the β-hydrogens by fluorine atoms, although the effect is not as linear as it is for the trifluoromethyl series, which is probably due to the competing β-C–H hyperconjugation effects for the mono- and difluoro-analogs.

4. EFFECT OF FLUORINE ON THE ACIDITY AND BASICITY

The successive bond polarizations due to the electron-withdrawing inductive effect (-I) of fluorine results in higher acidities of the neighboring acidic groups, and lower basicities of the neighboring amino groups. This inductive effect rapidly diminishes with increasing distance of the fluorine from the functional groups, and significant inductive effect is exerted only up to a distance of three to four carbons. This inductive bond polarization of the C–F bonds has a direct effect on the acidities of the hydrocarbons, alcohols, carboxylic acids, and phosphonic acids.

Due to the high electronegativity of fluorine, acidity of the organofluorine compounds is increased when fluorine is in close proximity to the acidic site; the closer the fluorine is to the acidic site, the higher is the acidity. Controlled modulation of the acidity of organic compounds by fluorine substitution is one of the key factors in the design of pharmaceutical candidates. For example, trifluoromethanesulfonamide ($CF_3SO_2NH_2$; $pK_a=5.8$) has remarkably higher binding affinity to carbonic anhyhdrase II as compared to methanesulfonamide ($CH_3SO_2NH_2$, $pK_a=10.5$), due to its fivefold increase in acidity. The increased acidity results in facile deprotonation of the sulfonamide moiety and thereby the stronger binding of the derived conjugate base with Zn(II) at the enzyme active site. The carbonic anhydrase inhibitory constants (K_i) for methane sulfonamide and trifluoromethanesulfonamide are on the order of 10^{-4} M and 10^{-9} M, respectively.[34] For a series of sulfonamides of varying acidity, it was found that there is a linear correlation between their pK_a and the corresponding enzyme inhibitory constant (pK_i) values; the lower the pK_a, the lower is the pK_i, that is, increased acidity results in higher binding constant to the carbonic anhydrase. These water-soluble carbonic anhydrase inhibitors are used for topical treatment of glaucoma.

Acetohydroxamic acid and trifluoroacetohydroxamic acid bind to the Zn^{2+} in human carbonic anhydrase II through their negatively charged nitrogen atoms (Figure 5). In case of the trifluoroacetohydroxamic acid, the CF_3 fluorine interacts with Zn^{2+}, forming a 5-membered chelate interaction. The fluorinated version of the hydroxamic acid has 10-fold enhanced affinity as

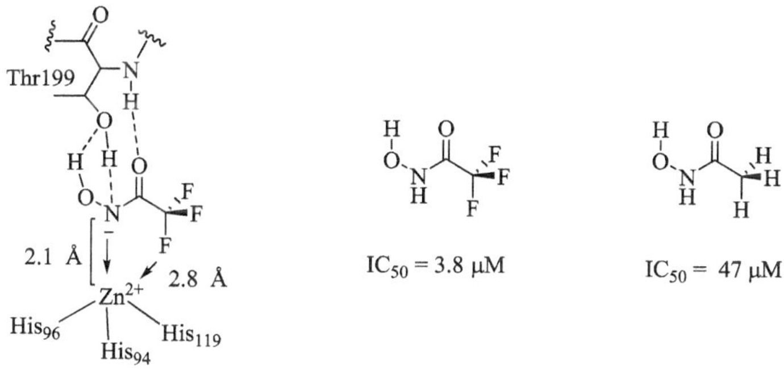

FIGURE 5 Schematic depiction of the binding of trifluoroacetohydroxamic acid to human carbonic anhyhdrase II, and structures of acetohydroxamic acid and trifluoroacetohydroxamic acid along with their enzyme inhibition constants.

compared to its nonfluorinated version, reflecting the relatively more fully ionized nitrogen, and therefore stronger zinc–nitrogen bond, in case of the fluorinated hydroxamic acid.[35] Thus, acidity (or basicity) of the key functional groups can be fine-tuned by the fluorines in order to modulate the enzyme–substrate binding affinities.

5. EFFECT OF FLUORINE ON THE METABOLIC STABILITY

Fluorine has the highest electronegativity of all the elements in the periodic table (Pauling electronegativity of F=4.0). However, its van der Waals radius (Bondi radii (Å): F=1.47, O=1.52, H=1.20; Table 3) is comparable to that of oxygen or hydrogen.

The C–F bond is broadly regarded as a bioisostere of either C–O or C–H bond, because most enzymes do not discriminate the C–F bonds from the corresponding C–H or C–O bonds; thus, monofluoro aromatic compounds are substrate mimetics of the corresponding nonfluorinated compounds in enzymatic reactions. For example, the cholesterol-lowering drug candidate SCH 48461 is metabolically relatively unstable, as the aryl rings are susceptible to enzymatic hydroxylation and other oxidative degradation reactions (Figure 6). The analogous compounds such as the US Food and Drug Administration (FDA)-approved ezetimibe, with fluorine substituents on aryl rings to prevent metabolic oxidation, have relatively improved pharmacokinetics and are more effective in lowering cholesterol levels than is the lead compound SCH 48461. Importantly, the target receptor Niemann-Pick C1-like transmembrane protein does not distinguish the aryl C–H from aryl C–F bonds in its binding to these compounds.[37]

Many drug candidates with improved metabolic stability are designed based on the strategic incorporation of fluorine at the oxidation sites. For example, BMS-422,461, a camptothecin analog, with a *gem*-difluoromethylenedioxy moiety, has a comparable antitumor activity to irinotecan (an anticancer

TABLE 3 Bondi's van der Waals radii (in Å)[36] and (Pauling Electronegativities in Parenthesis) of Fluorine and Other Biologically Abundant Elements

F	1.47 (4.0)	P	1.8 (2.1)
H	1.2 (2.1)	S	1.8 (2.5)
C	1.70 (2.5)	Cl	1.74 (3.0)
N	1.55 (3.0)	Br	1.85 (2.96)
O	1.52 (3.5)	I	1.98 (2.66)

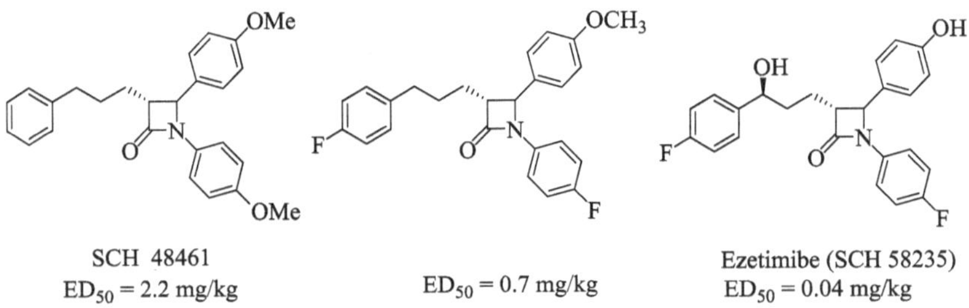

FIGURE 6 Structures of hypercholesterolemia drug ezetimibe and its precursor clinical candidates.

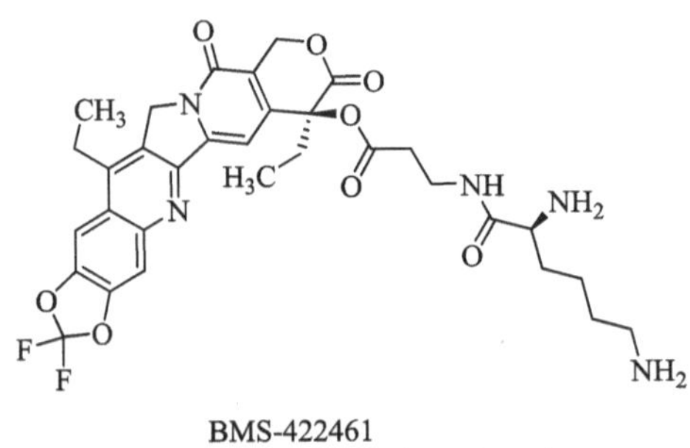

FIGURE 7 Structure of *gem*-difluoromethylenated camptothecin analog, BMS-422461.

camptothecin analog), with improved pharmacokinetics and reduced gastrointestinal toxicity (Figure 7).[38] Similar substitution of the methylenedioxy moiety by the *gem*-difluoromethylenedioxy moiety in benzo[1,3]dioxole derivatives, in general, attenuates the cytochrome-P450-enzymes-induced metabolic oxidation.[11]

The widely used antibacterial ciprofloxacin, a bacterial topoisomerase IV inhibitor (leading to the inhibition of the bacterial cell division), has two- to sevenfold increased potency, 70-fold increased cell permeability, and

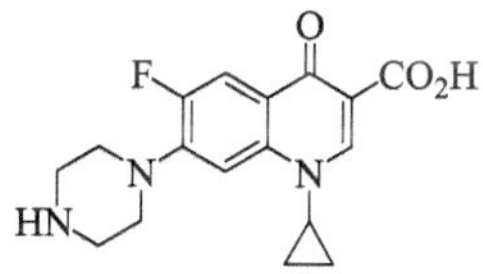

Ciprofloxacin

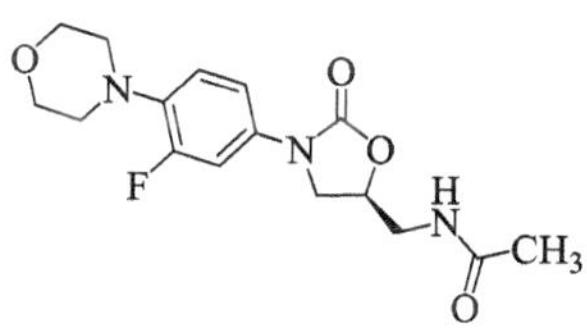

Linezolid

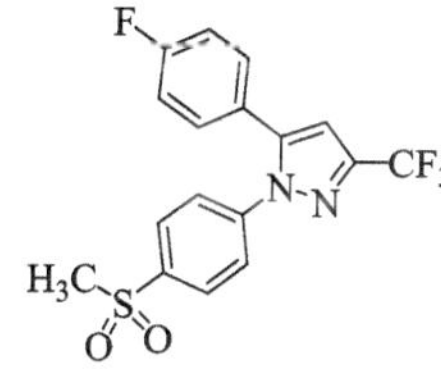

SC-58125
(plasma half life = 221 h)

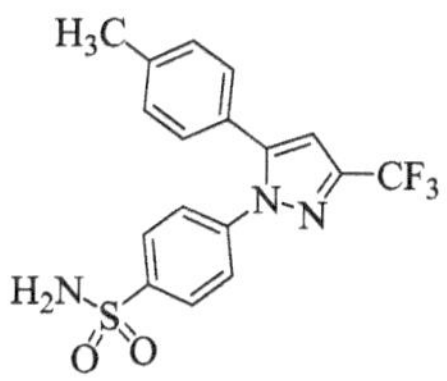

Celecoxib
(plasma half life = 3.5 h)

FIGURE 8 Structures of ciprofloxacin, linezolid, SC-57666, and celecoxib.

improved pharmacokinetic profile, as compared to its nonfluorinated analog (Figure 8).[39,40]

Similarly, the strategic incorporation of fluorine into the aromatic ring results in increased potency and efficacy in vivo for linezolid, which is an FDA-approved antibacterial drug to treat infections caused by vancomycin-resistant bacteria (Figure 8).[40] On the other hand, sometimes fluorine exerts undesirably high metabolic stability in drug candidates. For example, the plasma half-life of a drug candidate with a fluorophenyl moiety, SC-57666, an analog of celecoxib (a COX-2 inhibitor used for treatment of rheumatoid arthritis and osteoarthritis), is 221 h in rats.[41] Substitution of the aryl fluorine by a methyl group helped to decrease the metabolic stability to an acceptable level of 3.5 h for celecoxib. It is interesting to note that, in general, as shown by the above examples, a strategic substitution of aryl hydrogen by fluorine in the lead compounds can lead to superior drug candidates with improved bioavailability and potency.

The relatively greater C–F bond strength imparts greater metabolic stability to the organofluorine pharmaceuticals when the relatively labile C–H bond is replaced by the C–F bond. Further, due to the large electron-withdrawing inductive effect, the neighboring C–H bonds also are relatively polar and their oxidative stability is enhanced. The 1,25-dihydroxyvitamin D_3 is prone to hydroxylation at the C_{24} carbon by cytochrome P24 (24-hydroxylase) family of enzymes, which leads to its rapid oxidative degradation (Figure 9).[42] However, strategic replacement of the C_{24}-hydrogens by two fluorine atoms results in greater metabolic stability.[43] The nonsteroidal analogs of vitamin D_3—WY1113 and CD578—with two CF_3 groups on the side chain not only are metabolically stable to oxidative degradation, but also are superior agonists for the vitamin D receptor, as compared to their

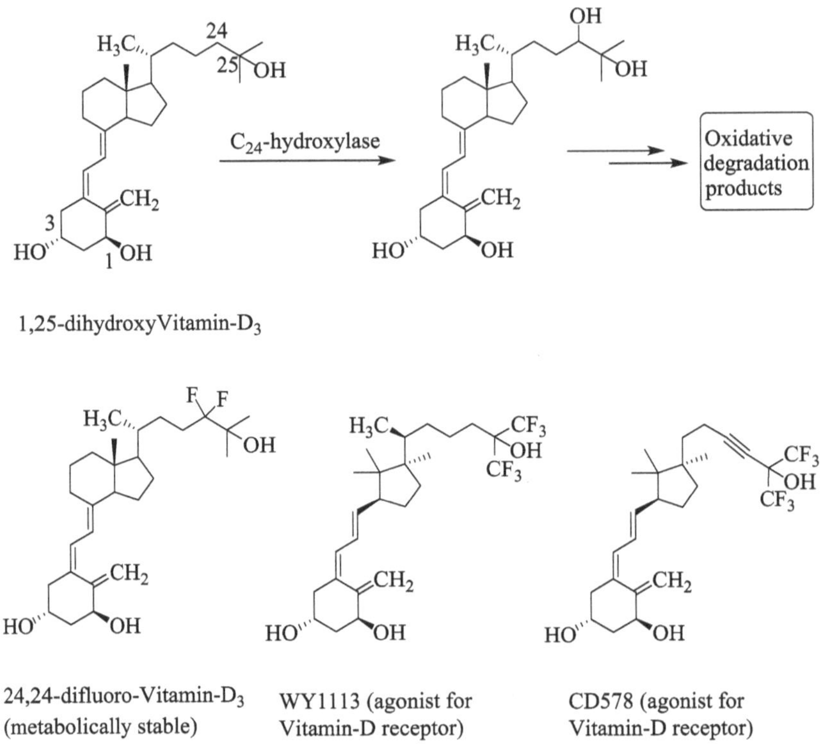

FIGURE 9 Structures of 1,25-dihydroxyvitamin D_3 and its metabolically stable analogs.

nonfluorinated analogs and 1,25-dihydroxyvitamin D_3.[44] In the case of the WY1113, the two CF_3 moieties on the C_{25} are able to inhibit the hydroxylation at C_{24}. Similarly, in case of CD578 the alkyne functionality is relatively resistant to metabolic oxidation due to the presence of the adjacent trifluoromethyl groups (Figure 9).

6. EFFECT OF FLUORINE ON THE BIOAVAILABILITY

Fluorine substitution in drug candidates, by virtue of its exerting increased lipophilicity to the molecules, aids in the tissue absorption and distribution. Further, due to the increased metabolic stability of the fluorinated drugs, their bioavailability is increased. The more lipophilic the drug candidate, the more is the effective concentration of the drug at the enzyme/receptor active sites. This is due to the fact that the active site regions of enzymes have relatively diminished polarity as compared to the cellular environment, and therefore the relatively lipophilic molecules have higher binding efficiency. Due to this high binding affinity of certain fluorinated drugs at the active sites, relatively lower drug doses may be required for their optimal efficacy. The lipophilicity of the drugs is normally measured as the logarithmic partition coefficient, log P, between

TABLE 4 Effect of Fluorine on the Acidity of Carboxylic Acids and Alcohols, and on the Basicity of Amines[11,45]

Amine	pK_a	Acid	pK_a	Alcohol	pK_a
$CH_3CH_2NH_3^+$	10.7	CH_3CO_2H	4.76	CH_3CH_2OH	15.9
$CH_2FCH_2NH_3^+$	9.0	CH_2FCO_2H	2.66	CF_3CH_2OH	12.4
$CHF_2CH_2NH_3^+$	7.3	CHF_2CO_2H	1.24	$(CF_3)_2CHOH$	9.3
$CF_3CH_2NH_3^+$	5.7	CF_3CO_2H	0.23	$(CF_3)_3COH$	5.4

n-octanol and water; the higher the log P, the higher is the lipophilicity of the compound. When fluorine is introduced closer to the amino groups, the amine basicity is decreased due to the strong electron-withdrawing inductive effect of the fluorine, so that the amines are less protonated and exist mostly in the neutral form and thus their log P values and their lipophilicities are relatively much higher than the corresponding nonfluorinated analogs. The effect of fluorine on the basicity of amines and acidity of carboxylic acids and alcohols is illustrated in Table 4. As can be seen from Table 4, trifluoroacetic acid is $10^{4.6}$ times as strong acid as acetic acid, 2,2,2-trifluoroethylammonium salts are 10^5 times less basic than ethylammonium salts, and perfluoro-*tert*-butyl alcohol is nearly as acidic as acetic acid.

The protein–ligand binding strengths can be fine-tuned by incorporation of fluorine atoms or fluoroalkyl groups on the ligands. Fluorine, due to its highest electronegativity of all the elements in the periodic table, is a weak H-bond acceptor. Nevertheless, protein–ligand X-ray structures reveal many instances in which fluorine engages in multipolar, H-bond type of interactions; for example, C–F··H–N, C–F··C=O, C–F··H–Cα interactions, involving protein backbones or side chains, are observed in many reported X-ray structures of the enzyme-substrate complexes. The hydrogen bonding type of interactions with fluorine in these cases is indicated by the short F–C(carbonyl) and F–C(α) carbon bond distances, close to a value of about 3.3 Å, which corresponds to the sum of the van der Waals distances of the interacting atoms.[46] Protein backbone amide carbonyl groups, positively charged side chains of arginine, glutamine, and guanidinium, serve as fluorophilic environments for the ligand at the enzyme active sites. Thus, drug efficacies can be fine-tuned by strategic incorporation of the fluorines into pharmaceutical candidates. However, the fluorine effect on the drug efficacy is not completely predictable, and in some instances fluorinated compounds may lead to reduced effectiveness.

There are numerous drug candidates with one or more fluorine atoms in the molecule, and nearly one-third of the currently marketed best-selling drugs are fluorinated molecules. Incorporation of fluorine into drug candidates generally

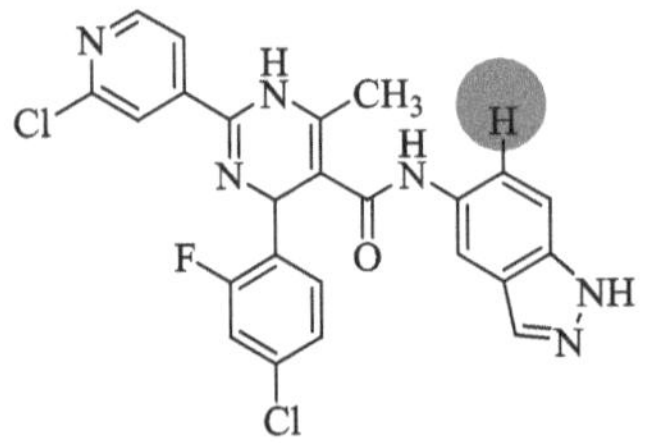

Indazole based ROCK1 Kinase inhibitor
$IC_{50} = 4$ nM
oral bioavalability = 7%

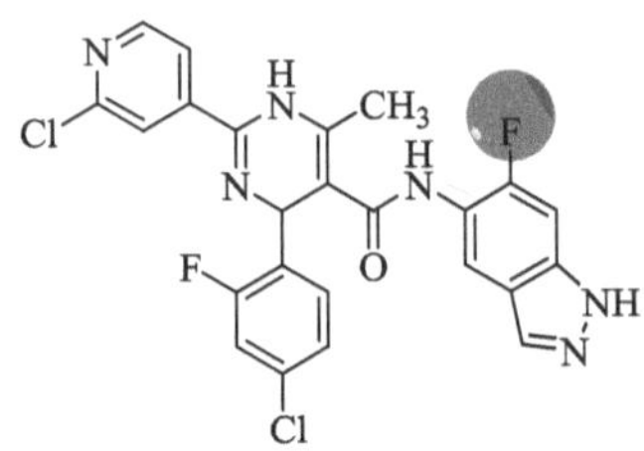

6-FluoroIndazole based ROCK1 kinase inhibitor
$IC_{50} = 7$ nM
oral bioavalability = 49%

FIGURE 10 Structure of a ROCK1 kinase inhibitor and its C6-fluorinated analog.

improves their potency, bioavailability, and metabolic stability. Moreover, lightly fluorinated molecules (e.g., CF, CF_2, and CF_3 groups) are, in general, nontoxic and mimic the corresponding nonfluorinated analogs in their stereoelectronic factors so that the drug-transporting receptors or enzymes have similar or, in some cases, improved affinities. Due to these inherent advantages, organofluorine compounds are significant components of many top-selling antibacterials, antivirals, antimalarial drugs, antihypercholesterolemia drugs, antidepressant drugs, and anticancer drugs. Detailed mechanistic understanding of the drug action and the effect of fluorine on the drug efficacy is still unclear in many instances, although much progress is rapidly being achieved through X-ray crystallographic and computational characterization of the protein–ligand complexes.

Strategic incorporation of fluorine into pharmaceutical candidates can substantially improve the drug potency. For example, in the series of quinolone antibiotics (ciprofloxacin-related compounds), a fluorine substitution at C_6 results in two-to sevenfold increase in potency, a 70-fold increase in cell permeability, improved pharmacokinetics, and attenuated plasma protein binding.[11] Further, incorporation of fluorine proximal to an amide moiety in several kinase inhibitors results in their improved oral bioavailability: Rho kinase (ROCK1) inhibitors are targeted toward cardiovascular diseases. The dihydropyrimidine-based ROCK1 inhibitors have poor bioavailability, whereas a fluorine substitution at C_6 retained the high ROCK1 inhibitory effect (IC_{50}=7 nM) of the nonfluorinated analog, and also improved bioavailability to 49% (Figure 10).[47]

Cathepsin K is a papain-like cysteine protease, predominantly expressed in osteoclasts. It is involved in the degradation of collagen, resulting in osteoporosis. The inhibitors of cathepsin K help in the suppression of collagenase activity of the cathepsin K, thereby enhancing bone formation.[48] Trifluoroethylamine is a hydrolytically stable isostere of peptide amide bond, with the C–CF_3 moiety mimicking the carbonyl group electronically, due to the high electronegativity of the CF_3 group. By replacing the hydrolytically unstable P2–P3 amide moiety of the dipeptide cathepsin K inhibitors by trifluorethylamine peptide

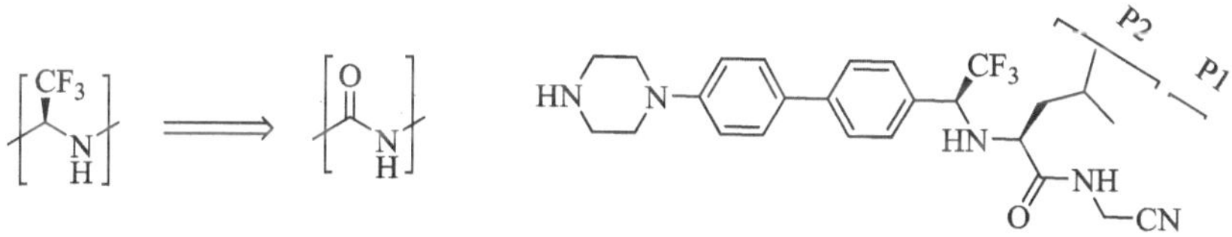

trifluoroethylamine
peptide isostere

Cathepsin K inhibitor
$IC_{50} = <0.005$ nM

FIGURE 11 Trifluoroethylamine-based peptide mimetic of cathepsin K dipeptide inhibitors.

isostere, 10- to 20-fold more potent inhibitors were developed.[49] The trifluoroethylamine residue resembles the tetrahedral transition state for the peptide hydrolysis, with enhanced binding affinity at the active centers of the enzymes.[50] The CF_3 group, by reducing the basicity of the amino group, allows the important N–H···O=C hydrogen bond to the Gly66 of cathepsin K. The strong electron-withdrawing effect of the trifluoromethyl group also suppresses the metabolic hydrolysis of the P1–P2 amide. This peptide mimetic is a ~5-pM inhibitor of cathepsin K, and is about 10,000-fold selective over other cathepsins.[49] Many of the cathepsin K inhibitors, including the trifluoroethylamine peptide isosteres, are currently in various stages of clinical development (Figure 11).[48]

7. POSITRON EMISSION TOMOGRAPHY TRACERS

The half-life of the ^{18}F isotope is the highest among all the positron emitters and thus ^{18}F-labeled compounds are ideally suited as PET tracers for disease diagnosis as well as for monitoring the disease progression. Despite the existence of numerous fluorinated pharmaceuticals, currently only four ^{18}F-radiolabeled PET tracers are approved by the FDA. [^{18}F]-2-Fluoro-2-deoxyglucose is used for monitoring the glucose metabolism in various diseased states, including Alzheimer disease and certain cancers. Three other ^{18}F-labeled compounds, florbetapir-^{18}F (^{18}F-AV-45), florbetaben-^{18}F (^{18}F-AV-1), and flutemetamol-^{18}F (^{18}F-GE067), have been approved by the FDA for monitoring the levels of amyloid beta peptide aggregation in neurodegenerative diseases (Figure 12). Many other ^{18}F-labeled PET tracers are in clinical trials, and it is expected that, with rapid advancement in the development of synthetic methods, almost all the fluorinated drugs would be modified to include the ^{18}F isotope for PET imaging. A separate chapter in this book (Chapter 7) is devoted to the synthesis and biomedical applications of ^{18}F-PET tracers.

8. ^{19}F NMR SPECTROSCOPIC TECHNIQUES FOR PROBING BIOCHEMICAL MECHANISMS

^{19}F NMR spectroscopy is uniquely suited for the characterization of the organofluorine compounds and their interactions with proteins and enzymes. ^{19}F has a

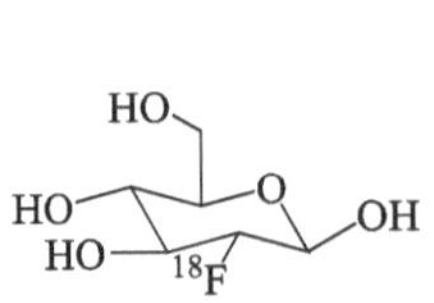

[^{18}F]-FDG

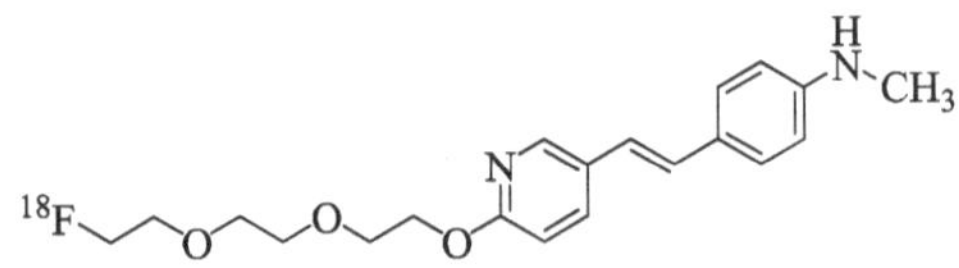

Florbetapir-^{18}F (^{18}F-AV45)

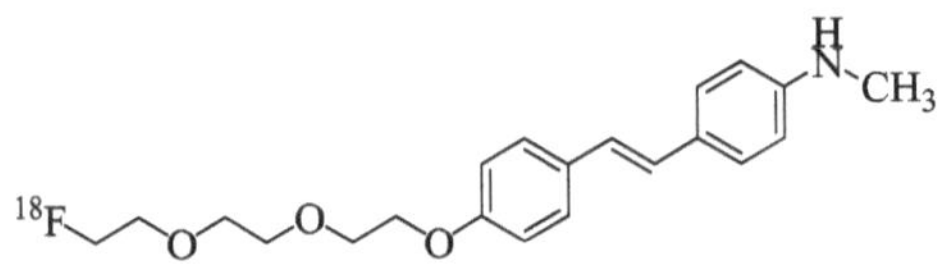

Florbetaben-^{18}F (^{18}FAV-1)

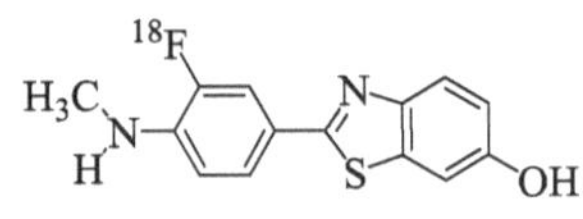

Flutemetamol-^{18}F (GE067)

FIGURE 12 FDA-approved ^{18}F-labeled PET tracers.

natural abundance of 100% and overall sensitivity of 83% relative to that of ^{1}H. The absence of the background ^{19}F NMR signals from the aqueous medium, combined with wide ^{19}F chemical shift range (typically over 400 ppm for organic compounds of biochemical interest), and high sensitivity of the δ^{19}F to its neighboring chemical environment are some of the principal reasons behind the extensive use of ^{19}F NMR spectroscopy in the study of biomacromolecules and their interactions with small molecule ligands. Thus, the ^{19}F NMR is now widely used for studying biochemical reaction mechanisms in chemical biology and in the area of drug design.

The surge in the applications of the fluorinated amino acids and the corresponding peptides in biology and medicine is spurred by advances in NMR techniques for their characterization and quantification in the protein environments; for example, the protein–ligand (drug) interactions and protein folding could be conveniently monitored by the ^{19}F NMR spectroscopy.[51] Using high-field NMR spectrometers, it is now possible to detect ^{19}F signals of the fluorinated amino acids or peptides at the micromolar concentration. Further, as mentioned above, whereas the ^{1}H NMR spectra of the probing molecules in biological media are complicated by the occurrence of spurious signals from other organic compounds that are ubiquitously present in the medium, the ^{19}F NMR spectra are free of such interfering signals as there are no other endogenous fluorinated molecules in the biological media, except for the presence of the products, if any, of the probing molecules. The fluorinated amino acids, or fluorinated peptide mimetic molecules, show distinct ^{19}F NMR signals, well resolved from the signals from the other fluorinated molecules, due to the relatively large chemical shift dispersion (>400-ppm spectral window for ^{19}F NMR as compared to about 10 ppm for ^{1}H NMR). The ^{19}F NMR studies of the site-specific fluorinated amino acids into peptides or proteins gives information on structural and conformational dynamics based on the δ^{19}F and line widths.[52]

TABLE 5 $\delta^{19}F$ for Selected Organic Compounds ($\delta^{19}F(CFCl_3) = 0$)

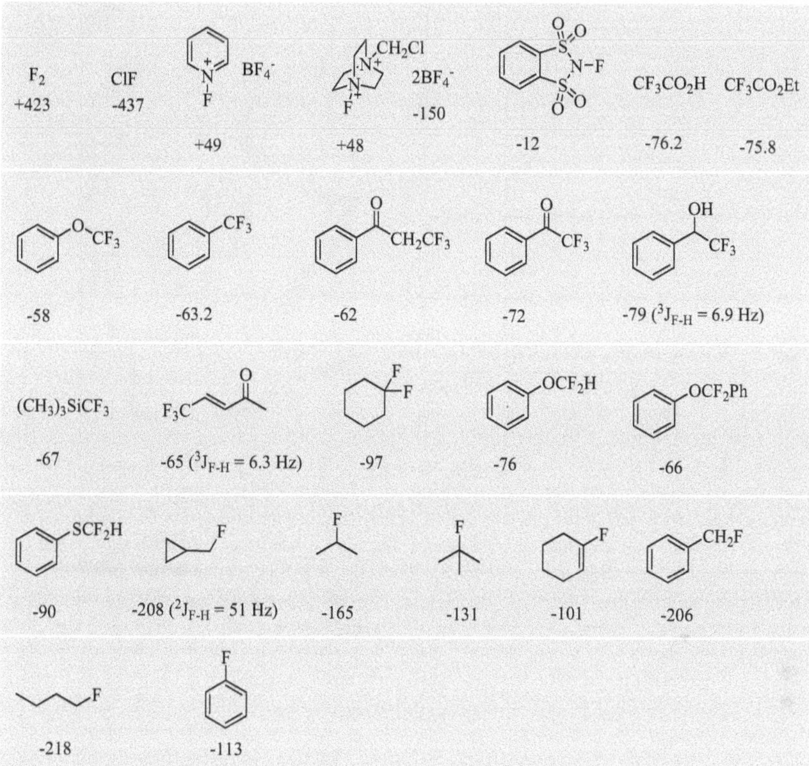

In general, the $\delta^{19}F$ of fluorines in CF_3 groups are relatively more downfield (deshielded) than those for the CF_2, and the $\delta^{19}F$ for the monofluorinated compounds is relatively more shielded as compared to that of the *gem*-difluoro compounds. The $\delta^{19}F$ for the CF_3 group ranges from −52 to −87, that for the CF_2 ranges from −80 to −180, and that for the monofluorinated compounds range from −70 to −238 ($\delta^{19}F$: aryl-F ~ vinyl-F >> sp^3 C–F). Table 5 summarizes the $\delta^{19}F$ values for some of these compounds (taken from the Dolbier's monograph book on ^{19}F NMR).[45] The local environment (neighboring groups as well as the surrounding medium) around the fluorine affects its $\delta^{19}F$. Through the analysis of the X-ray structures of the protein-fluorinated ligand complexes from the Protein Data Bank (PDB) and Cambridge Structural Database, Vulpetti and coworkers have hypothesized that local environment of the fluorines, as determined by ^{19}F NMR and obtainable by computations, also is relevant for the protein–substrate recognition mechanism.[53,54] Their observations suggest that the relatively shielded fluorines (i.e., those with relatively

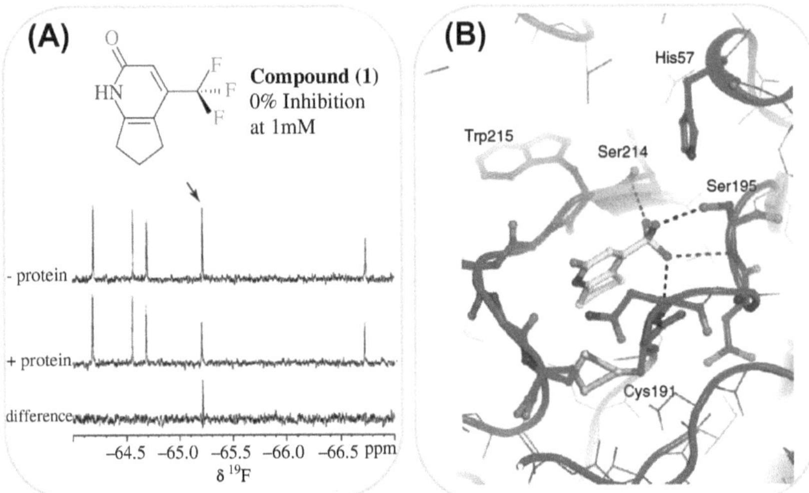

FIGURE 13 (A): ^{19}F NMR-based binding assay for a local environment of fluorine library of fluorine compounds, each of which is associated with a single ^{19}F signal, against bovine trypsin. The signal intensity of one of the peaks is diminished in the presence of the protein, which is clearly evident in the difference spectrum (at the bottom). The diminished intensity of this signal indicates that the trifluoromethyl-substituted compound is relatively strongly bound at the enzyme active site. (B): High-resolution X-ray structure (PDB: 3NK8) of bovine trypsin complexed to the above ligand; the hydrogen bonding interactions of the fluorine (green) to the Ser195, Ser214, and Cys191 residues are shown as dashed lines. (For interpretation of the references to color in this figure legend, the reader is referred to the online version of this book.) *Adapted from Ref. 53. Copyright 2012, Elsevier.*

increased electron density) tend to have hydrogen bonding interactions (F···H–O or F···H–N) with the surrounding amino acid residues in proteins, whereas the relatively deshielded fluorines are in close contact with hydrophobic side chains (i.e., hydrophobic interactions), and orthogonal multipolar interactions with the backbone carbonyl groups.

8.1 ^{19}F NMR in Fragment-Based Drug Screening

Fragment-based drug screening using ^{19}F NMR has gained popularity in pharmaceutical companies, including Novartis, Abbott, and AstraZeneca.[53] Fluorinated compounds that bind to a protein can be identified by a reduction in the intensity (or sometimes complete disappearance of the signal) of their ^{19}F absorptions, when exposed to the proteins (or enzymes) of interest. Libraries of many compounds can be screened in a single run to identify the lead compounds that have potential therapeutic applications. This ^{19}F NMR-based ligand screening method is able to detect even relatively weak-affinity ligands, as in the case of a trifluoromethyl-substituted pyridine-2-one derivative (Figure 13), which is inactive in an enzymatic assay at 1 mM concentration.[53] The binding mode of this ligand to the active site of trypsin could be inferred through its X-ray structure.

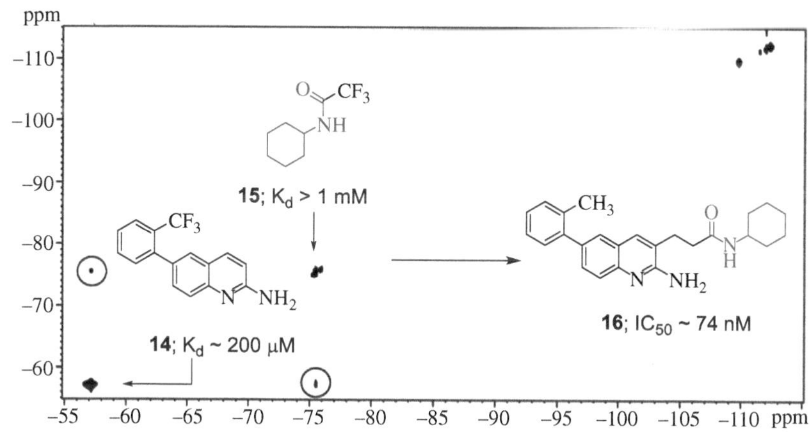

FIGURE 14 Schematic illustration of ^{19}F–^{19}F (^{1}H broadband decoupled) NOE spectrum for a mixture of six CF_3-containing ligands; the cross-peaks at about 76 ppm (circled) indicate the spatial proximity of two of these ligands (compounds **14** and **15**). Conjugating these two ligands gives the lead compound **16** with an improved IC_{50} value of 74 nM (cf. 200 μM and 1 mM for the ligands **14** and **15**, respectively). *Adapted with permission from Ref. 55. Copyright 2012, American Chemical Society.*

The trifluoromethyl group has hydrogen bonding interactions with the neighboring amino acids, Ser214, His57, and Cys191, which accounts for its diminished ^{19}F signal intensity. Thus screened and validated compounds through ^{19}F NMR method can be used as "spy molecules" for performing competitive binding assays for other potential ligands, and for measuring their binding constants.

The spatial proximity of two fluorinated compounds (from a library of several compounds) at the enzyme active site can be inferred through a ^{19}F–^{19}F NOE spectrum (Figure 14).[55] For example, in a β-secretase (BACE-1) assay using a library of trifluoromethylated compounds, two such compounds were identified that interact with the active site and that are spatially close to each other at the active site. By conjugating these two fragments (and with the exclusion of their fluorine substituents), a potent BACE-1 inhibitor was identified, which has a superior binding affinity (IC_{50} ~74 nM) to the enzyme as compared to either of the two precursor ligands (Figure 14).[55]

8.2 ^{19}F NMR in the Study of Protein Dynamics

The ^{19}F NMR technique can be used to monitor the protein dynamics in the in-cell environments; that is the NMR can be recorded in the cell suspension prior to lysis.[56] The ^{19}F NMR for the in-cells show relatively broadened peaks as compared to the purified protein in solution (for the lysates), which show relatively sharp peaks. This broadening for the ^{19}F signals obtained before lysis provides local structural and conformational dynamics. Site-specific incorporation of *p*-trifluoromethylphenylalanine (tfmF) in green fluorescent protein (GFP) at positions 39 (Y39tfmF) and 221 (F221tfmF) allows the

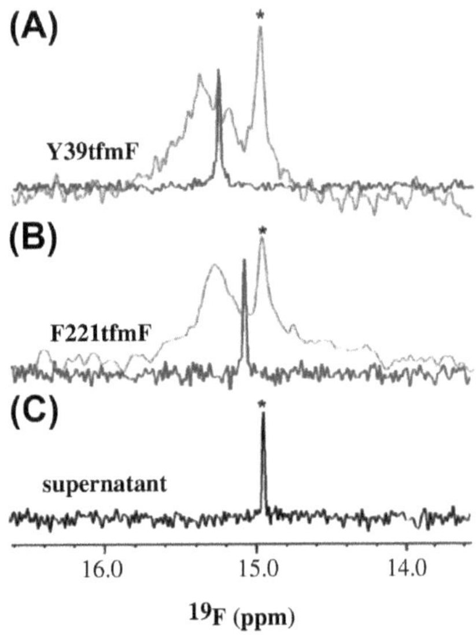

FIGURE 15 ^{19}F NMR spectra of 4-tfmF-labeled GFP cells (green) and the purified protein in solution (blue) (A and B), and the supernatants collected after the in-cell experiments (C); the asterisk indicates the residual tfmF that is not incorporated into the protein. The relatively more broadened signal for Y39tfmF as compared to that of F221tmF indicates that the side chain at 39 is relatively less flexible than the side chain at position 221. (For interpretation of the references to color in this figure legend, the reader is referred to the online version of this book.) *Adapted with permission from Ref. 56. Copyright 2010, American Chemical Society.*

study of the conformational dynamics in cells. Both these lines are relatively broad with line widths at half height ($\Delta\nu_{1/2}$) of about 0.4 ppm, whereas the corresponding absorptions for the purified protein in solution have narrower line widths of about 0.1 ppm (Figure 15). The relatively broader line width for the Y39tfM GFP in cells as compared to that of the F221tfmF version shows that the side chain at position 221 is relatively more mobile than the side chain at 39.

Because of the rapid advances in solid-state NMR techniques, including cross polarization magic angle spinning and Carr-Purcell-Meiboom-Gill-based pulse sequences, it is now possible to study the structural details of the membrane-associated polypeptides.[57–60] Due to the strong dipolar coupling of ^{19}F, ^{19}F–^{19}F distances of up to 15–20 Å can be measured using solid-state NMR.[61] In situ solid-state ^{19}F NMR can be used to investigate protein conformational changes, binding of the substrate at the active site, and protein–ligand interactions. It does not require any additional workup, nor does it involve any loss of compounds. Using this in situ ^{19}F NMR experiment, protein–ligand interactions can be conveniently probed to characterize protein conformational changes and to investigate the enzyme-catalyzed reactions. For example, this method has been used to monitor the auto-phosphorylation of the 3,5-difluorotyrosine residue, which was genetically incorporated into the activation loop of the bacterial protein tyrosine kinase Etk.[62] The ^{19}F signal for the phosphorylated tyrosine residue is shifted downfield by 12 ppm with respect to that of the 3,5-difluorotyrosine (F2Y) residue, and is well resolved from that of the unreacted substrate. The newly arising signal at $\delta^{19}F$ −122.30 disappears after dephosphorylation using tyrosine phosphatase enzyme, showing that it is indeed due to the

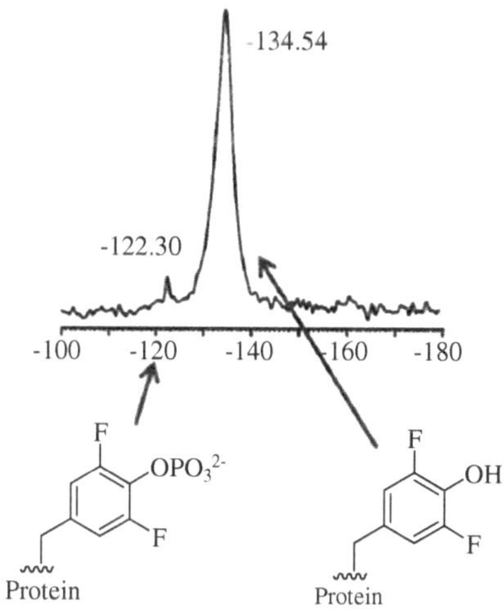

FIGURE 16 ^{19}F solid-state NMR spectrum of the partially phosphorylated Etk-574-F2Y (trifluoroacetic acid was used as an internal standard), showing signals for unreacted and phosphorylated difluorotyrosine residues. *Adapted with permission from Ref. 62. Copyright 2013, Wiley-VCH Verlag GmbH and CO. KGaA, Weinheim.*

phosphorylated residue (Figure 16). A separate chapter (Chapter 4) is devoted to various biological and medicinal applications of fluorinated noncanonical peptides and proteins.

8.3 ^{19}F NMR in the Study of Enzyme Mechanisms

Mechanism-based inactivation of glycosidases has implications in the treatment of Gaucher disease, diabetes, and influenza. Withers and coworkers have pioneered several 2-fluoro-2-deoxyglycoside (C_2–F glycoside) derivatives as substrate mimetics for inactivating the glycosidase enzymes.[63–65] The mechanism of these reactions involves double displacement of the leaving group at the anomeric carbon, mediated by the glycosidase enzymes. In the presence of the 2-fluorine substituent such nucleophilic substitutions are disfavored, because the β-fluorine atom strongly destabilizes the oxocarbenium intermediate due to its strong electron-withdrawing inductive effect. Despite the deactivating effect of the adjacent fluorine on the anomeric carbocation, the first step of the nucleophilic substitution reactions for the C_2–F glycosides is achievable using a good leaving group at the anomeric carbon, that with a $pK_a \leq 4.0$.[66,67] However, the second step of the hydrolysis is dramatically attenuated for the C_2–F glycosides (but not for the nonfluorinated glycoside), because of the poor leaving group effect of the glutamate (Glu233), so that the C_2–F glycosides act as suicide inactivators of the enzymes (Figure 17).

^{19}F NMR spectroscopy was used to monitor the formation of the covalent adduct of a β-glycosidase with 2-fluoro-2-deoxy-β-cellobioside **2**.[68] The *Cellulomonas fimi*-derived glucanase, a retaining glycosidase, forms a covalent adduct **2** by reacting with 2′,4′-dinitrophenyl 2-fluoro-2-deoxy-β-cellobioside **1**, which is assisted by protonation of the leaving group (2,4-dinitrophenol) through a general acid catalysis provided by Glu127

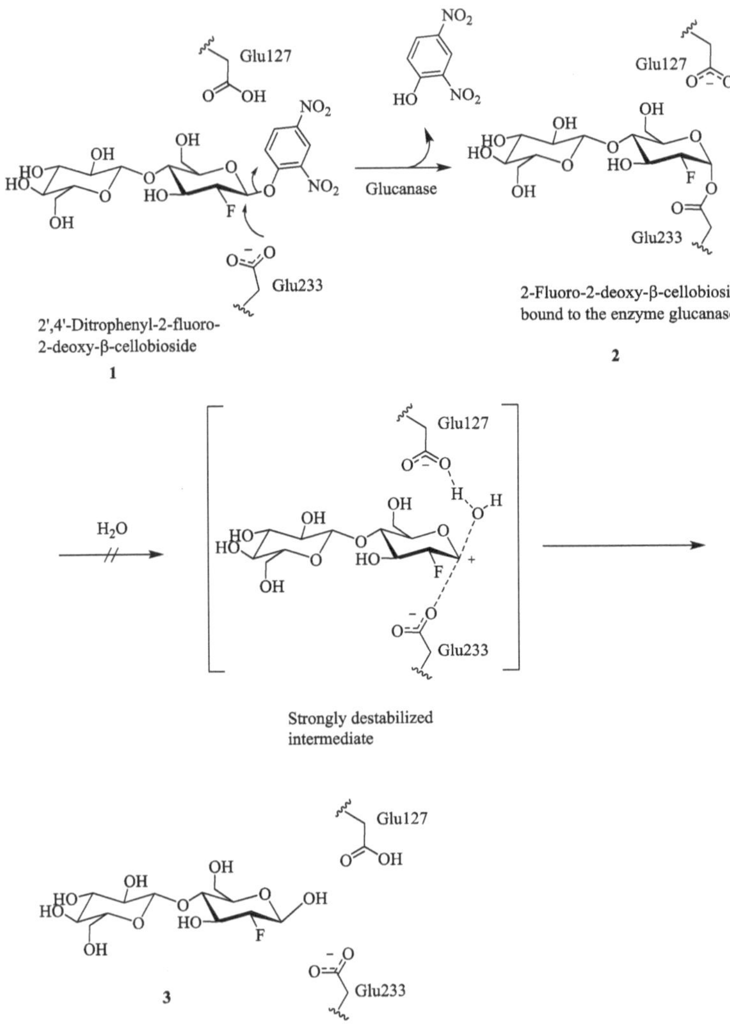

FIGURE 17 Mechanism-based inactivation of 2-fluoro-2-deoxy-β-cellobioside by a retaining β-glycosidase (glucanase); the second step of the hydrolysis is suppressed due to the strong electron-withdrawing effect of the C_2–F, so that the enzyme is irreversibly inactivated.

(Figure 18). This covalent adduct appears as a broad peak at $\delta^{19}F$ −195.4, downfield of the starting compound **1** ($\delta^{19}F = -198.1$) and its partially hydrolyzed product **3** ($\delta^{19}F = -198.4$) (Figure 18). This covalent adduct is inert to further hydrolysis reaction as the intermediate oxocarbenium ion would be strongly destabilized by the electron-withdrawing inductive effect of the C_2–F. An X-ray crystal structure for the covalent adduct was obtained at a resolution of 1.8 Å, which further supports the S_N2 type of mechanism for this enzymatic reaction (Figure 18).[68]

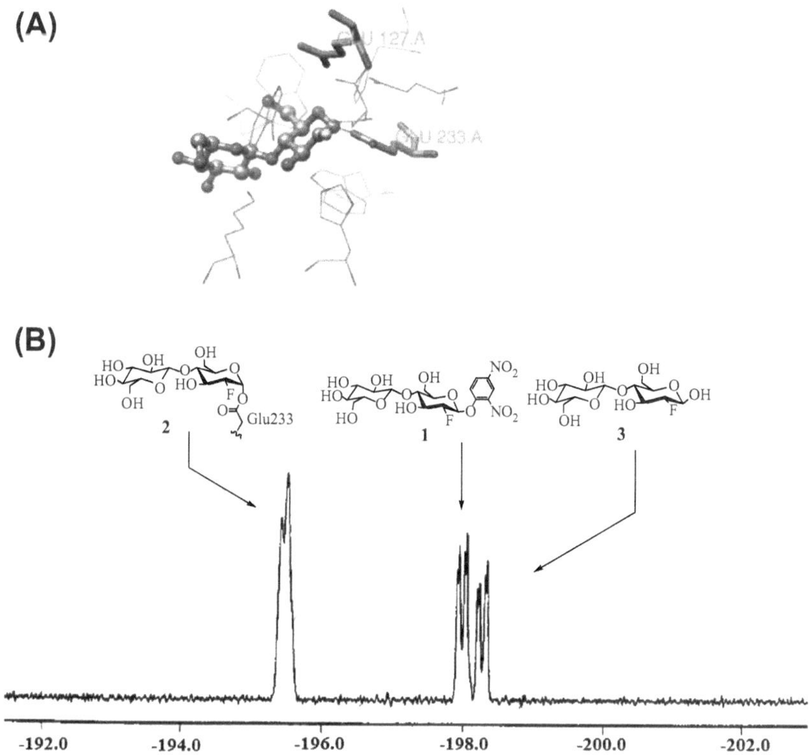

FIGURE 18 (A) an X-ray structure of the covalent adduct of the enzyme, 2-fluoro-2-deoxy-β-cellobioside (**2**) (oxygens are depicted in red, nitrogens in blue, and fluorine in green; the structure was generated using UCSF Chimera software; PDB 1EXP); (B) ^{19}F NMR spectrum of the products resulting from the reaction of *Cellulomonas fimi*-derived glucanase with 2′,4′-dinitrophenyl 2-fluoro-2-deoxycellobioside (**1**); the broad peak at δ^{19}F −195.4 was assigned to the covalent adduct **2**, and the absorptions at δ^{19}F −198.1 and −198.4 were assigned to the unreacted compound **1** and its partially hydrolyzed product **3**, respectively. (For interpretation of the references to color in this figure legend, the reader is referred to the online version of this book.) *Adapted with permission from Ref. 68. Copyright 1996, Nature Publishing Group.*

8.4 ^{19}F Magnetic Resonance Imaging (MRI)

^{19}F NMR spectra are uncomplicated by the absence of other background signals in contrast to that of ^{1}H NMR spectroscopy. Due to the high sensitivity of the δ^{19}F on the local chemical environment, e.g., pH-dependent conformational changes, the MRI signal intensities are useful for disease diagnosis at early stages.[69,70]

9. SUMMARY AND OUTLOOK

Organofluorine compounds, because of their unique physicochemical and chemical reactivity characteristics are widely used as pharmaceuticals, materials, and agrochemicals. Many drug candidates, with enhanced metabolic and oxidative stability, are designed by strategic incorporation of fluorine into the lead pharmaceutical candidates. Fluorinated pharmaceuticals have high metabolic stability, enhanced lipophilicity, and membrane permeability, and exhibit enhanced binding affinity to the targeted proteins as compared to their nonfluorinated analogs. Acidity, basicity, and lipophilicity of potential pharmaceutical compounds can be modulated by strategic incorporation of fluorine. Due to the relatively longer half-life of the ^{18}F isotope, ^{18}F-PET is emerging as an alternative to the ^{11}C-labeled PET imaging. ^{19}F NMR is well suited for probing biochemical mechanisms, and is used in the fragment-based drug discovery. In situ ^{19}F NMR also provides information on protein–ligand interactions and protein dynamics in the in-cell environments.

REFERENCES

1. Filler, R., Kobayashi, Y., Yagupolskii, L. M., Eds. *Organofluorine Compounds in Medicinal Chemistry and Biomedical Applications; Stud. Org. Chem.*, Vol. 48, Elsevier: Amsterdam, 1993.
2. Kirsch, P. *Modern Fluoroorganic Chemistry: Synthesis, Reactivity, Applications*, 2nd ed.; Wiley-VCH, 2013.
3. Ojima, I., Ed. *Fluorine in Medicinal Chemistry and Chemical Biology;* Wiley: New York, 2009.
4. Uneyama, K. *Organofluorine Chemistry;* Blackwell Publishing Ltd.: Oxford, UK, 2006; p. 339.
5. Vlassenko, A. G.; Benzinger, T. L. S.; Morris, J. C. PET Amyloid-beta Imaging in Preclinical Alzheimer's Disease. *Biochim. Biophys, Acta, Mol. Basis Dis.* **2012,** *1822,* 370–379.
6. Zhang, W.; Arteaga, J.; Cashion, D. K.; Chen, G.; Gangadharmath, U.; Gomez, L. F.; Kasi, D.; Lam, C.; Liang, Q.; Liu, C.; Mocharla, V. P.; Mu, F.; Sinha, A.; Szardenings, A. K.; Wang, E.; Walsh, J. C.; Xia, C.; Yu, C.; Zhao, T.; Kolb, H. C. A Highly Selective and Specific PET Tracer for Imaging of Tau Pathologies. *J. Alzheimer's Dis.* **2012,** *31,* 601–612.
7. Dalvit, C.; Vulpetti, A. Intermolecular and Intramolecular Hydrogen Bonds Involving Fluorine Atoms: Implications for Recognition, Selectivity, and Chemical Properties. *Chem. Med. Chem.* **2012,** *7,* 262–272.
8. Schneider, H.-J. Hydrogen Bonds with Fluorine. Studies in Solution, in Gas Phase and by Computations, Conflicting Conclusions from Crystallographic Analyses. *Chem. Sci.* **2012,** *3,* 1381–1394.
9. Howard, J. A. K.; Hoy, V. J.; O'Hagan, D.; Smith, G. T. How Good is Fluorine as a Hydrogen Bond Acceptor? *Tetrahedron* **1996,** *52,* 12613–12622.
10. Dunitz, J. D.; Taylor, R. Organic Fluorine Hardly Ever Accepts Hydrogen Bonds. *Chem. – Eur. J.* **1997,** *3,* 89–98.
11. Meanwell, N. A. Synopsis of Some Recent Tactical Application of Bioisosteres in Drug Design. *J. Med. Chem.* **2011,** *54,* 2529–2591.
12. Morgenthaler, M.; Schweizer, E.; Hoffmann-Roder, A.; Benini, F.; Martin, R. E.; Jaeschke, G.; Wagner, B.; Fischer, H.; Bendels, S.; Zimmerli, D.; Schneider, J.; Diederich, F.; Kansy, M.; Muller, K. Predicting and Tuning Physicochemical Properties in Lead Optimization: Amine Basicities. *ChemMedChem.* **2007,** *2,* 1100–1115.
13. Ruzziconi, R.; Spizzichino, S.; Mazzanti, A.; Lunazzi, L.; Schlosser, M. The biphenyl-monitored effective size of unsaturated functional or fluorinated ortho substituents. *Org. Biomol. Chem.* **2010,** *8,* 4463–4471.
14. Wolf, C.; Koenig, W. A.; Roussel, C. Influence of substituents on the rotational energy barrier of atropisomeric biphenyls - studies by polarimetry and dynamic gas chromatography. *Liebigs Ann.* **1995,** 781–786.
15. Leroux, F. Atropisomerism, biphenyls, and fluorine: A comparison of rotational barriers and twist angles,. *ChemBioChem* **2004,** *5,* 644–649.

16. Prakash, G. K. S.; Wang, F.; Rahm, M.; Zhang, Z.; Ni, C.; Shen, J.; Olah, G. A. The Trifluoromethyl Group as a Conformational Stabilizer and Probe: Conformational Analysis of Cinchona Alkaloid Scaffolds. *J. Am. Chem. Soc.* **2014,** *136,* 10418–10431.
17. Upton, T. G.; Kashemirov, B. A.; McKenna, C. E.; Goodman, M. F.; Prakash, G. K. S.; Kultyshev, R.; Batra, V. K.; Shock, D. D.; Pedersen, L. C.; Beard, W. A.; Wilson, S. H. α, β-Difluoromethylene Deoxynucleoside 5'-Triphosphates: A Convenient Synthesis of Useful Probes for DNA Polymerase β Structure and Function. *Org. Lett.* **2009,** *11,* 1883–1886.
18. Shakhmin, A.; Jones, J.-P.; Bychinskaya, I.; Zibinsky, M.; Oertell, K.; Goodman, M. F.; Prakash, G. K. S. Preparation of fluorinated RNA nucleotide analogs potentially stable to enzymatic hydrolysis in RNA and DNA polymerase assays. *J. Fluorine Chem.* **2014,** *167,* 226–230.
19. Prakash, G. K. S.; Zibinsky, M.; Upton, T. G.; Kashemirov, B. A.; McKenna, C. E.; Oertell, K.; Goodman, M. F.; Batra, V. K.; Pedersen, L. C.; Beard, W. A.; Shock, D. D.; Wilson, S. H.; Olah, G. A. Synthesis and biological evaluation of fluorinated deoxynucleotide analogs based on bis-(difluoromethylene)triphosphoric acid. *Proc. Natl. Acad. Sci. U S A.* **2010,** *107,* 15693–15698.
20. Cai, C.; Yi, W.-B.; Zhang, W.; Shen, M.-G.; Hong, M.; Zeng, L.-Y. Fluorous Lewis acids and phase transfer catalysts. *Mol. Divers.* **2009,** *13,* 209–239.
21. Lai, C.-Z.; Reardon, M. E.; Boswell, P. G.; Buehlmann, P. Cation-coordinating properties of perfluoro-15-crown-5. *J. Fluorine Chem.* **2010,** *131,* 42–46.
22. Irandoust, M.; Daraei, H.; Rostamian, R. Proton NMR probing of stoichiometry and thermodynamic data for the complexation of Na^+ and Li^+ ions with 15-Crown-5 in acetonitrile-nitrobenzene mixtures. *Polyhedron* **2011,** *30,* 1207–1212.
23. Filler, R.; Saha, R. Fluorine in medicinal chemistry: a century of progress and a 60-year retrospective of selected highlights. *Future Med. Chem.* **2009,** *1,* 1370.
24. Banks, R. E., Smart, B. E., Tatlow, J. C., Eds. *Organofluorine Chemistry: Principles and Commercial Applications;* Plenum: *New York*, 1994.
25. Chambers, R. D., Ed. *Organofluorine Chemistry: Techniques and Synthons;* Top. Curr. Chem, *193*, , 1997.
26. Yamamoto, H. *Organofluorine Compounds: Chemistry and Applications;* Springer-Verlag: New York, 2000.
27. O'Hagan, D. Understanding organofluorine chemistry. An introduction to the C-F bond. *Chem. Soc. Rev.* **2008,** *37,* 308–319. Refrository Exact.
28. Yamamoto, H. In *Organofluorine Compounds: Chemistry and Applications;* . Springer-Verlag, 2000.
29. Dolbier, W. R., Jr.; Palmer, K. W. Effect of terminal fluorine substitution on the Cope rearrangement: boat versus chair transition state. Evidence for a very significant fluorine steric effect. *J. Am. Chem. Soc.* **1993,** *115,* 9349–9350.
30. Griffith, G. A.; Hillier, I. H.; Moralee, A. C.; Percy, J. M.; Roig, R.; Vincent, M. A. Interplay of Structure and Reactivity in a Most Unusual Furan Diels-Alder Reaction. *J. Am. Chem. Soc.* **2006,** *128,* 13130–13141.
31. Marshall, P.; Schwartz, M. Computational Study of C-H Bond Strengths in Polyfluoroalkanes. *J. Phys. Chem. A* **1997,** *101,* 2906–2909.
32. Zachariah, M. R.; Westmoreland, P. R.; Burgess, D. R. F., Jr.; Tsang, W.; Melius, C. F. BAC-MP4 Predictions of Thermochemical Data for C1 and C2 Stable and Radical Hydrofluorocarbons and Oxidized Hydrofluorocarbons. *J. Phys. Chem.* **1996,** *100,* 8737–8747.
33. Wood, G. P. F.; Moran, D.; Jacob, R.; Radom, L. Bond Dissociation Energies and Radical Stabilization Energies Associated with Model Peptide-Backbone Radicals. *J. Phys. Chem. A* **2005,** *109,* 6318–6325.

34. Maren, T. H.; Conroy, C. W. A new class of carbonic anhydrase inhibitor. *J. Biol. Chem.* **1993,** *268,* 26233–26239.
35. Scolnick, L. R.; Clements, A. M.; Liao, J.; Crenshaw, L.; Hellberg, M.; May, J.; Dean, T. R.; Christianson, D. W. Novel Binding Mode of Hydroxamate Inhibitors to Human Carbonic Anhydrase II. *J. Am. Chem. Soc.* **1997,** *119,* 850–851.
36. Bondi, A. van der Waals volumes and radii. *Journal of Physical Chemistry* **1964,** *68,* 441–451.
37. Rosenblum, S. B.; Huynh, T.; Afonso, A.; Davis, H. R., Jr; Yumibe, N.; Clader, J. W.; Burnett, D. A. Discovery of 1-(4-fluorophenyl)-(3R)-[3-(4-fluorophenyl)-(3S)- hydroxypropyl]-(4S)-(4-hydroxyphenyl)-2-azetidinone (SCH 58235): A designed, potent, orally active inhibitor of cholesterol absorption. *J. Med. Chem.* **1998,** *41,* 973–980.
38. Rose, W. C.; Marathe, P. H.; Jang, G. R.; Monticello, T. M.; Balasubramanian, B. N.; Long, B.; Fairchild, C. R.; Wall, M. E.; Wani, M. C. Novel fluoro-substituted camptothecins: in vivo antitumor activity, reduced gastrointestinal toxicity and pharmacokinetic characterization. *Cancer Chemother. Pharmacol.* **2006,** *58,* 73–85.
39. Johannes, M.; Oberbillig, T.; Hoffmann-Roeder, A. Synthesis of fluorinated Thomsen-Friedenreich antigens: direct deoxyfluorination of αGalNAc-threonine *tert*-butyl esters. *Org. Biomol. Chem.* **2011,** *9,* 5541–5546.
40. Senn, H. M.; O'Hagan, D.; Thiel, W. Insight into Enzymatic C-F Bond Formation from QM and QM/MM Calculations, *J. Am. Chem. Soc.* **2005,** *127,* 13643–13655.
41. Penning, T. D.; Talley, J. J.; Bertenshaw, S. R.; Carter, J. S.; Collins, P. W.; Docter, S.; Graneto, M. J.; Lee, L. F.; Malecha, J. W.; Miyashiro, J. M.; Rogers, R. S.; Rogier, D. J.; Yu, S. S.; Anderson, G. D.; Burton, E. G.; Cogburn, J. N.; Gregory, S. A.; Koboldt, C. M.; Perkins, W. E.; Seibert, K.; Veenhuizen, A. W.; Zhang, Y. Y.; Isakson, P. C. Synthesis and Biological Evaluation of the 1,5-Diarylpyrazole Class of Cyclooxygenase-2 Inhibitors: Identification of 4-[5-(4-Methylphenyl)-3-(trifluoromethyl)-1H-pyrazol-1-yl]benzenesulfonamide (SC-58635, Celecoxib). *J. Med. Chem.* **1997,** *40,* 1347–1365.
42. Bikle, D. D. Vitamin D Metabolism, Mechanism of Action, and Clinical Applications. *Chem. Biol. (Oxford, U. K.)* **2014,** *21,* 319–329.
43. Tanaka, Y.; Wichmann, J. K.; De Luca, H. F.; Kobayashi, Y.; Ikekawa, N. Metabolism and binding properties of 24,24-difluoro-25-hydroxyvitamin D3. *Arch. Biochem. Biophys.* **1983,** *225,* 649–655.
44. Eelen, G.; Verlinden, L.; Bouillon, R.; De Clercq, P.; Munoz, A.; Verstuyf, A. CD-ring modified vitamin D3 analogs and their superagonistic action. *J. Steroid Biochem. Mol. Biol.* **2010,** *121,* 417–419.
45. Dolbier, W. R. *Guide to Fluorine NMR for Organic Chemists*. John Wiley & Sons, Inc., New York (N.Y.), **2009**.
46. Paulini, R.; Mueller, K.; Diederich, F. Orthogonal multipolar interactions in structural chemistry and biology. *Angew. Chem., Int. Ed.* **2005,** *44,* 1788–1805.
47. Sehon, C. A.; Wang, G. Z.; Viet, A. Q.; Goodman, K. B.; Dowdell, S. E.; Elkins, P. A.; Semus, S. F.; Evans, C.; Jolivette, L. J.; Kirkpatrick, R. B.; Dul, E.; Khandekar, S. S.; Yi, T.; Wright, L. L.; Smith, G. K.; Behm, D. J.; Bentley, R.; Doe, C. P.; Hu, E.; Lee, D. Potent, Selective and Orally Bioavailable Dihydropyrimidine Inhibitors of Rho Kinase (ROCK1) as Potential Therapeutic Agents for Cardiovascular Diseases. *J. Med. Chem.* **2008,** *51,* 6631–6634.
48. Helali, A. M.; Iti, F. M.; Mohamed, I. N. Cathepsin K Inhibitors: A Novel Target but Promising Approach in the Treatment of Osteoporosis. *Curr. Drug Targets* **2013,** *14,* 1591–1600.
49. Black, W. C.; Bayly, C. I.; Davis, D. E.; Desmarais, S.; Falgueyret, J.-P.; Leger, S.; Li, C. S.; Masse, F.; McKay, D. J.; Palmer, J. T.; Percival, M. D.; Robichaud, J.; Tsou, N.; Zamboni, R. Trifluoroethylamines as amide isosteres in inhibitors of cathepsin K. *Bioorg. Med. Chem. Lett.* **2005,** *15,* 4741–4744.
50. Sani, M.; Volonterio, A.; Zanda, M. The trifluoroethylamine function as peptide bond replacement,. *ChemMedChem* **2007,** *2,* 1693–1700.

51. Neerathilingam, M.; Greene, L. H.; Colebrooke, S. A.; Campbell, I. D.; Staunton, D. Quantitation of protein expression in a cell-free system: Efficient detection of yields and ^{19}F NMR to identify folded protein. *J. Biomol. NMR* **2005,** *31,* 11–19.
52. Chen, H.; Viel, S.; Ziarelli, F.; Peng, L. ^{19}F NMR: a valuable tool for studying biological events. *Chem. Soc. Rev.* **2013,** *42,* 7971–7982.
53. Vulpetti, A.; Dalvit, C. Fluorine local environment: from screening to drug design. *Drug Discov. Today* **2012,** *17,* 890–897.
54. Dalvit, C.; Ko, S. Y.; Vulpetti, A. Application of the rule of shielding in the design of novel fluorinated structural motifs and peptidomimetics. *J. Fluorine Chem* **2013,** *152,* 129–135.
55. Jordan, J. B.; Poppe, L.; Xia, X.; Cheng, A. C.; Sun, Y.; Michelsen, K.; Eastwood, H.; Schnier, P. D.; Nixey, T.; Zhong, W. Fragment Based Drug Discovery: Practical Implementation Based on ^{19}F NMR Spectroscopy. *J. Med. Chem.* **2012,** *55,* 678–687.
56. Li, C.; Wang, G.-F.; Wang, Y.; Creager-Allen, R.; Lutz, E. A.; Scronce, H.; Slade, K. M.; Ruf, R. A. S.; Mehl, R. A.; Pielak, G. J. Protein ^{19}F NMR in Escherichia coli. *J. Am. Chem. Soc.* **2010,** *132,* 321–327.
57. Tkachenko, A. N.; Radchenko, D. S.; Mykhailiuk, P. K.; Afonin, S.; Ulrich, A. S.; Komarov, I. V. Design, Synthesis, and Application of a Trifluoromethylated Phenylalanine Analogue as a Label to Study Peptides by Solid-State ^{19}F NMR Spectroscopy. *Angew. Chem., Int. Ed.* **2013,** *52,* 6504–6507.
58. Ulrich, A. S.; Wadhwani, P.; Duerr, U. H. N.; Afonin, S.; Glaser, R. W.; Strandberg, E.; Tremouilhac, P.; Sachse, C.; Berditchevskaia, M.; Grage, S. *Solid-state ^{19}F-nuclear magnetic resonance analysis of membrane-active peptides.* CRC Press (Boca Raton, Fla); **2006,** 215–236; *Chem. Abstr.* 2006:189119.
59. Ulrich, A. S. Solid state ^{19}F NMR methods for studying biomembranes. *Prog. Nucl. Magn. Reson. Spectrosc.* **2005,** *46,* 1–21.
60. Koch, K.; Afonin, S.; Ieronimo, M.; Berditsch, M.; Ulrich, A. S. Solid-state ^{19}F-NMR of peptides in native membranes. *Top. Curr. Chem.* **2012,** *306,* 89–118.
61. Hong, M.; Schmidt-Rohr, K. Magic-Angle-Spinning NMR Techniques for Measuring Long-Range Distances in Biological Macromolecules. *Acc. Chem. Res.* **2013,** *46,* 2154–2163.
62. Li, F.; Shi, P.; Li, J.; Yang, F.; Wang, T.; Zhang, W.; Gao, F.; Ding, W.; Li, D.; Li, J.; Xiong, Y.; Sun, J.; Gong, W.; Tian, C.; Wang, J. A Genetically Encoded ^{19}F NMR Probe for Tyrosine Phosphorylation. *Angew. Chem., Int. Ed.* **2013,** *52,* 3958–3962.
63. Rempel, B. P.; Tropak, M. B.; Mahuran, D. J.; Withers, S. G. Tailoring the Specificity and Reactivity of a Mechanism-Based Inactivator of Glucocerebrosidase for Potential Therapeutic Applications. *Angew. Chem., Int. Ed.* **2011,** *50,* 10381–10383. S10381/10381.
64. Tysoe, C.; Withers, S. G. Fluorinated Mechanism-Based Inhibitors: Common Themes and Recent Developments. *Curr. Top. Med. Chem.* **2014,** *14,* 865–874.
65. Goddard-Borger, E. D.; Wennekes, T.; Withers, S. G. Molecular probes Getting lucky in the lysosome. *Nat. Chem. Biol.* **2010,** *6,* 881–883.
66. Rempel, B. P.; Withers, S. G. Covalent inhibitors of glycosidases and their applications in biochemistry and biology. *Glycobiology* **2008,** *18,* 570–586.
67. Street, I. P.; Kempton, J. B.; Withers, S. G. Inactivation of a β-glucosidase through the accumulation of a stable 2-deoxy-2-fluoro-α-D-glucopyranosyl-enzyme intermediate: a detailed investigation. *Biochemistry* **1992,** *31,* 9970–9978.
68. White, A.; Tull, D.; Johns, K.; Withers, S. G.; Rose, D. R. Crystallographic observation of a covalent catalytic intermediate in a β-glycosidase. *Nat. Struct. Biol.* **1996,** *3,* 149–154.
69. Ruiz-Cabello, J.; Barnett, B. P.; Bottomley, P. A.; Bulte, J. W. M. Fluorine (^{19}F) MRS and MRI in biomedicine. *NMR Biomed.* **2011,** *24,* 114–129.
70. Gaye, B.; Adejare, A. Fluorinated molecules in the diagnosis and treatment of neurodegenerative diseases. *Future Med. Chem.* **2009,** *1,* 821–833.

Chapter 2

Fluorinated Compounds in Enzyme-Catalyzed Reactions

Chapter Outline

1. INTRODUCTION

Fluorine is the most abundant element among the halogens in the Earth's crust, yet less than 20 organofluorine compounds have been isolated from natural products. This is because only a limited number of fluorinated organic compounds are derived from biosynthetic pathways, although there are about a million synthetically derived organofluorine compounds.[1–4] The naturally occurring organofluorine compounds (i.e., the compounds derived from biosynthetic pathways) that are isolated to date include fluoroacetic acid (isolated from the South African plants, *Dichapetalum cymosum* and *Dichapetalum braunii*, and the soil-dwelling bacterium *Streptomyces cattleya*),[1,4] fluoroacetone (from the plant source *Acacia georginae*),[5] (2R, 3R)-2-fluorocitric acid (from *S. cattleya* and from the commercial tea leaves (<30 ppm; nontoxic levels)),[6]

Organofluorine Compounds in Biology and Medicine. http://dx.doi.org/10.1016/B978-0-444-53748-5.00002-2

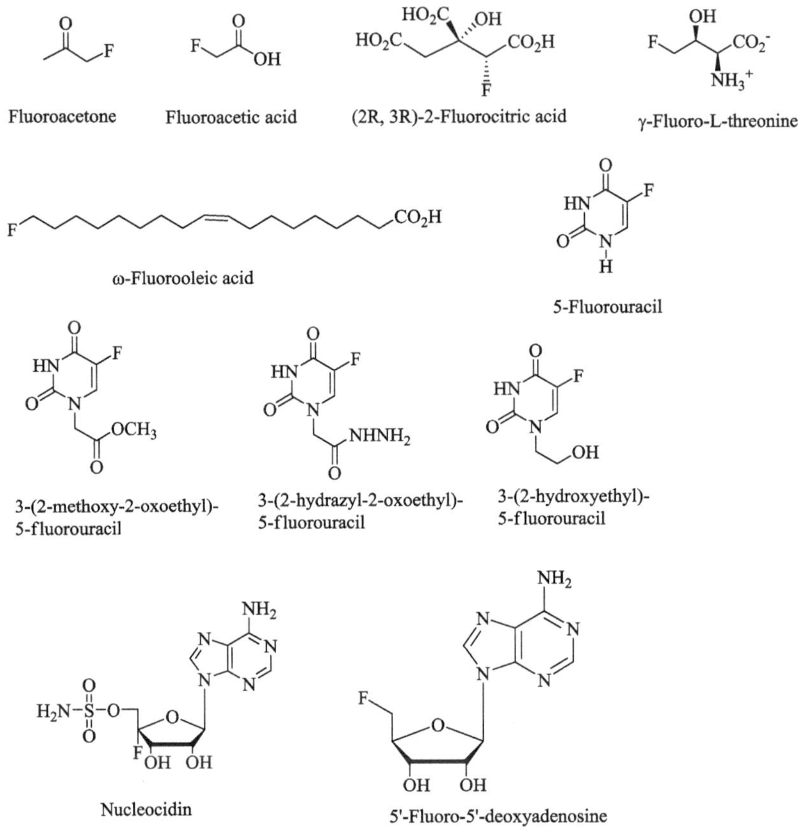

FIGURE 1 Naturally occurring organofluorine compounds; some bacterial species, and some plants have evolved mechanisms for the biosynthesis of the fluorinated organic compounds. Fluoroacetic acid is especially highly toxic to humans as it shuts down the citric acid cycle.

γ-fluoro-L-threonine (from *S. cattleya*),[4] ω-fluorooleic acid (from the seeds of *Dichapetalum toxicarium*),[7] nucleocidin (a fluorinated adenosine derivative),[8,9] and 5′-fluoro-5′-deoxyadenosine (from *S. cattleya*).[3,10] Interestingly, 5-fluorouracil, a popular anticancer drug, and its derivatives have been isolated from the sponge *Phakellia fusca*, collected from the South China Sea (Figure 1).[11]

The mechanism of the biosynthesis of organofluorine compounds, especially that of the C–F bond formation is not clear-cut to date, although significant progress has been made in elucidating the biosynthetic pathways for the fluoroacetate and 4-fluorothreonine, which are isolated from the bacterium *S. cattleya*.[4,10,12–20] The only known fluorinating enzyme that has been well characterized to date is the fluorinase, which catalyzes the nucleophilic substitution reaction of S-adenosylmethionine (SAM) by fluoride anion to form the 5′-fluoro-5′-deoxyadenosine.[21] This classic enzymatic reaction follows the bimolecular nucleophilic substitution (S_N2) mechanism, as supported by the

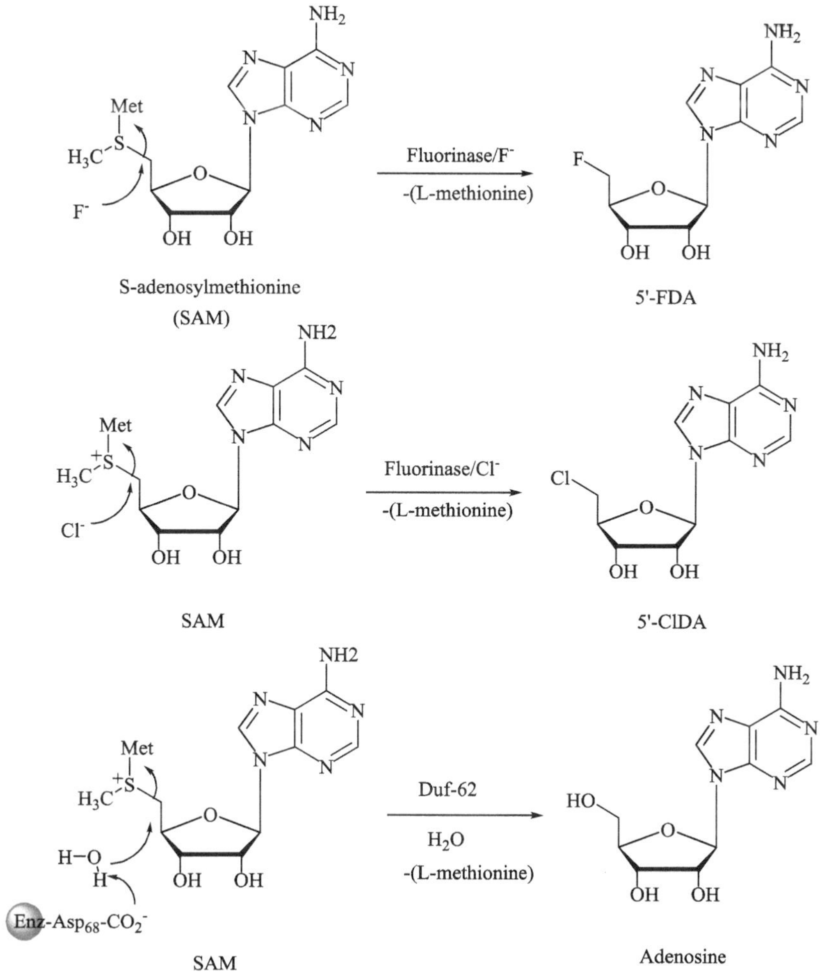

FIGURE 2 Fluorinase and Duf-62-catalyzed S_N2 fluorination, chlorination, and hydroxylation of SAM; X-ray crystallographic evidence strongly supports the S_N2 mechanism for these reactions.

X-ray crystallographic evidence (Figure 2). The 5′-fluoro-5′-deoxyadenosine formed in this reaction is the starting material for the biosynthesis of fluoroacetate and other fluorinated amino acids such as fluorothreonine.

Interestingly, fluorinase also catalyzes the conversion of SAM to the 5′-chloro-5′-deoxyadenosine, presumably proceeding through a similar S_N2 reaction mechanism.[21] The fluorinase enzyme has some homology to the proteins that are expressed by the *duf-62* gene. The duf-62 proteins are involved in the hydrolytic cleavage of the SAM to form the adenosine.[21] The active nucleophile in this hydrolytic reaction is the hydroxide anion, derived from water, at the active site, involving a general base catalysis mediated by Asp68.

2. BACTERIAL METABOLISM OF THE FLUOROACETATE

In order to circumvent the lethal action of the fluorocitrate intermediate, an aconitase inhibitor derived from fluoroacetyl coenzyme A (CoA) in the citric acid cycle, the soil-dwelling bacterium *S. cattleya* has evolved a mechanism, whereby it produces fluoroacetyl-CoA thioesterase that has over 10^4 times as high catalytic efficiency in the hydrolysis of the fluoroacetyl-CoA as that for the hydrolysis of the cellularly abundant analog, acetyl-CoA.[22,23]

This enormous selectivity in the enzymatic hydrolysis is ascribed to a change in the rate-determining step from that of acetyl-CoA; the enzymatic deprotonation at the fluorine-bearing carbon of the fluoroacetyl-CoA is accelerated due to the relatively higher acidity of this hydrogen (Figure 3). The acetyl-CoA, on the other hand, is hydrolyzed rather slowly through the acyl nucleophilic substitution in a nonenzymatic, uncatalyzed process. The enzymatic α-deprotonation of fluoroacetyl-CoA forms the enolate-CoA that upon spontaneous expulsion of the CoA forms fluoroketene as an intermediate. The highly reactive fluoroketene intermediate rapidly hydrolyzes to form the fluoroacetic acid. *S. cattleya*'s resistance to the toxicity of the fluoroacetate that it produces from the inorganic fluoride source is thus due to the temporal control of the production of fluoroacetyl-CoA rather than that of the discrimination of the fluorinated versus nonfluorinated molecules in the citric acid cycle in the host's metabolic machinery. Using chiral (R)- and (S)-[^{2}H]fluoroacetyl-CoA substrates, Weeks and coworkers demonstrated an enantioselective primary deuterium kinetic isotope effect of 1.7 for the (R)-^{2}H isomer (but not observed for the (S)-^{2}H isomer).[23] These results show that the selectivity of the fluoroacetyl-CoA substrate over the acetyl-CoA in the enzyme catalyzed reactions is not only due to

(A)

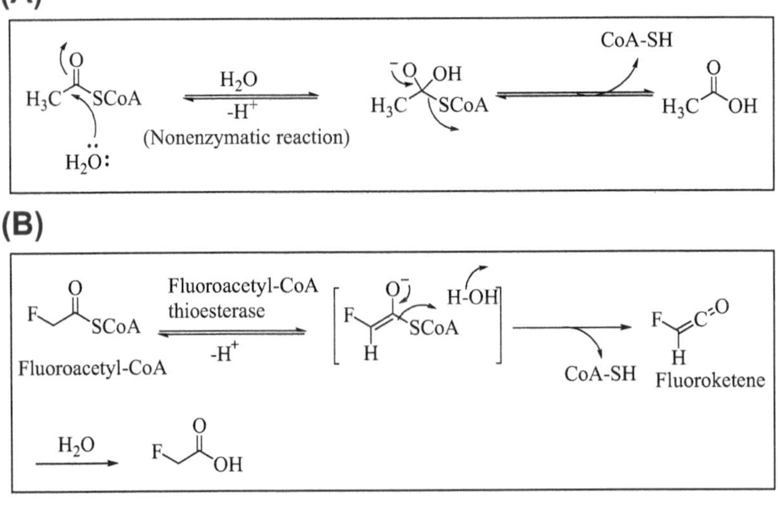

FIGURE 3 Nonenzymatic hydrolysis of acetyl-CoA (A), and fluoroacetyl-CoA thioesterase-catalyzed hydrolysis of the fluoroacetyl-CoA (B); the nonenzymatic hydrolysis of acetyl-CoA is 10^4 times slower than that of the enzymatic hydrolysis of the fluoroacetyl-CoA.

the enhanced acidity of the α-hydrogen, but also due to the molecular recognition afforded by the fluorine at the enzyme active site.

3. BIOSYNTHESIS OF FLUOROACETATE AND FLUOROTHREONINE

Streptomyces cattleya is the only known bacterium that incorporates an inorganic fluoride source (NaF) to produce fluoroacetate and fluorothreonine.[10,24,25] The bacterial growth is unaffected by the presence or absence of the inorganic fluoride source (Figure 4). ^{19}F nuclear magnetic resonance (NMR) monitoring of the culture supernatant showed time-dependent rise in the production of fluoroacetate and fluorothreonine showing the occurrence of two independent routes for their biosynthesis.[25]

Figure 5 shows the stepwise formation of the fluoroacetate and γ-fluorothreonine starting from the SAM. The fluorinase-catalyzed nucleophilic substitution (S_N2) reaction of SAM with the fluoride anion results in the formation of the 5′-deoxy-5′-fluoroadenosine, which upon enzymatic deadenylative phosphorylation (S_N2), followed by isomerization catalyzed by 5-fluoro-5-deoxyribose-1-phosphate (5-FDRP) isomerase gives 5-fluoro-5-deoxyribulose-1-phosphate. The latter compound then forms either fluoroacetate or fluorothreonine, catalyzed by fluoroacetaldehyde dehydrogenase or fluorothreonine transaldolase, respectively.[25] Thus, the biosynthesis of fluoroacetate is independent of that of fluorothreonine, although they are both derived from a common intermediate, fluorodeoxyribulose-1-phosphate.

Based on the X-ray structural evidence on the fluorinase catalyzed reactions it was inferred that the C–F bond formation significantly precedes the departure of the leaving group methionine[21]; the C–S bond distance of 3.4 Å for the departing methionine and the C–F bond distance of 1.3 Å for the newly forming carbon–fluorine bond indicates an early transition state, that is, the C–F bond formation is nearly complete just before the departure of the leaving

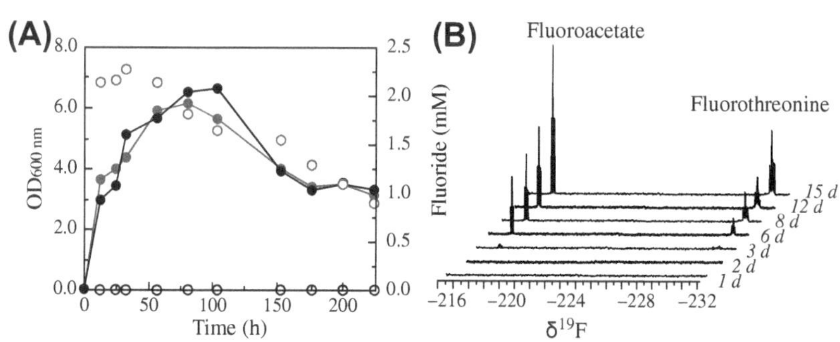

FIGURE 4 Analysis of the fluorine response in *S. cattleya*: (A) growth of the bacterial cultures in the presence (red circles) and absence (black circles) of added 2 mM NaF; (B) ^{19}F NMR monitoring of the culture supernatant medium for the formation of fluoroacetate and fluorothreonine (OD = optical density). *Adapted with permission from Ref. 25. Copyright 2012, American Chemical Society.*

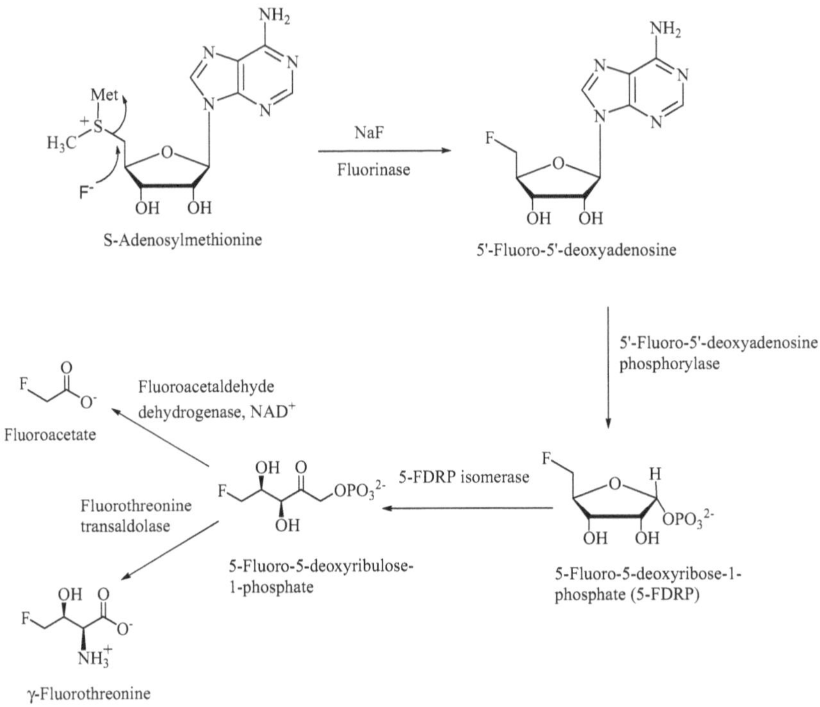

FIGURE 5 Biosynthesis of fluoroacetate and γ-fluorothreonine.

group methionine. Serine-158, aspartic acid-16, and threonine-80 play a key role in the fluoride activation at the active center of the fluorinase (Figures 6 and 7).

The DFT/CHARMM (combined Density-functional theory and the force field CHARMM) calculations of the transition state at the active site further confirm the proposed S_N2 reaction mechanism.[26] These calculations indicate that the activation energy for the S_N2 reaction at the active site by fluoride anion (53 kJ/mol) is 49 kJ/mol more favorable than that in solution (92 kJ/mol). The major role of the enzyme in this reaction is in the positioning of substrates for bringing them into closer proximity.

The DFT/CHARMM calculations of the transition state for the enzymatic nucleophilic substitution reaction of the SAM are also in accordance with the occurrence of an early transition state. Thus, the role of the enzyme in the fluorination of SAM is mainly targeted toward the substrate organization at the transition state, desolvation of the bound water molecules, and activation of the fluoride anion nucleophile through hydrogen bonds.[21] In other words, the solvated water molecules around the fluoride anion are replaced by protein hydrogen bonds so that fluoride anion is relatively more nucleophilic toward the nucleophilic substitution of methionine in SAM (Figure 8).

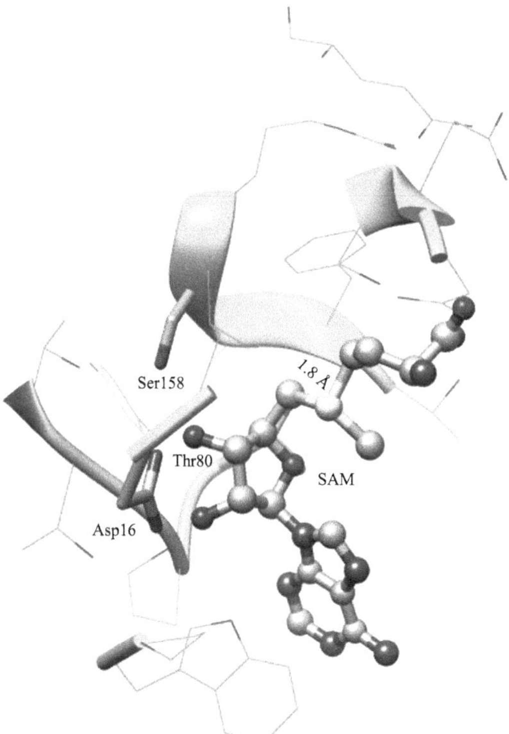

FIGURE 6 X-ray crystal structure of SAM (shown as ball and stick: red, oxygen; blue, nitrogen; yellow, sulfur) bound to the active site of the fluorinase (before the start of reaction); the key amino acid residues Asp16, Ser158, and Thr80 involved in the fluoride activation are shown as sticks (the structure was created using UCSF Chimera software; PDB: 1rqp).

4. ENZYMATIC DEFLUORINATION

Enzymatic dehalogenation of halogen-containing organic compounds is a facile process for their detoxification. However, these processes are rarely observed for the organofluorine compounds. As an exceptional case, haloacetate dehalogenase enzyme, found in the soil *Pseudomonas* sp. and the soil fungus *Fusarium solani*, has high specificity for cleaving the C–F bond of fluoroacetate and that of other organofluorine compounds. This enzyme is stable from 15 °C to 55 °C, and is inhibited by excess inorganic fluoride sources.[27]

5. MECHANISM-BASED ENZYME INHIBITORS

In most cases, the C–F bond is isosteric with the corresponding C–H bond and the enzymes are unable to distinguish the fluorinated substrate versus the nonfluorinated compound. Similarly, the CF_3 and the CF_2 groups are isosteric with the isopropyl and ethereal oxygen, respectively. The CF_2 group is also considered to be isosteric and isopolar with respect to the carbonyl group. This bioisostere concept is utilized in the design of various mechanism-based enzyme inhibitors. The bioisosteric organofluorine compounds mimic the substrates in binding to

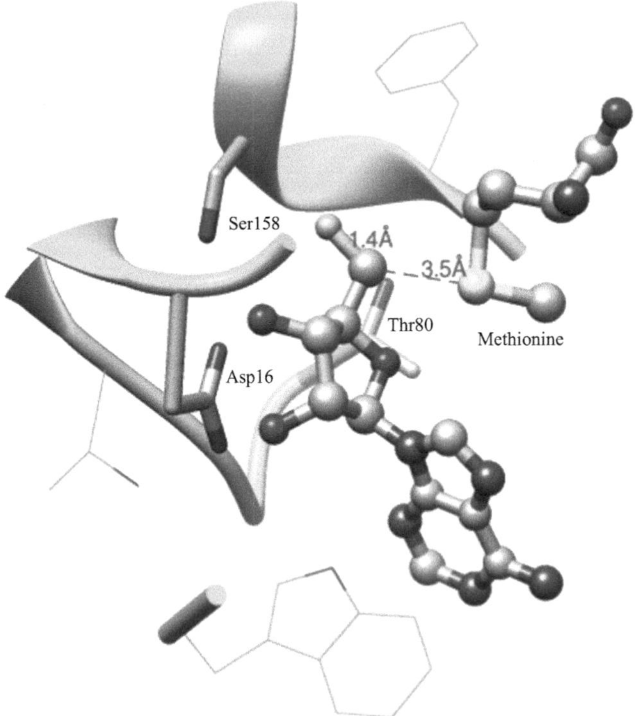

FIGURE 7 X-ray crystal structure of the S_N2 reaction products of SAM at the fluorinase active site; the 5′-fluoro-5′-deoxyadenosine and the leaving group methionine are shown in ball and stick representations (red, oxygen; blue, nitrogen; green, fluorine; yellow, sulfur). The key amino acid residues Asp16, Ser158, and Thr80 involved in fluoride activation are shown as sticks (the structure was created using UCSF Chimera software; PDB: 1rqr).

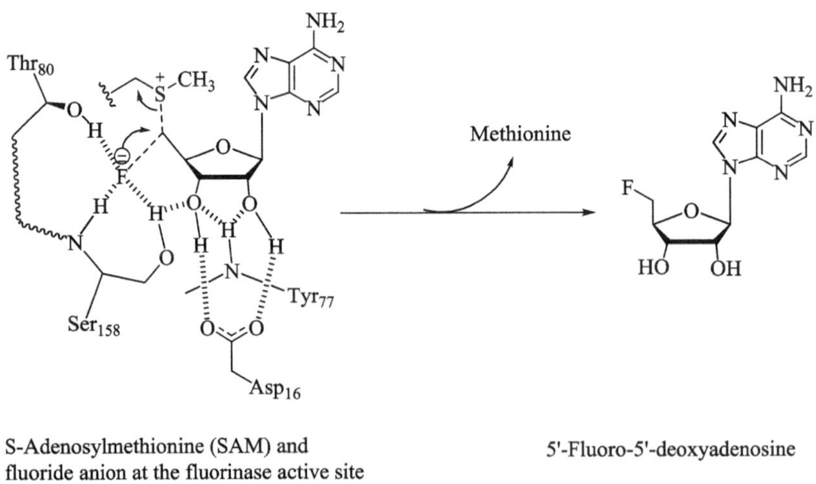

FIGURE 8 A schematic view of the fluorinase-catalyzed fluorination of SAM; the Thr80, Asp16, Tyr77, and Ser158 are the main amino acids responsible for the fluoride anion activation (through hydrogen bonding interactions).[21]

the enzyme active site, and in some cases, there is a greater preference for the fluorinated mimetics for binding to the active site. However, after binding to the active site, the mechanisms of enzyme action may be entirely different for the fluorinated compounds; in case of reversible inhibitors, the fluorinated compounds would have relatively high binding energy to the enzyme active site so that they effectively compete with the corresponding nonfluorinated substrates and thereby block the enzyme active site. The irreversible inhibitors, also called as suicide substrates (or suicide inhibitors), on the other hand, may react with an active site residue, forming a covalent adduct so that the enzyme cannot be regenerated. These mechanism-based inhibitor concepts are instrumental in the development of numerous therapeutic candidates.[28]

5.1 Irreversible Inhibitors (Suicide Inhibitors)

In the case of irreversible inhibitors, the inhibitor is covalently bound to the active site residues of the enzyme and therefore the enzyme is permanently inactivated. The suicide inhibitor resembles the substrate in its structural features as well as in its mechanism of action during the initial stages of enzyme inhibition. However, unlike the natural substrate, the suicide inactivator chemically reacts with an active site residue and either binds to the active site through a covalent linkage or forms a secondary metabolite that has high affinity to the active site residues of the enzyme, so that the enzyme is irreversibly inactivated. About 39 drugs that exert their therapeutic action by forming covalent adducts with proteins were approved by US Food and Drug Administration.[29] The irreversible enzyme inhibition may be due to acyl nucleophilic substitution reactions (e.g., aspirin, orlistat), aromatic nucleophilic substitution reactions (e.g., cladribine), S_N2 reactions (e.g., phenoxybenzamine), nucleophilic addition reactions (e.g., vildagliptin), or Michael additions (e.g., canertinib, exemestane, ethacrynic acid). The nonselective binding of these drugs to proteins other than the targeted proteins or enzymes results in adverse side effects in many cases. Further, the covalent binding of these drugs or drug candidates may arise through reactions at functionally distinct nucleophilic sites on the proteins, further complicating the safety of these drugs when used at doses higher than the recommended values; for example, the Michael additions of epidermal growth factor receptor (EGFR) inhibitors, such as canertinib and neratinib (HKI-272), may take place either through the sulfhydryl groups or through the ε-amino group of lysine residues.[29,30] The Pfizer group, Dahal and coworkers, proposed that a daily covalent binding body burden value of 10 mg or less should be the upper threshold when designing new covalent drugs.[29]

5.1.1 Suicide Inactivation of Acetylcholinesterase

Acetylcholinesterase is involved in the hydrolysis of acetylcholine, an essential neurotransmitter of the central nervous system, into choline (Figure 9). This enzyme catalyzes the hydrolysis of the neurotransmitter acetylcholine at neuronal synapses, and at neuromuscular junctions, at the end of the signaling process.

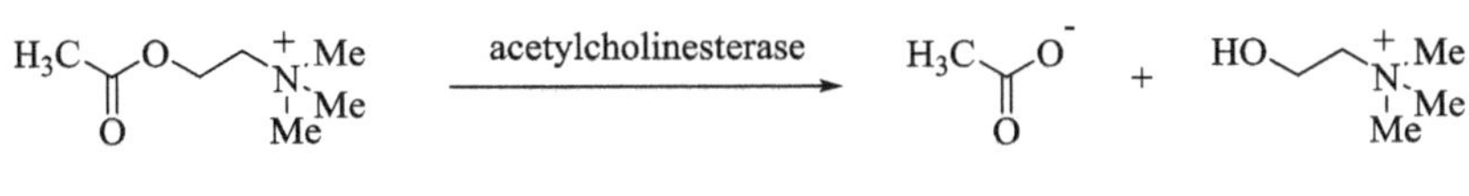

FIGURE 9 Acetylcholinesterase-catalyzed hydrolysis of acetylcholine.

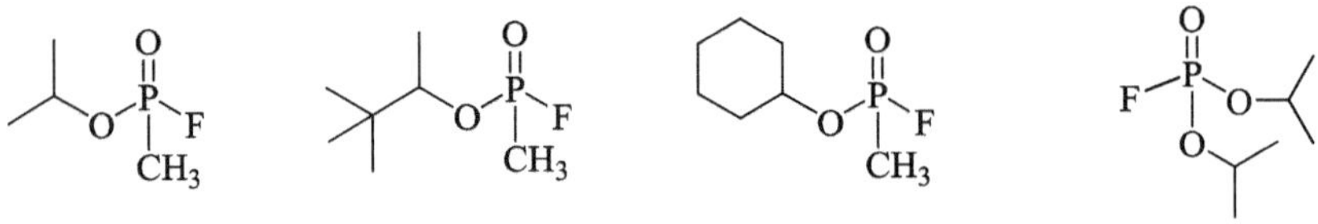

FIGURE 10 Structures of some fluorophosphonate-based nerve agents.

In certain neurological disorders such as Alzheimer's disease (AD), acetylcholinesterase is overactivated in the synapses so that levels of acetylcholine in the brains is significantly diminished, which leads to weakened neurotransmission and thereby memory loss and other adverse effects. The cholinesterase enzyme may also be, at least partly, responsible for the buildup of amyloid β plaques and the neurofibrillary tangles in AD brains. Some of the current therapeutics for AD, such as donepezil and galantamine, are therefore focused on controlled, reversible inhibition of acetylcholinesterase. Through reversible inhibition of acetylcholinesterase, the neurotransmitter acetylcholine is retained for relatively longer periods of time, and therefore may have beneficial effects on the memory retention, at least in the short term.

On the other hand, the fluorophosphonate-based nerve gas agents, sarin (isopropyl methylfluorophosphonate), soman (pinacolyl methylfluorophosphonate), cyclosarin (methyl cyclohexylfluorophosphonate), and DFP (diisopropylfluorophosphonate, a widely used simulant for the nerve gas agents), irreversibly inactivate acetylcholinesterase by phosphonylating the serine residue at the active site (Figure 10). The irreversible inhibition of acetylcholinesterase thus results in the overaccumulation of neurotransmitter acetylcholine at the synaptic cleft, leading to the overstimulation of the cholinergic receptors, and ultimately seizures, respiratory arrest, and death.[31]

The organophosphates also inhibit other enzymes, such as chymotrypsin, which have serine residues at their active sites. Even though fluorine is one of the poorest leaving groups in aliphatic nucleophilic substitution reactions, it is an excellent leaving group in nucleophilic acyl substitution reactions because the departure of the fluoride in the second step is facilitated by intramolecular neighboring group (O^-) participation. Further dealkylation (in the case of sarin, soman, and cyclosarin, which can form a relatively stable secondary carbocation-like intermediate), aided by the general acid-general base catalysis at the active site, leads to the formation of the phosphonate adduct to the Ser200 (Figures 11 and 12). This covalent adduct is quite stable under biological conditions, so that the enzyme is irreversibly inactivated. Exposure to these nerve

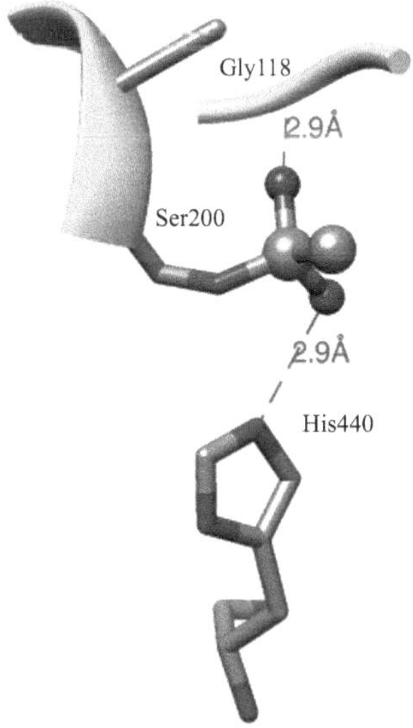

FIGURE 11 Expanded view of the X-ray structure of the covalent adduct of acetylcholinesterase with methylphosphonate, derived from the nerve gases soman, sarin, or cyclosarin; the hydroxyl group of the Ser200 undergoes S_N2 type of reaction with these compounds displacing the fluoride anion, and subsequent general acid-general base-catalyzed dealkylation leads to the formation of the phosphonate adduct. The phosphonate moiety (ball and stick) has hydrogen bonding interactions with His440, Gly118, and other neighboring amino acid residues (UCSF Chimera software; PDB 1CFJ).

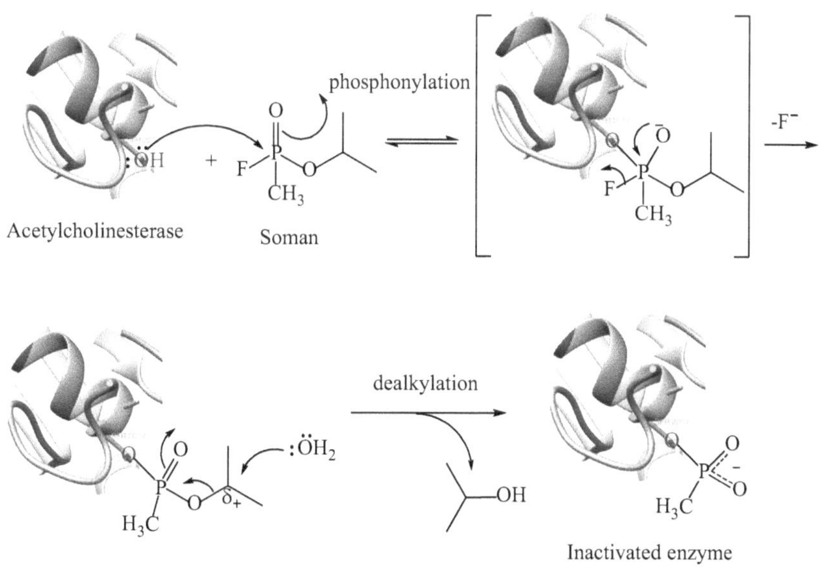

FIGURE 12 Mechanism-based suicide inactivation of cholinesterase by the nerve gas, soman; soman has relatively higher affinity for the active site of the acetylcholinesterase as compared to that of acetylcholine, and at the active site it undergoes rapid nucleophilic substitution reaction by the Ser200 hydroxyl group to form a covalently bound adduct. Further dealkylation of this adduct aided by the general acid-general base catalysis results in the irreversibly inactivated (aged) enzyme.

gases has fatal consequences and, even upon exposure to minute quantities, immediate symptoms such as memory impairment, headache, nausea, respiratory depression, severe muscle contractions, and paralysis are manifested.

Using a [^{18}F]F$^-$-labeled organophosphate tracer, *p*-nitrophenyl(2-fluoroethyl) methylphosphonate (**4**; a hybrid of nerve gas VX (*O*-Ethyl S-(2-diisopropylaminoethyl) methylphosphonothioate and insecticide paraoxon), and PET imaging, the organophosphate-adducted acetylcholinesterase was probed in vivo within rats by evaluation of its brain penetration and cerebral distribution (Figure 13). This PET tracer effectively localizes to cerebral regions that express acetylcholinesterase target. The formation of the enzyme–organophosphate adduct was inferred through pharmacokinetic properties of the tracer.[32]

The ^{18}F-labeled *p*-nitrophenyl(2-fluoroethyl)methylphosphonate **4** was synthesized through an S_N2 reaction of the methylphosphonic acid *p*-nitrophenyl ester **3** with the ^{18}F-labeled 2-[^{18}F]fluoroethyl tosylate (**2**).[32] The latter radiolabeled fluoroethyltosylate was synthesized by the S_N2 reaction of the 1,2-ethaneditosylate **1** with [^{18}F]F$^-$ under phase transfer conditions using Kryptofix–2.2.2 (Figure 14).

Anticholinergic drugs, such as atropine, scopolamine, and pralidoxime chloride are generally used to reactivate the acetylcholinesterase in the affected cases (Figure 15).[31] These antidotes, at large concentrations, hydrolyze the serine phosphate esters to release the reactivated cholinesterase enzyme. Because of their bioisosteric resemblance with acetylcholine, these compounds have strong binding

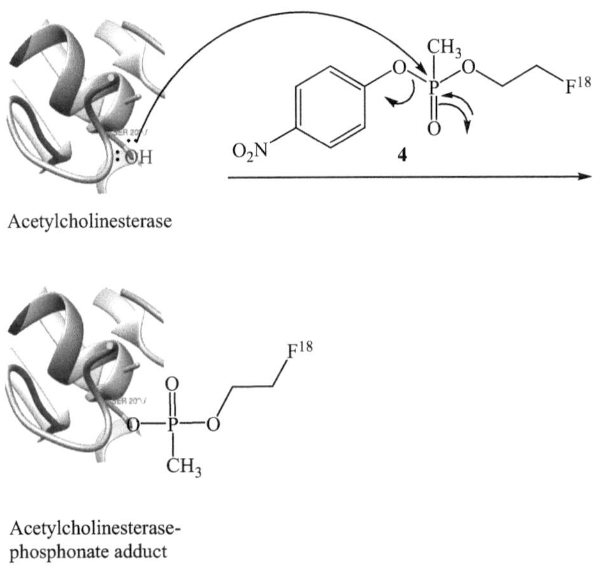

FIGURE 13 Formation of the acetylcholinesterase-^{18}F-labeled phosphonate adduct in vivo within the cerebral region of the rats.

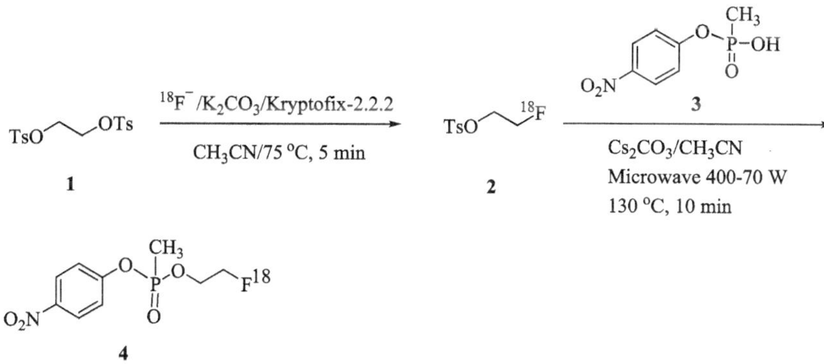

FIGURE 14 Radiosynthesis of the ^{18}F-labeled *p*-nitrophenyl β-fluoroethoxymethylphosphonate.

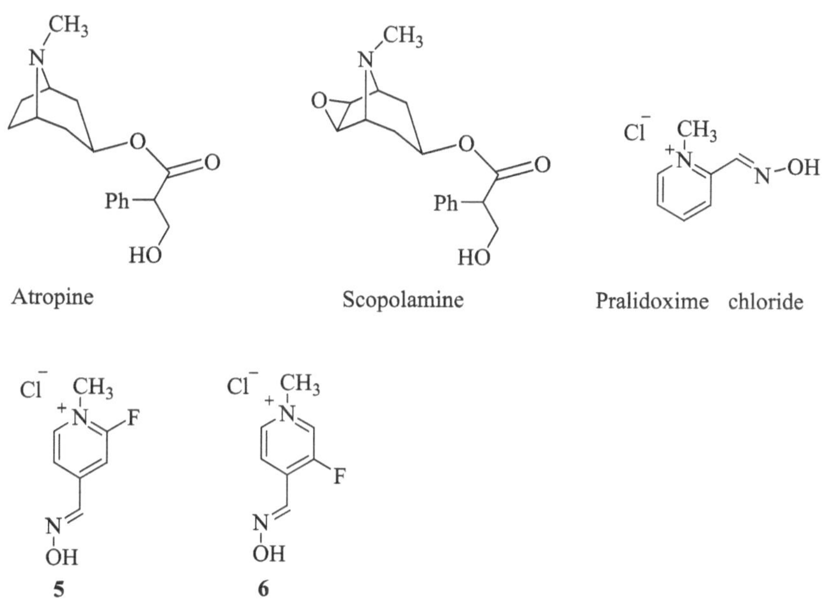

FIGURE 15 Structures of some of the mechanism-based acetylcholinesterase reactivators (atropine, scopolamine, pralidoxime chloride, and the monofluoro derivatives of 4-formylpyridinium oximes).

interactions at the cholinesterase active site, and thereby reactivate the cholinesterase through displacement of the enzyme-bound phosphonate substrates.

Pralidoxime chloride and its isomeric compound *N*-methyl-4-pyridinium oxime react with the enzyme-bound phosphonate ester by the acyl nucleophilic substitution mechanism to regenerate the active acetylcholinesterase enzyme (Figure 16). The oxime hydroxyl moiety of pralidoxime and its derivatives, and the hydroxyl moieties of atropine and scopolamine, compete more effectively

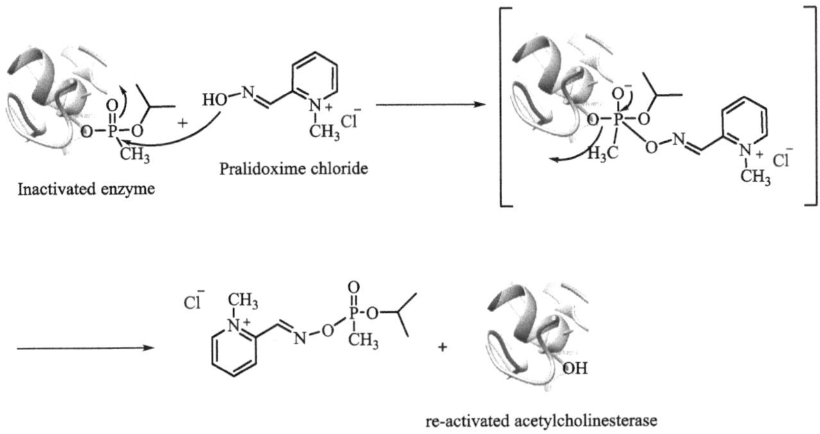

FIGURE 16 Mechanism for the reactivation of the cholinesterase by Pralidoxime; Pralidoxime and other anticholinergic drugs undergo acyl nucleophilic substitution with the enzyme-bound phosphonate esters to release the reactivated acetylcholinesterase.

in the acyl nucleophilic substitution reaction of the enzyme-bound phosphonate ester as compared to the serine moiety, and so these compounds act as mechanism-based reactivators of the acetylcholinesterase.

In order to improve the blood–brain barrier (BBB) capability of these compounds, fluorination of the pyridinium ring was explored.[31] The fluoro substituents impart increased lipophilicity to the molecules and thereby increase the BBB. Both the 2-fluoro and 3-fluoro analogs of the 4-formyl-*N*-methylpyridinium oximes (**5** and **6**) have relatively higher BBB permeability, and have better inhibitory effect on the acetylcholinesterase than their nonfluorinated analog, pralidoxime chloride. The 2-fluoroanalog has relatively more lipophilicity and BBB characteristic than the 3-fluoroanalog, suggesting that the fluorine adjacent to the pyridinium moiety helps to decrease the polarity of the pyridinium moiety, and thereby enhances its lipophilicity.

5.2 Suicide Inactivation of Ornithine Decarboxylase

Ornithine decarboxylase inhibition by α-difluoromethylornithine (DFMO) is a classic example of suicide inhibition, and mechanistic understanding gained from this reaction is the basis of designing many other suicide inhibitors as potential drug candidates. Ornithine decarboxylation, catalyzed by ornithine decarboxylase, produces putrescine, a precursor for the polyamines, which are essential for cell division (*vide infra*); targeting this enzyme, for example, proved to be beneficial in the treatment of the African sleeping sickness.

The African sleeping sickness is a disease found primarily in tropical and subtropical regions, and is devastating to millions of the affected people. It is caused by a protozoan pathogen *Trypanosoma brucei*, or by a related species, *Trypanosoma gambiense*.[33] The ornithine decarboxylase in these strains of pathogens is a long-lasting protein and is not readily regenerated in the parasites, whereas in the human host cells it is rapidly recycled. Taking advantage of this vulnerability of the parasite, DFMO was developed as a leading pharmaceutical, which serves as an excellent mechanism-based inhibitor.

Pyridoxal phosphate (PLP) is a cofactor for the ornithine decarboxylase. The initial step of the decarboxylation of ornithine involves the formation of an imine intermediate (Schiff base **8**) through reaction of its α-amino group with the carbonyl group of the pyridoxal 5′-phosphate (**7**). This Schiff base is rapidly hydrolyzed to regenerate the PLP and the diamine putrescine. Putrescine is the required diamine for the biosynthesis of the polyamines, spermine ($NH_2(CH_2)_3$–NH–$(CH_2)_4$–NH–$(CH_2)_3NH_2$) and spermidine ($NH_2(CH_2)_4$–NH–$(CH_2)_3NH_2$), which are essential components for deoxyribonucleic acid (DNA) packaging during the cell cycle (Figure 17). Whereas in humans the ornithine decarboxylase has a rapid turnover, in the *Trypanosoma* parasite it is a long-lived protein and is not regenerated once it is inactivated.

However, in the presence of the suicide inactivator DFMO, the Schiff base **10** preferentially undergoes elimination of the fluoride ion to form a terminal fluoroolefin derivative **11** as a secondary metabolite. Due to the extended conjugation

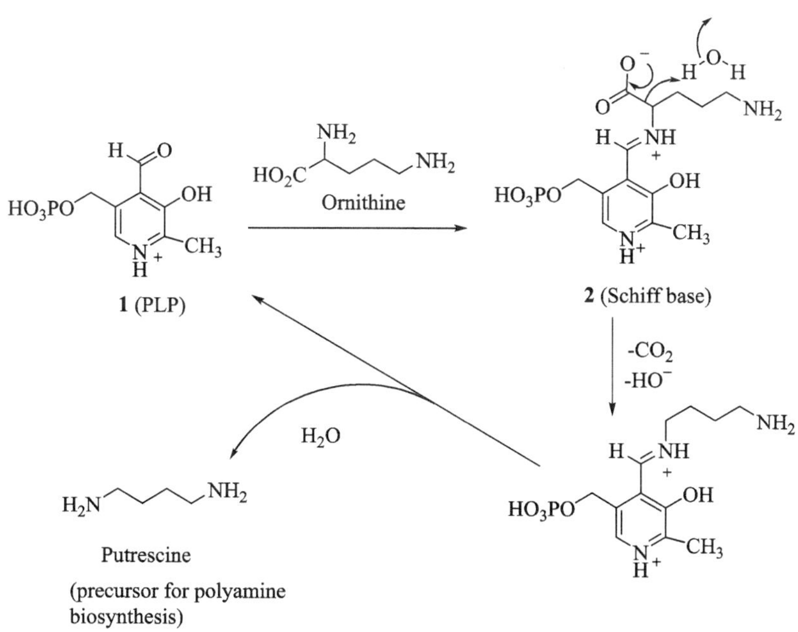

FIGURE 17 Ornithine-decarboxylase-catalyzed biosynthesis of putrescine, an essential starting compound for the biosynthesis of polyamines; the Schiff base formed from the PLP and ornithine is decarboxylated and hydrolyzed at the enzyme active site to give the putrescine; large amounts of polyamines are required for DNA packaging during cell cycle, and therefore the enzyme inactivation by DFMO results in permanent inhibition of the cell cycle in the *Trypanosoma* sp.

with the pyridinium ion, the terminal fluoroolefin moiety of the intermediate **11** is highly reactive with the nucleophilic thiol residues at the enzyme active site (e.g., cysteine residues), and forms Michael adducts. Michael addition of the active site residues, for example, the thiol group of Cys360, to compound **11**, followed by rapid elimination of the fluoride from the Michael adduct **12** (through reverse polarization of the π-electrons) gives the enzyme inhibitor adduct **13**. The enzyme is thereby irreversibly inactivated (Figure 18). Although this inactivation of ornithine decarboxylase occurs both in human cells and in *Trypanosoma* sp., human cells are capable of rapid regeneration of the enzyme, while those of *Trypanosoma* sp. are unable to express the protein; this results in permanent cell cycle arrest in *Trypanosoma* sp. and thereby the disease is suppressed.

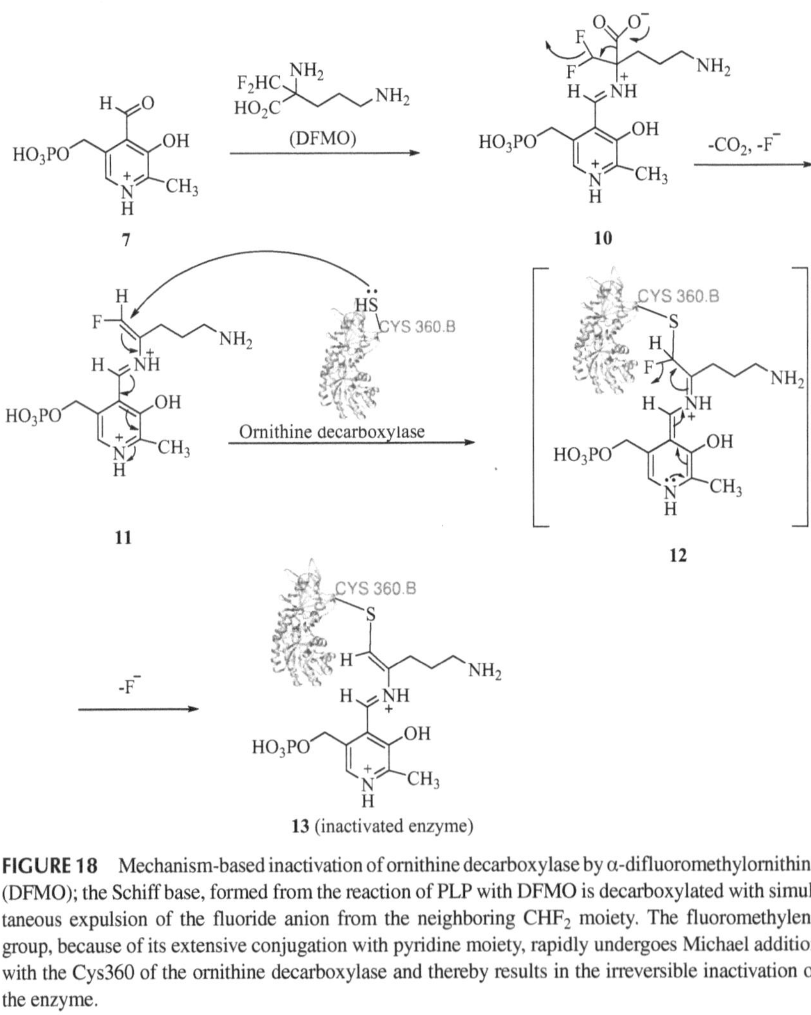

FIGURE 18 Mechanism-based inactivation of ornithine decarboxylase by α-difluoromethylornithine (DFMO); the Schiff base, formed from the reaction of PLP with DFMO is decarboxylated with simultaneous expulsion of the fluoride anion from the neighboring CHF_2 moiety. The fluoromethylene group, because of its extensive conjugation with pyridine moiety, rapidly undergoes Michael addition with the Cys360 of the ornithine decarboxylase and thereby results in the irreversible inactivation of the enzyme.

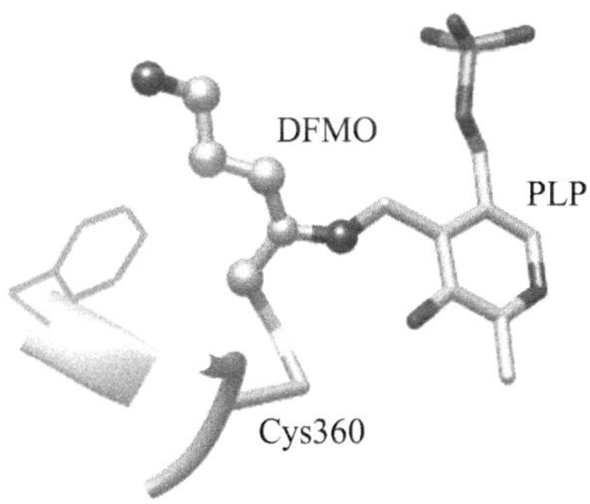

FIGURE 19 The X-ray structure for the Schiff base **13**, derived from the reaction of pyridoxal 5′-phosphate (PLP) and DFMO, covalently bound to the ornithine decarboxylase (*T. brucei*) through the Cys360 sulfhydryl group (UCSF Chimera software; PDB 2TOD).

The structure of the enzyme-bound Schiff base was confirmed by a single-crystal X-ray structure determination, which shows that the fluoromethylene carbon of the Schiff base is covalently bound to the sulfhydryl group of the Cys360 (Figure 19).[34]

Plasmodium parasite, which is the cause of malaria, also heavily depends on the polyamines for its cell cycle. Encouraged by the suicide inactivation of ornithine decarboxylase by DFMO that resulted in a lasting cure for African sleeping sickness, similar studies have been directed to control the polyamine metabolism in the *Plasmodium* sp., and this approach led to a moderate success in the eradication of the *Plasmodium* parasite.[35] Antineoplastics of similar structural characteristics, based on their polyamine depletion abilities, are currently being developed at various levels of success.[36]

5.3 Irreversible Inactivation of Thymidylate Synthase by 5-Fluoro-2′-Deoxyuridine 5′-Monophosphate (FdUMP)

The DNA nucleotide 2′-deoxythymidine 5′-monophosphate (dTMP) is biosynthesized from 2′-deoxyuridine 5′-monophosphate, mediated by thymidylate synthase (TS). In this process the Cys146 thiol group at the enzyme active site activates the uracil moiety toward nucleophilic attack on the N_5,N_{10}-methylene carbon of the N_5,N_{10}-methylenetetrahydrofolate (**14**) to give adduct **15**. Subsequent intramolecular 1,3-hydride shift (with the release of the 7,8-dihydrofolate **17**) followed by E1cb-type expulsion of the Cys146 thiol group results in the formation of dTMP, and the enzyme active site is regenerated for further catalytic conversions (Figure 20).

Inhibition of the deoxythymidine biosynthesis has significance in the treatment of cancer and many other protozoan-derived diseases, such as African sleeping sickness; the protozoa *T. brucei* lacks alternate nucleotide biosynthetic pathways so that it is deactivated by the pharmacological inhibition of the thymidylate synthase. Through a mechanism-based inhibition (suicide inhibition) strategy, 5-Fluoro-2′-deoxyuridine 5′-monophosphate (FdUMP) acts as an irreversible inhibitor of the thymidylate synthase.[37] The initial steps of reaction of the FdUMP with the enzyme active site are similar to

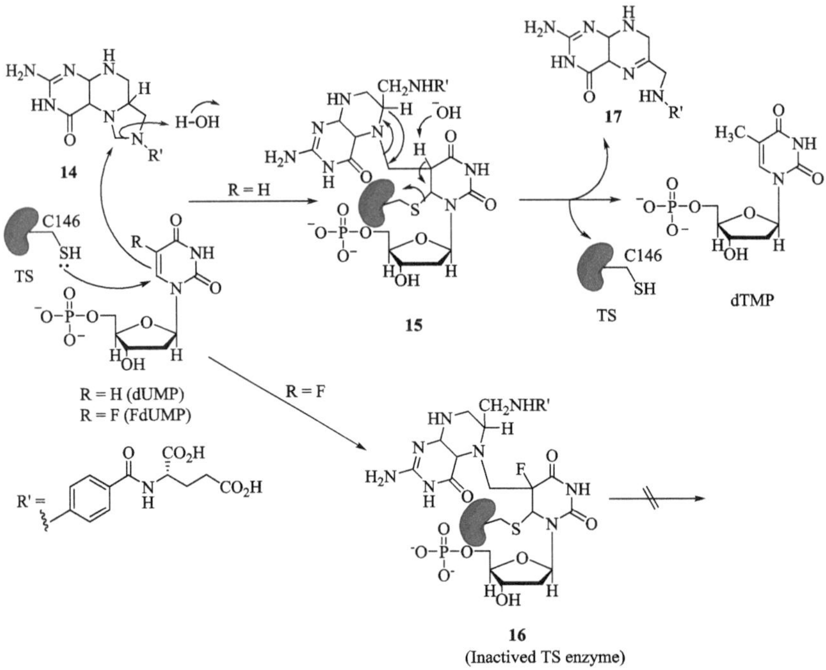

FIGURE 20 Mechanism-based irreversible inactivation of the thymidylate synthase (TS) by 5-Fluoro-2′-deoxyuridine 5′-monophosphate (FdUMP); biosynthesis of 2′-deoxythymidylic acid (dTMP) from 2′-deoxyuridine 5′-monophosphate (dUMP) involves thymidylate-synthase-catalyzed reaction of dUMP with N_5,N_{10}-methylenetetrahydrofolate (**14**) to give the intermediate enzyme-adduct **15**; the latter adduct is converted to dTMP and the enzyme is regenerated through a general base-catalyzed E1cB mechanism; However, this process is not possible in case of the adduct for FdUMP, resulting in the irreversible inactivation of the enzyme.

those for the deoxyuridine 5′-monophosphate.[37] The cysteine-thiol-assisted nucleophilic displacement of the N_5,N_{10}-methylenetrahydrofolate by FdUMP proceeds as expected to give the enzyme-adduct **16** (Figure 20). However, due to the strong electron-withdrawing inductive effect of the fluorine, the 1,3-hydride shift, a key step for the incorporation of the 5-methyl group, is disfavored, and, at the same time, enzyme regeneration through the E1cB-type elimination of the cysteine146-thiol group is not possible due to the lack of the C_5–H in the latter enzyme-adduct. Thus, the enzyme is irreversibly inactivated by the FdUMP through this covalent bond formation to the cysteine residue at the active site.

5-Fluorouracil, the biosynthetic precursor of FdUMP, is one of the first anticancer drugs and the above-mentioned insights into the enzyme mechanisms laid the foundation for the design and development of many novel drugs with improved efficacy and reduced cytotoxicity as compared to that of fluorouracil.[38–40]

5.4 Block Effect on Enzyme Inhibition: Aconitase Inhibition in the Citric Acid Cycle

The citric acid cycle (also called as Krebs cycle or tricarboxylic acid cycle) takes place in the mitochondria and is an integral part for the generation of adenosine triphosphate (ATP). In the citric acid cycle, 36 mol of ATP are formed from a single glucose molecule, and reduced nicotinamide adenosine diphosphate and other organics are also formed, which serve as intermediates for the biosynthesis of amino acids (e.g., glutamate is synthesized from α-ketoglutaric acid, an intermediate in the citric acid cycle).

Fluoroacetate is the starting compound for the biosynthesis of the fluorinated compounds in certain soil bacteria. However, fluoroacetyl-CoA (**18**), derived from the fluoroacetate, also competes with acetyl-CoA in its citrate-synthase-catalyzed reaction with oxaloacetate to form (2R, 3R)-2-fluorocitrate (**20**), which undergoes aconitase-catalyzed dehydration to give the 2(R)-2-fluoro-*cis*-aconitic acid (**21**) (in analogy to the normal citric acid cycle, in which *cis*-aconitic acid is formed from acetyl-CoA). Whereas the hydration of *cis*-aconitic acid gives isocitric acid, the next intermediate in the normal citric acid cycle, **21** undergoes S_N2' type of hydroxylation–defluorination to give (R)-hydroxy-*trans*-aconitic acid (**22**), which no longer can participate in the citric acid cycle and, moreover, it acts as an irreversible inhibitor of the aconitase enzyme, resulting in shutting down of the citric acid cycle, and thereby exerting cytotoxicity in humans and other mammals.[41] In order to establish the mechanism of the aconitase inhibition by the secondary metabolite **22**, a single-crystal X-ray structure determination was undertaken. The single-crystal structure of this enzyme–inhibitor complex revealed the tight binding of **22** to the aconitase at the active site.[41] His101, His147, and His167 have strong hydrogen bonding interactions with the substrate carboxylate moieties, and the resulting protonated His101 also forms hydrogen bond to the inhibitor hydroxyl group, which, in turn, forms a hydrogen bond to Asp165 (Figure 21). Thus the enzyme inhibitory action of the toxic metabolite **22** is not through its covalent binding to the amino acid residues at the enzyme active site, but rather through its tight fitting in, and thereby blocking, the active site.[41]

Importantly, the 2(S)-2-fluoro-*cis*-aconitic acid, (enantiomer of **21**) would not be able to undergo the similar S_N2' type of hydroxylation–defluorination and therefore would not be able to inhibit the aconitase enzyme. The citrate-synthase-catalyzed reaction of oxaloacetate with Z-enolate-CoA (*Z*-**19**), followed by aconitase catalyzed dehydration would give the 2(S)-2-fluoro-*cis*-aconitic acid, whereas *E*-enolate-CoA (*E*-**19**) gives predominantly **21**. The citrate synthase converts the fluoroacetyl-CoA (**18**), stereoselectively (98:2 diastereoselectivity) to *E*-**19**, resulting in the cytotoxicity associated with the fluoroacetate (Figure 22). Using ab initio quantum mechanics/molecular mechanics modeling Mulholland and coworkers[42] have rationalized the high stereoselectivity of the citrate synthase (i.e., the stereoselective formation of the *E*-**19**): the computed enzyme–substrate complex for the *E*-**19** is about 2 kcal/mol more stable than that for *Z*-**19**, which is also in accord with the estimated $\Delta(\Delta G^{\ddagger})$ value based on

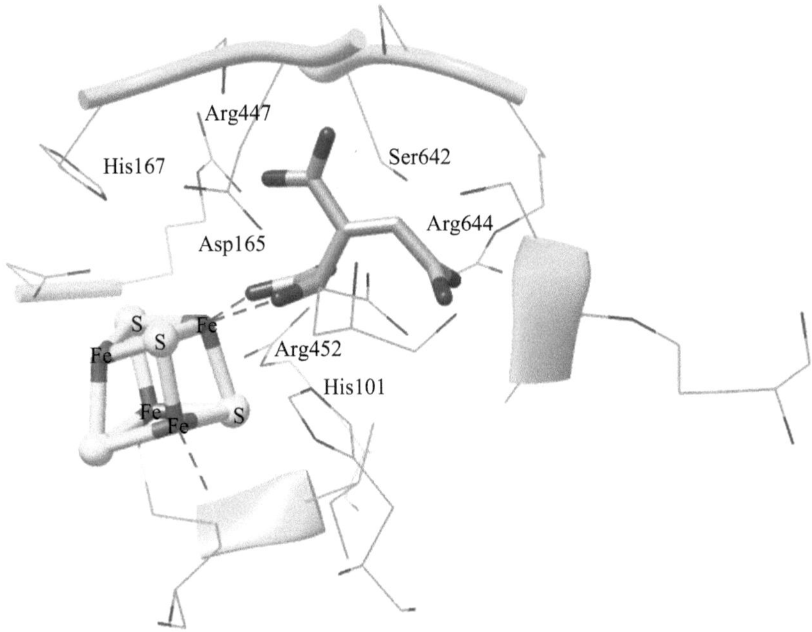

FIGURE 21 Single-crystal X-ray structure of 4-hydroxy-(*E*)-aconitate (**22**; shown as stick model with oxygens in red) bound to the active site of aconitase at 2.05 Å resolution (generated using UCSF Chimera software; pdb:1FGH); The metabolite **22** binds to the active site very tightly but not through covalent binding, and is stabilized by extensive hydrogen bonding interactions with the His101, His147, and His167 residues; thus, it inactivates the aconitase and shuts down the citric acid cycle.[41]

the experimentally observed ~98:2 diastereomeric ratio of the 2(R)- and 2(S)-2-fluorocitrates.

6. REVERSIBLE INHIBITORS

The reversible inhibitors, in most cases, bind to the enzyme active site with relatively greater affinity as compared to that of their native substrate analogs. The enzyme inactivation is not irreversible in these cases, as the bound inhibitor is not covalently bound at the active site, and is reversibly released from the active site. Most of the reversible inhibitors are developed based on the concept of transition state analog inhibition, that is, the enzymes have greater affinity toward the substrates that resemble the transition states, as compared to the native substrate molecules in the ground state. The reversible inhibition of serine proteases is an example of this type of inhibitors.

6.1 Reversible Transition State Analog Inhibitors

Peptidyl trifluoromethyl ketones are examples of reversible, transition state analog inhibitors of serine proteases. The strong electron-withdrawing inductive

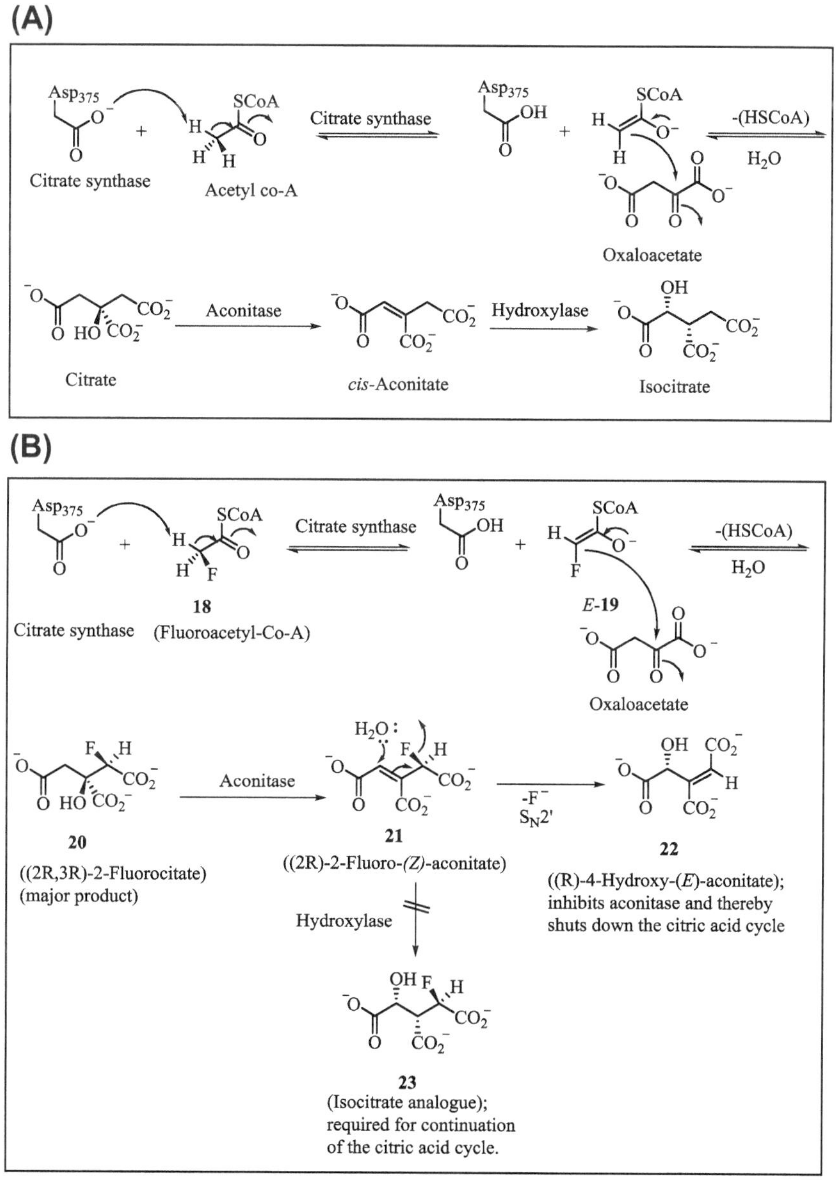

FIGURE 22 (A) Conversion of acetyl-CoA to citrate in the citric acid cycle and (B) conversion of fluoroacetyl-CoA (**18**) to (R)-4-hydroxy-(*E*)-aconitate (**22**), which inhibits aconitase and thereby shuts down the citric acid cycle.

effect of the trifluoromethyl group renders the carbonyl carbon in trifluoromethyl ketones highly electrophilic, so that the trifluoromethyl ketones in aqueous solutions exist virtually completely in the hydrated form. In the hydrate form the trifluoromethyl ketones resemble the transition state structures for the

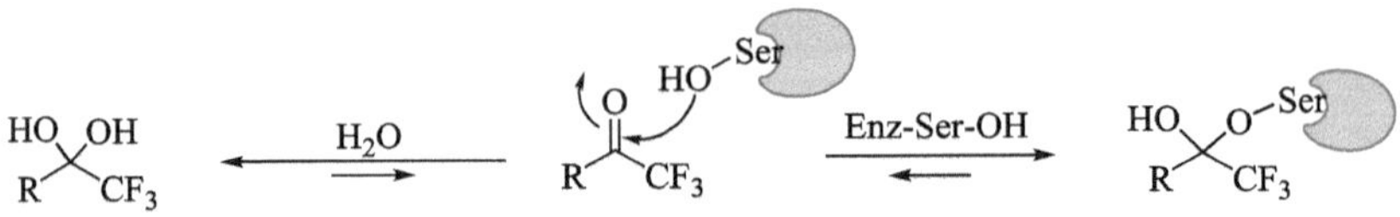

FIGURE 23 Enzyme inactivation by the trifluoromethyl ketones; the trifluoromethyl ketones exist mostly as their hydrates, and appropriate peptide trifluoromethyl ketone hydrates are transition state analog inhibitors of serine proteases. Due to the high electrophilicity of the trifluoromethyl ketones, they rapidly react with the serine hydroxyl groups at the active site and thereby act as reversible inhibitors.

peptide hydrolysis, and therefore have relatively higher binding affinity to the protease active sites. At the active sites, the hydrates of the peptide trifluoromethyl ketones reversibly form the corresponding hemiacetals by reacting with the serine hydroxyl group. Because the serine residues are reversibly blocked, through the formation of the trifluoromethyl ketone hemiacetals, the enzyme activity of the serine proteases is reduced (Figure 23). The corresponding difluoromethyl ketones have relatively lower concentration of the hydrate form, and accordingly show relatively higher K_i values (Figure 25).

^{19}F NMR studies indicate that the AcLeuPheCF$_3$ dipeptide trifluoromethyl ketone hydrate reversibly reacts with the serine hydroxyl group to form the corresponding hemiacetal; the δ^{19}F for the serine-bound hemiacetal is downfield shifted with respect to that of the hydrate form by 2 ppm, and is relatively broad as compared to that of the hydrate form, as expected for the reversible equilibration (Figure 24).[43]

K_i values for the trifluoromethyl ketone AcLeuPheCF$_3$ (**24**) and its difluoromethyl (**25**), monofluoromethyl (**26**), and methyl ketone (**27**) analogs for chymotrypsin were measured as 1.2, 25, 200, and 500 μM, respectively (Figure 25).[43] This observation is consistent with the correspondingly lower concentration of the hydrate forms across this series (i.e., **24** is the most hydrated and **27** is the least hydrated in this series). Related oligopeptide trifluoromethyl ketones are also effective inhibitors of the human cytomegalovirus protease[44] and human leukocyte elastase,[45] which are serine-based proteases.

6.2 Reversible Inhibitors for Human and Plasmodium Arginases

DFMO and its structural analogs such as 2-amino-6-borono-2-(difluoromethyl) hexanoic acid (FABH) and 2-amino-6-borono-2-methylhexanoic acid (MABH) form reversible complexes with human arginase I (HAI) and *Plasmodium falciparum* arginase (PFA) enzymes (Figure 26).[46] At the active site of HAI, the boronic acid side chains in the MABH an FABH undergo nucleophilic attack by the metal-bridged hydroxide anion to form the tetrahedral boronate anion that mimics the tetrahedral transition state for the catalytic transformation of arginine into L-ornithine and urea.

The X-ray structures of the plasmodium arginase complexes PFA-MABH and PFA-FABH do not show the presence of the tetrahedral boronate anions, unlike

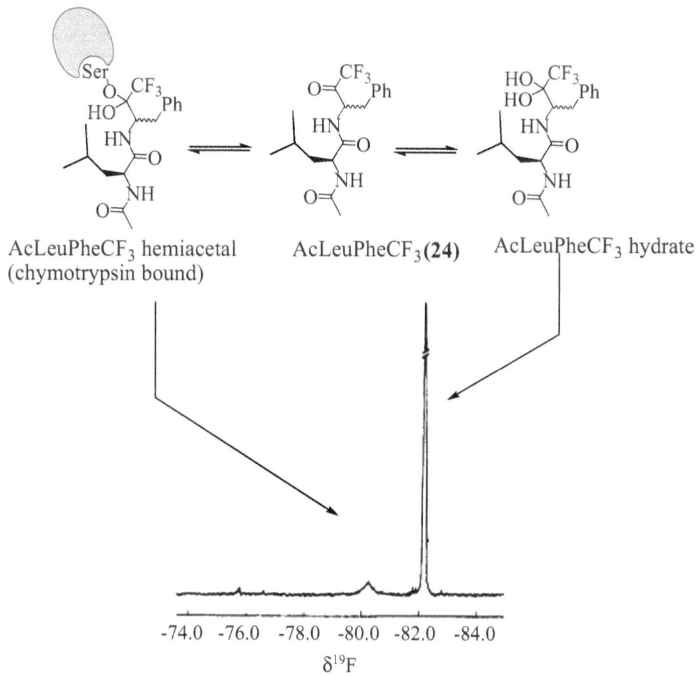

FIGURE 24 ^{19}F NMR spectrum of chymotrypsin (1 mM) reversibly bound to the $AcLeuPheCF_3$ inhibitor (4 mM); the downfield and broadened shift for the enzyme-bound inhibitor reflects reversible equilibration of the enzyme-bound hemiacetal with the ketone hydrate; the spectrum was referenced relative to the $CFCl_3$ ($\delta^{19}F(CFCl_3)=0$). *Adapted with permission from Ref. 43. Copyright 1986, American Chemical Society.*

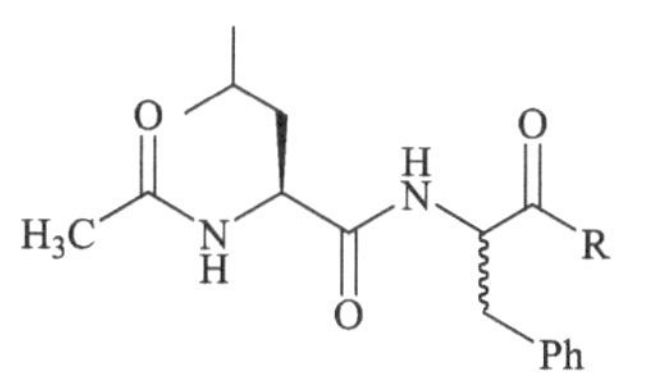

24; R = CF_3; K_i = 1.2 μM
25; R = CF_2H; K_i = 25 μM
26; R = CFH_2; K_i = 200 μM
27; R = CH_3; K_i = 500 μM

FIGURE 25 Chymotrypsin inhibitory constants (K_i) for the AcLeuPhe ketones; the higher the electrophilicity of the carbonyl group, the lower are the K_i values. *Data taken from Ref. 43.*

their HAI complexes. Thus, these inhibitors do not bind as transition state analog inhibitors to PFA. The different mechanisms of the inhibitor binding to the HAI and PFA suggest the possibility of differential inhibition of the human and parasite arginase enzymes. Selective inhibition of arginase, especially the differential inhibition of the parasite and the human arginase, would lead to more potent drug candidates that may be useful for the treatment of diseases such as asthma, cerebral malaria, and atherosclerosis caused by aberrant activity of arginase.

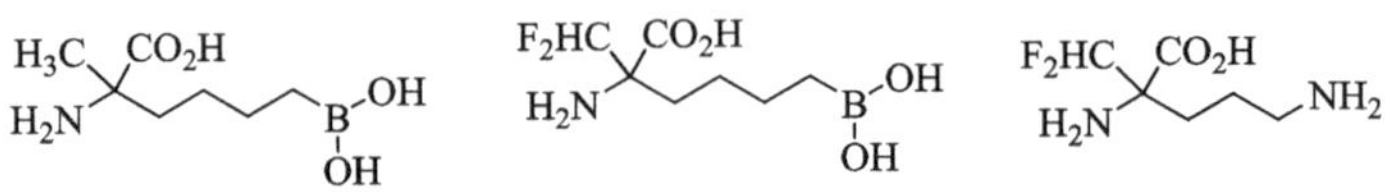

FIGURE 26 Structures of MABH, FABH, and DFMO, the reversible inhibitors of HAI.

6.3 Serotonin Reuptake Inhibitors

Depression and anxiety disorders, which may be related to the imbalance of the neurotransmitter concentrations in the neuronal cells and synapses, have huge societal impact. Cell to cell signaling is achieved by neurotransmitters, which are released from the neuronal axons into the synapse. In order to prevent continued stimulation of the postsynaptic cells, these neurotransmitters have to be retransported back into the neuronal cells, mediated by a class of monoamine transporter proteins. The imbalance of the concentrations of the neurotransmitters in the neuronal cells and the synapse results in various pathologies associated with neurotransmission. In the case of depression-related diseases, serotonin transporters are overly active, depleting the serotonin (5-hydroxytryptamine (Figure 27)) rapidly from the synapses. Selective blocking of the serotonin reuptake proteins by inhibitors is therefore a desired target. By selectively blocking the serotonin reuptake proteins these drugs are expected to elevate the concentration of serotonin in the synapse spatiotemporally (i.e., in time and space) and alleviate the symptoms of depression and anxiety disorders. These selective, reversible serotonin reuptake inhibitors exhibit their inhibitory action by temporarily blocking the active site, and thus they are reversible inhibitors by virtue of being bioisosteric with serotonin.

Only during the past two decades, progress has been made in identifying the selective serotonin reuptake inhibitors (SSRIs). Prozac (fluoxetine; used as racemic form), a SSRI, is currently a blockbuster drug, and is used for the treatment of depression. It is widely used all over the world, such that trace amounts of this drug are often observed in the municipal effluents.[47] Interestingly, both enantiomers of Prozac are active inhibitors of the serotonin transporter. Escitalopram and paroxetine are examples of other fluorine derived antidepressants, which act as SSRIs (Figure 27).

Specificity of these serotonin transporters to SSRI depends mostly on the interaction of the fluorine with the halogen-binding site of the protein, which is composed of leucine, glycine, tyrosine, and phenylalanine residues. These amino acid residues are involved in the binding of the antidepressant SSRI drugs as well as in the selective binding to serotonin. Thus, fluoxetine and escitalopram are competitive, reversible inhibitors of serotonin reuptake proteins. Figure 28 shows the active site binding of fluoxetine to a leucine transporter (LeuT), of the bacterium *Aquifex aeolicus*, which has similar primary sequence of amino acids and active site residues as those of the human serotonin reuptake proteins.[48] In the hydrophobic binding site of fluoxetine–LeuT complex, the trifluoromethyl fluorines have C–H-dipolar interactions

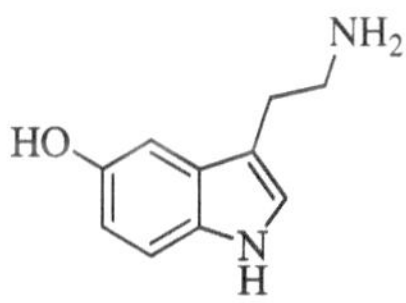

Serotonin, a neurotransmitter

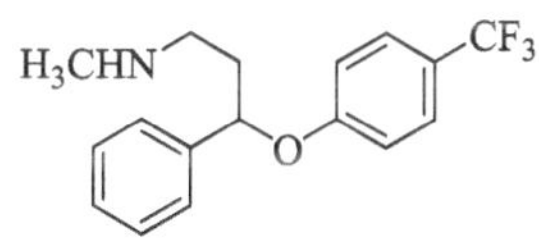

Fluoxetine (Prozac)

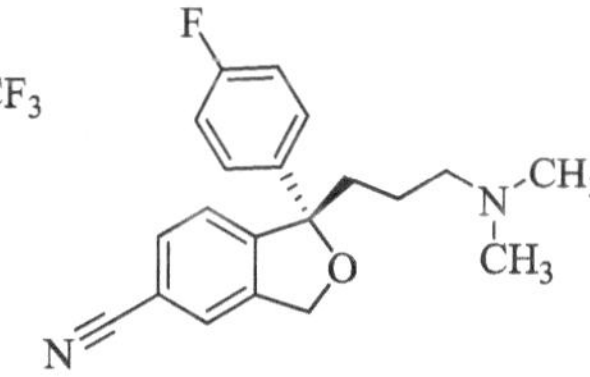

S-Escitalopram (S-Citalopram)

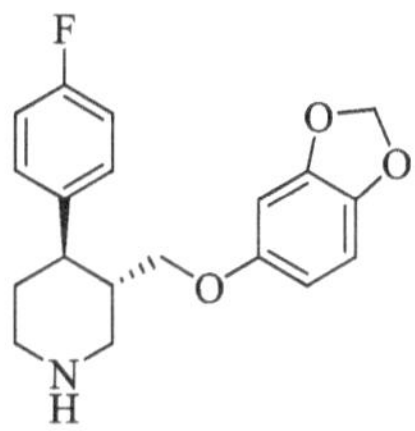

Paroxetine

FIGURE 27 Structures of serotonin and some fluorinated serotonin reuptake inhibitors.

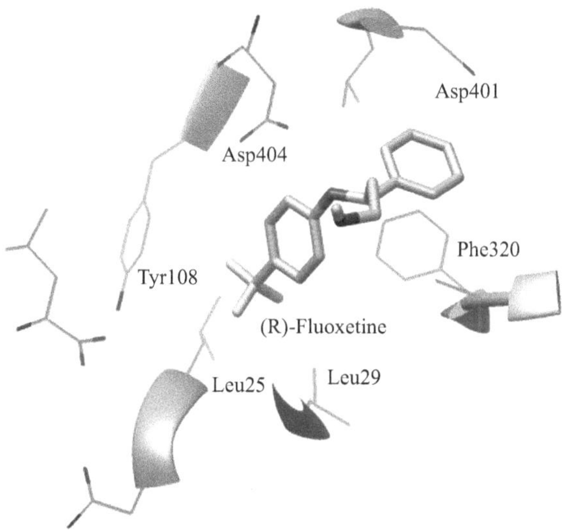

FIGURE 28 (R)-fluoxetine bound to LeuT at the halogen-binding pocket (HBP); the structure was created using UCSF Chimera software; PDB: 3GWV.

with Leu29, Tyr108, Leu25, and Gly26 residues, as shown by the C–H---F distances of 3.5 Å.

Fluoxetine is metabolized by *N*-demethylation to produce norfluoxetine by cytochrome P450 isoenzymes in the human liver; the metabolite norfluoxetine is also a pharmacologically active SSRI inhibitor. Racemic, as well as the enantiomers of fluoxetine are reversible inhibitors of cytochrome P450 enzymes CYP2D6, CYP2C19, CYP3A4, and CYP2C9.[34]

7. SUMMARY AND OUTLOOK

Fluorinase enzyme is the only known enzyme that catalyzes the fluorination reactions, and it is involved in the S_N2 type of fluorination reaction of SAM to give 5′-fluoro-5′-deoxyadenosine, an intermediate for the biosynthesis of fluoroacetate and fluorothreonine. Fluorinase plays a key role in the substrate organization at the transition state, desolvation of the bound water molecules, and activation of the fluoride anion for the nucleophilic substitution reaction.

Organofluorine compounds have important applications in the design of enzyme inhibitors. Mechanism-based irreversible inactivation of acetylcholinesterase is achieved by fluorophosphonate-based compounds, including pesticides and nerve gas agents, whereas in certain neurological disorders such as AD, reversible inhibition of acetylcholinesterase is required. The reversible binding of the trifluoromethyl ketone hydrates to serine proteases can be demonstrated by ^{19}F NMR experiments. Serotonin reuptake inhibitors, such as fluoxetine, S-escitalopram, and paroxetine (used to treat depression and anxiety disorders), exert their enzyme inhibitory effect through reversible binding at the active sites. Mechanism-based inactivation of enzymes is useful in the treatment of a variety of diseases. For example, suicide inactivation of ornithine decarboxylase by DFMO prevents polyamine synthesis in *Trypanosoma* pathogens, and therefore is useful for treating African sleeping sickness. Similarly, irreversible inactivation of thymidylate synthase by 5-Fluoro-2′-deoxyuridine 5′-monophosphate (FdUMP) results in inhibition of the deoxythymidine biosynthesis; the FdUMP is used in the treatment of cancers and African sleeping sickness. Fluorinated compounds, as enzyme inhibitors, therefore, have significant potential as therapeutics.

REFERENCES

1. O'Hagan, D.; Perry, R.; Lock, J. M.; Meyer, J. J. M.; Dasaradhi, L.; Hamilton, J. T. G.; Harper, D. B. High Levels of Monofluoroacetate in *Dichapetalum braunii*. *Phytochemistry* **1993,** *33,* 1043–1045.
2. O'Hagan, D.; Harper, D. B. Fluorine-Containing Natural Products. *J. Fluorine Chem.* **1999,** *100,* 127–133.
3. Dong, C.; Huang, F.; Deng, H.; Schaffrath, C.; Spencer, J. B.; O'Hagan, D.; Naismith, J. H. Crystal Structure and Mechanism of a Bacterial Fluorinating Enzyme. *Nature* **2004,** *427,* 561–565.
4. Sanada, M.; Miyano, T.; Iwadare, S.; Williamson, J. M.; Arison, B. H.; Smith, J. L.; Douglas, A. W.; Liesch, J. M.; Inamine, E. Biosynthesis of Fluorothreonine and Fluoroacetic Acid by the Thienamycin Producer, *Streptomyces cattleya*. *J. Antibiot.* **1986,** *39,* 259–265.
5. Peters, R. A.; Shorthouse, M. Identification of a Volatile Constituent Formed by Homogenates of *Acacia georginae* Exposed to Fluoride. *Nature* **1971,** *231,* 123–124.
6. Lauble, H.; Kennedy, M.C.; Emptage, M.H.; Beinert, H.; Stout, C.D. The Reaction of Fluorocitrate with Aconitase and the Crystal Structure of the Enzyme-Inhibitor Complex. *Proc. Natl. Acad. Sci. U.S.A.* **1996,** *93,* 13699–13703.

7. Peters, R. A.; Hall, R. J. Fluorine Compounds in Nature; the Distribution of Carbon–Fluorine Compounds in Some Species of *Dichapetalum*. *Nature* **1960,** *187,* 573–575.
8. Morton, G. O.; Lancaster, J. E.; Van Lear, G. E.; Fulmor, W.; Meyer, W. E. Structure of Nucleocidin. III. Revised Structure. *J. Am. Chem. Soc.* **1969,** *91,* 1535–1537.
9. Maruyama, T.; Ikejiri, M.; Izawa, K.; Onishi, T. Synthesis and Biological Activity of Fluorinated Nucleosides. In *Fluorine in Medicinal Chemistry and Chemical Biology;* Ojima, I., Ed.; Wiley, 2009; pp 165–198.
10. Zhao, C.; Li, P.; Deng, Z.; Ou, H.-Y.; McGlinchey, R. P.; O'Hagan, D. Insights into Fluorometabolite Biosynthesis in *Streptomyces cattleya* DSM46488 Through Genome Sequence and Knockout Mutants. *Bioorg. Chem.* **2012,** *44,* 1–7.
11. Xu, X.-H.; Yao, G.-M.; Li, Y.-M.; Lu, J.-H.; Lin, C.-J.; Wang, X.; Kong, C.-H. 5-Fluorouracil Derivatives from the Sponge *Phakellia fusca*. *J. Nat. Prod.* **2003,** *66,* 285–288.
12. Deng, H.; Ma, L.; Bandaranayaka, N.; Qin, Z.; Mann, G.; Kyeremeh, K.; Yu, Y.; Shepherd, T.; Naismith, J. H.; O'Hagan, D. Identification of Fluorinases from *Streptomyces* Sp MA37, *Nocardia brasiliensis*, and *Actinoplanes* Sp N902-109 by Genome Mining. *ChemBioChem* **2014,** *15,* 364–368.
13. Zhao, C.; Li, P.; Deng, Z.; Ou, H. Y.; McGlinchey, R. P.; O'Hagan, D. Insights into Fluorometabolite Biosynthesis in *Streptomyces cattleya* DSM46488 through Genome Sequence and Knockout Mutants. *Bioorg. Chem.* **2012,** *44,* 1–7.
14. Zhu, X.; Robinson, D. A.; McEwan, A. R.; O'Hagan, D.; Naismith, J. H. Mechanism of Enzymatic Fluorination in *Streptomyces cattleya*. *J. Am. Chem. Soc.* **2007,** *129,* 14597–14604.
15. Cobb, S. L.; Deng, H.; Hamilton, J. T.; McGlinchey, R. P.; O'Hagan, D.; Schaffrath, C. The Identification of 5'-fluoro-5-deoxyinosine as a Shunt Product in Cell Free Extracts of *Streptomyces cattleya*. *Bioorg. Chem.* **2005,** *33,* 393–401.
16. Murphy, C. D.; Schaffrath, C.; O'Hagan, D. Fluorinated Natural Products: the Biosynthesis of Fluoroacetate and 4-fluorothreonine in *Streptomyces cattleya*. *Chemosphere* **2003,** *52,* 455–461.
17. Dong, C.; Deng, H.; Dorward, M.; Schaffrath, C.; O'Hagan, D.; Naismith, J. H. Crystallization and X-ray Diffraction of 5'-fluoro-5'-deoxyadenosine Synthase, a Fluorination Enzyme from *Streptomyces cattleya*. *Acta Crystallogr. D. Biol. Crystallogr.* **2003,** *59,* 2292–2293.
18. Schaffrath, C.; Cobb, S. L.; O'Hagan, D. Cell-Free Biosynthesis of Fluoroacetate and 4-fluorothreonine in *Streptomyces cattleya*. *Angew. Chem. Int. Ed. Engl.* **2002,** *41,* 3913–3915.
19. Murphy, C. D.; O'Hagan, D.; Schaffrath, C. Identification of a PLP-dependent Threonine Transaldolase: A Novel Enzyme Involved in 4-Fluorothreonine Biosynthesis in *Streptomyces cattleya*. *Angew. Chem. Int. Ed. Engl.* **2001,** *40,* 4479–4481.
20. Murphy, C. D.; Moss, S. J.; O'Hagan, D. Isolation of an Aldehyde Dehydrogenase Involved in the Oxidation of Fluoroacetaldehyde to Fluoroacetate in *Streptomyces cattleya*. *Appl. Environ. Microbiol.* **2001,** *67,* 4919–4921.
21. Deng, H.; O'Hagan, D. The Fluorinase, the Chlorinase and the Duf-62 Enzymes. *Curr. Opin. Chem. Biol.* **2008,** *12,* 582–592.
22. Weeks, A. M.; Chang, M. C. Y. Catalytic Control of Enzymatic Fluorine Specificity. *Proc. Natl. Acad. Sci. U.S.A.* **2012,** *109,* 19667–19672. S19667/19661-S19667/19633.
23. Weeks, A. M.; Keddie, N. S.; Wadoux, R. D. P.; O'Hagan, D.; Chang, M. C. Y. Molecular Recognition of Fluorine Impacts Substrate Selectivity in the Fluoroacetyl-CoA Thioesterase FlK. *Biochemistry* **2014,** *53,* 2053–2063.

24. O'Hagan, D.; Deng, H. Enzymatic Fluorination and Biotechnological Developments of the Fluorinase. *Chem. Rev* **2014,** Ahead of Print; http://dx.doi.org/10.1021/cr500209t, *Chem. Abstr.* 2014:1616926.
25. Walker, M. C.; Deng, H.; Temporal and Fluoride Control of Secondary Metabolism Regulates Cellular Organofluorine Biosynthesis. *ACS Chem. Biol.* **2012,** *7,* 1576–1585.
26. Senn, H. M.; O'Hagan, D.; Thiel, W. Insight into Enzymatic C–F Bond Formation from QM and QM/MM Calculations. *J. Am. Chem. Soc.* **2005,** *127,* 13643–13655.
27. Walker, J. R. L.; Lien, B. C. Metabolism of Fluoroacetate by a Soil *Pseudomonas* sp. and *Fusarium solani*. *Soil. Biol. Biochem.* **1981,** *13,* 231–235.
28. Mishanina, T. V.; Corcoran, J. M. Kohen; A. Substrate activation in flavin-dependent thymidylate synthase. *J. Am. Chem. Soc.* **2014,** *136,* 10597–10600.
29. Dahal, U. P.; Obach, R. S.; Gilbert, A. M. Benchmarking In vitro Covalent Binding Burden as a Tool to Assess Potential Toxicity Caused by Nonspecific Covalent Binding of Covalent Drugs. *Chem. Res. Toxicol.* **2013,** *26,* 1739–1745.
30. Wang, J.; Li-Chan, X. X.; Atherton, J.; Deng, L.; Espina, R.; Yu, L.; Horwatt, P.; Ross, S.; Lockhead, S.; Ahmad, S.; Chandrasekaran, A.; Oganesian, A.; Scatina, J.; Mutlib, A.; Talaat, R. Characterization of HKI-272 Covalent Binding to Human Serum Albumin. *Drug. Metab. Dispos.* **2010,** *38,* 1083–1093.
31. Mercey, G.; Verdelet, T.; Renou, J.; Kliachyna, M.; Baati, R.; Nachon, F.; Jean, L.; Renard, P.-Y. Reactivators of Acetylcholinesterase Inhibited by Organophosphorus Nerve Agents. *Acc. Chem. Res.* **2012,** *45,* 756–766.
32. James, S. L.; Ahmed, S. K.; Murphy, S. T.; Braden, M. R.; Belabassi, Y.; Van Brocklin, H. F.; Thompson, C. M.; Gerdes, J. M. A Novel Fluorine-18 Β-Fluoroethoxy Organophosphate Positron Emission Tomography Imaging Tracer Targeted to Central Nervous System Acetylcholinesterase. *ACS Chem. Neurosci.* **2014,** *5,* 519–524.
33. Heby, O.; Persson, L.; Rentala, M. Targeting the Polyamine Biosynthetic Enzymes: A Promising Approach to Therapy of African Sleeping Sickness, Chagas' Disease, and Leishmaniasis. *Amino Acids* **2007,** *33,* 359–366.
34. Grishin, N. V.; Osterman, A. L.; Brooks, H. B.; Phillips, M. A.; Goldsmith, E. J. X-ray structure of ornithine decarboxylase from Trypanosoma brucei: The native structure and the structure in complex with α-difluoromethylornithine. *Biochemistry* **1999,** *38,* 15174–15184.
35. Müller, I. B.; Das Gupta, R.; Lüersen, K.; Wrenger, C.; Walter, R. D. Assessing the Polyamine Metabolism of *Plasmodium falciparum* as Chemotherapeutic Target. *Mol. Biochem. Parasitol.* **2008,** *160,* 1–7.
36. Marton, L. J.; Pegg, A. E. Polyamines as Targets for Therapeutic Intervention. *Annu. Rev. Pharmacol. Toxicol.* **1995,** *35,* 55–91.
37. Gmeiner, W. H. Novel Chemical Strategies for Thymidylate Synthase Inhibition. *Curr. Med. Chem.* **2005,** *12,* 191–202.
38. Longley, D. B.; Harkin, D. P.; Johnston, P. G. 5-fluorouracil: Mechanisms of Action and Clinical Strategies. *Nat. Rev. Cancer* **2003,** *3,* 330–338.
39. Pinedo, H. M.; Peters, G. F. J. Flourouracil: Biochemistry and Pharmacology. *J. Clin. Oncol.* **1988,** *6,* 1653–1664.
40. Rustum, Y. M.; Harstrick, A.; Cao, S.; Vanhoefer, U.; Yin, M. B.; Wilke, H.; Seeber, S. Thymidylate Synthase Inhibitors in Cancer Therapy: Direct and Indirect Inhibitors. *J. Clin. Oncol.* **1997,** *15,* 389–400.
41. Lauble, H.; Kennedy, M. C.; Emptage, M. H.; Beinert, H.; Stout, C. D. The reaction of fluorocitrate with aconitase and the crystal structure of the enzyme-inhibitor complex. *Proc. Natl. Acad. Sci. U. S. A.* **1996,** *93,* 13699–13703.

42. van der Kamp, M. W.; McGeagh, J. D.; Mulholland, A. J. "Lethal Synthesis" of Fluorocitrate by Citrate Synthase Explained through QM/MM Modeling. *Angew. Chem., Int. Ed.* **2011,** *50,* 10349–10351.
43. Imperiali, B.; Abeles, R. H. Inhibition of Serine Proteases by Peptidyl Fluoromethyl Ketones. *Biochemistry* **1986,** *25,* 3760–3767.
44. Ogilvie, W.; Bailey, M.; Poupart, M.-A.; Abraham, A.; Bhavsar, A.; Bonneau, P.; Bordeleau, J.; Bousquet, Y.; Chabot, C.; Duceppe, J.-S.; Fazal, G.; Goulet, S.; Grand-Maitre, C.; Guse, I.; Halmos, T.; Lavallee, P.; Leach, M.; Malenfant, E.; O'Meara, J.; Plante, R.; Plouffe, C.; Poirier, M.; Soucy, F.; Yoakim, C.; Deziel, R. Peptidomimetic Inhibitors of the Human Cytomegalovirus Protease. *J. Med. Chem.* **1997,** *40,* 4113–4135.
45. Warner, P.; Green, R. C.; Gomes, B.; Strimpler, A. M. Non-peptidic Inhibitors of Human Leukocyte Elastase. 1. The Design and Synthesis of Pyridone-Containing Inhibitors. *J. Med. Chem.* **1994,** *37,* 3090–3099.
46. Ilies, M.; Di Costanzo, L.; Dowling, D. P.; Thorn, K. J.; Christianson, D. W. Binding of α,α-Disubstituted Amino Acids to Arginase Suggests New Avenues for Inhibitor Design. *J. Med. Chem.* **2011,** *54,* 5432–5443.
47. Vidal-Dorsch, D. E.; Bay, S. M.; Greenstein, D. J.; Baker, M. E.; Hardiman, G.; Reyes, J. A.; Kelley, K. M.; Schlenk, D. Biological Responses of Marine Flatfish Exposed to Municipal Wastewater Effluent. *Environ. Toxicol. Chem.* **2014,** *33,* 583–591.
48. Zhou, Z.; Zhen, J.; Karpowich, N. K.; Law, C. J.; Reith, M. E. A.; Wang, D.-N. Antidepressant Specificity of Serotonin Transporter Suggested by Three LeuT-SSRI Structures. *Nat. Struct. Mol. Biol.* **2009,** *16,* 652–657.

Chapter 3

Synthetic Methods

Chapter Outline

1. INTRODUCTION

Organofluorine compounds have proved to be of enormous value in the areas of medicinal chemistry, materials chemistry, and agrochemicals. Nearly one-third of the blockbuster drugs are fluorinated compounds, and the focus increasingly is on the use of chiral drugs.[1] Accordingly, numerous novel synthetic methodologies have been developed for efficient preparation of organofluorine compounds in the past few decades. Depending on the nature of the reactive intermediates, these reactions may be classified as nucleophilic, electrophilic, and radical fluorinations. Transition-metal-ion-catalyzed fluorinations (or perfluoroalkylations) have been introduced for the selective fluorination of

Organofluorine Compounds in Biology and Medicine. http://dx.doi.org/10.1016/B978-0-444-53748-5.00003-4

aromatic as well as aliphatic compounds, and there is emerging interest in this area for developing milder, safer, and regio- and stereoselective fluorinations.[2–7]

Direct fluorination using elemental fluorine has been tamed by using dilute solutions of fluorine in inert gases. Alternatively, using microreactors it is now possible to perform selective fluorination or perfluorination of a wide variety of hydrocarbons, although use of elemental fluorine, due to its inherent hazards, is limited to select dedicated laboratories (vide infra).[8,9] The electrophilic nature of this reaction is shown by the relatively high reactivity of the electron-rich tertiary C–H bonds over the secondary or primary C–H bonds under controlled conditions. Through direct fluorination, the synthesis of 9(α)-F steroids, which serve as intermediates for other pharmaceutically interesting corticosteroids, has been achieved (vide infra). Relatively stable and commercially available electrophilic reagents such as xenon difluoride (XeF_2), acetyl hypofluorite (CH_3CO_2F), and Selectfluor, which are synthesized using the elemental fluorine, can be substituted for reactions that require the use of the elemental fluorine. These reagents also provide convenient access to the preparation of the ^{18}F-labeled compounds for use as tracers in positron emission tomography (PET). For example, (6-[^{18}F]fluoro-L-3-(3,4-dihydroxyphenyl) alanine) (6-[^{18}F]fluoro-L-DOPA) has been synthesized from the corresponding arylboronic ester using [^{18}F]Selectfluor.[10] Chapter 7 in this book outlines the synthesis and applications of various PET tracers.

Hydrogen fluoride and its amine complexes, especially pyridinium polyhydrogen fluoride (PPHF; Olah's reagent), have been widely used for the convenient fluorination of alkenes, alkynes, alcohols and oxidative desulfuration–fluorination, and so forth.[11,12] Several selective reagents for the deoxyfluorination of alcohols and carbonyl compounds have been developed, which include diethylaminosulfur trifluoride (DAST), Deoxo-Fluor, Fluolead, and XtalFluor-E or XtalFluor-M (vide infra). These reagents are commercially available, and are used for the synthesis of a wide variety of organofluorine compounds.

Trifluoromethyl (CF_3) and difluoromethylene (CF_2) groups are incorporated into many of the pharmaceuticals in order to increase their bioavailability, lipophilicity, and metabolic stability (see Chapter 5). Accordingly, this area has been very active for many years. Nucleophilic trifluoromethylations of carbonyl compounds, conveniently performed using the Ruppert–Prakash reagent CF_3SiMe_3 is widely used for synthesis of trifluoromethylated alcohols and trifluoromethyl ketones (vide infra).[13–20] Electrophilic trifluoromethylation using Togni's hypervalent iodine(III)-CF_3 reagent and Umemoto's S-trifluoromethyldibenzothiophenium (or oxonium, selenonium, and telluronium) salts can be used for transformations such as α-trifluoromethylation (or perfluoroalkylation) of carbonyl compounds, *S*-trifluoromethylations (e.g., in the preparation of S-CF_3-methionine), and *O*- and *N*-trifluoromethylation of alcohols and amines (vide infra).[21–28]

There is an emerging interest in the development of pentafluorosulfanyl-substituted pharmaceuticals due to their favorable physicochemical and electronic properties (e.g., relatively higher lipophilicity than the CF_3-substituted compounds and enhanced thermal stabilities). SF_5 analogs of pharmaceuticals such as fluoxetine,[29] mefloquine,[30] and cinacalcet[31] have been synthesized and tested for their biological activity (vide infra). Although no SF_5-containing drugs are currently approved by the US Food and Drug Administration (FDA) to date, it is likely that they will complement the other fluorinated drugs.

There are several excellent monographs and reviews covering the synthesis of organofluorine compounds, and the interested reader is referred to this literature.[13,15,32–39] Because of the enormity of the publications in this area, this chapter addresses only the synthetic methodologies as relevant to the synthesis of the fluorinated pharmaceuticals. Detailed synthetic procedures for the relevant pharmaceuticals are provided in the subsequent chapters in this book.

2. FLUORINATION BY ELEMENTAL FLUORINE

2.1 C–H Fluorination

Elemental fluorine (bp = −188 °C) is rarely used in the electrophilic fluorination reactions of organic compounds because of its extremely violent reactivity and toxicity. Due to the very low F–F bond energy of F_2 (38 kcal/mol), coupled with the high C–F bond energies, about 110 kcal/mol, the fluorination of alkanes by elemental fluorine is highly exothermic, and these reactions usually proceed highly exothermally to give tars under uncontrolled reaction conditions. Somewhat convenient techniques for the fluorination of hydrocarbons (including perfluorination of alkyl chlorides, ethers, and esters) have been developed that involve: (1) continuous flow aerosol fluorination, in which mixtures of NaF and hydrocarbons are transferred as aerosols into the fluorination chamber; (2) the use of CoF_3 as the fluorinating agent; and (3) the use of fluorine gas in microfluidic reactors.[40–45]

Electrophilic fluorination of alkanes and electrophilic additions to alkenes can be achieved in moderate yields by the use of a dilute solution of fluorine in nitrogen (~2–10%), at relatively low temperatures (<0 °C). These electrophilic fluorinations, in some cases, are regio- and stereoselective (Figure 1). The relatively electron-rich tertiary C–H bond is preferentially fluorinated in the presence of the primary and secondary hydrogens, with high regio- and stereoselectivity. For example, fluorination of the (*Z*)-4-*tert*-butylcyclohexyl *p*-nitrobenzoate gives predominantly the (*E*)-4-fluoro-4-*tert*-butylcyclohexyl *p*-nitrobenzoate, in yields of up to 83%.[46] This high stereoselectivity rules out the involvement of the free radical intermediates in these reactions, and these fluorinations most likely proceed via a pentacoordinated, 2-electron-3-centered intermediate (Eqn (1)).

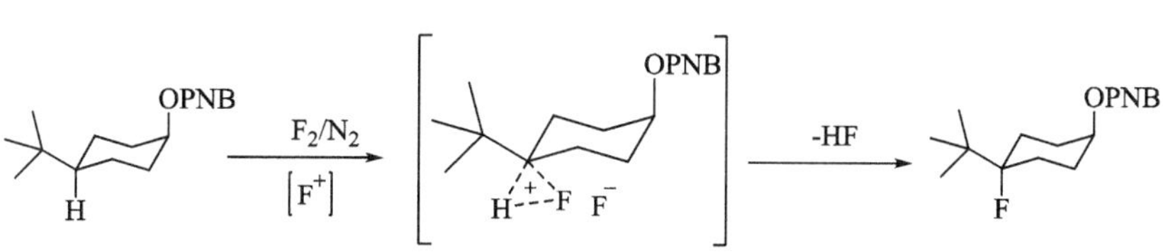

OPNB = *p*-nitrobenzoate

(1)

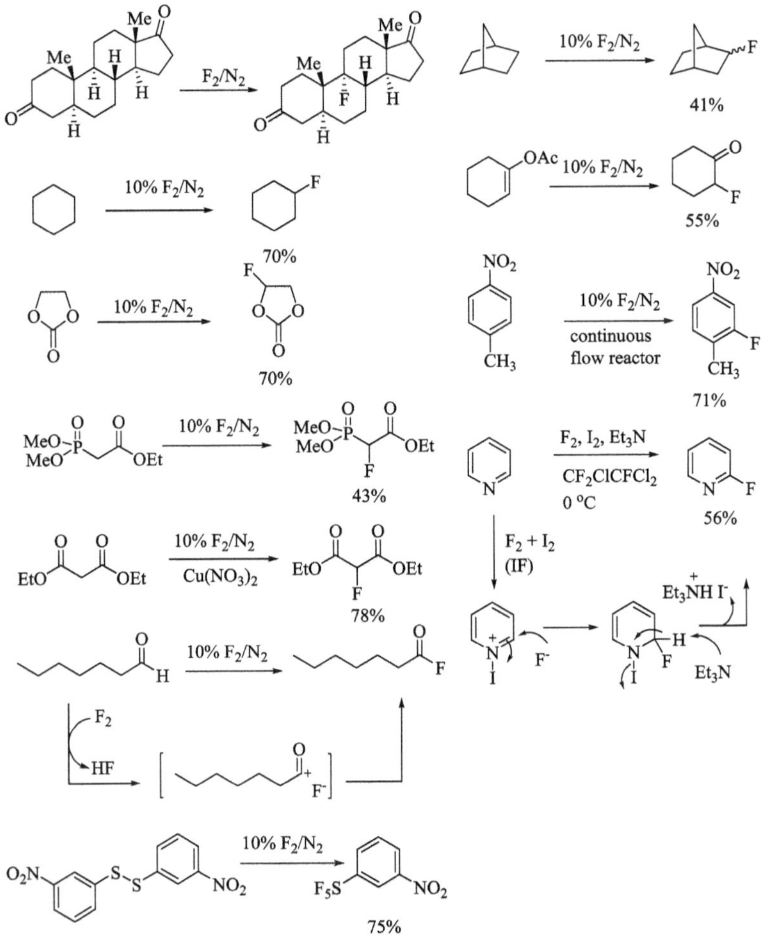

FIGURE 1 Electrophilic fluorination of selected hydrocarbons by elemental fluorine.

Relatively more electron-rich C–H bonds react with the elemental fluorine (as dilute solution in nitrogen or helium) with high regioselectivity, as illustrated by the selective fluorination of the 9(α)-H of the steroids; this selective fluorination of electron-rich C–H bonds further supports the electrophilic nature of these fluorination reactions.[46] The 9(α)-F steroids are important intermediates in the preparation of the pharmaceutically relevant fluorinated corticosteroids.

Microfluidic reactor design has been applied for the fluorination of a variety of organic compounds, under operationally convenient reaction temperatures (operating temperatures are in the range of 0–5 °C).[42] This tamed direct electrophilic fluorination reaction is broadly applicable for the exhaustive fluorination of ethers, esters, and selective fluorination of aliphatic, aromatic, and heterocyclic organic compounds. For example, using this technique, the monofluorination of cyclic and

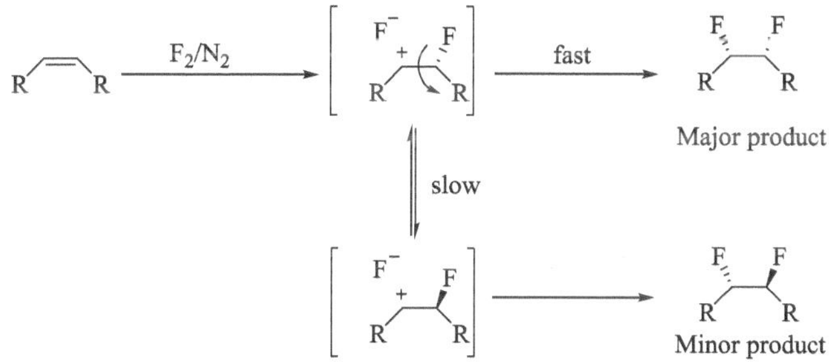

FIGURE 2 Fluorination of alkenes by elemental fluorine.

acyclic hydrocarbons, fluorination of enol acetates, conversion of aldehydes to acyl fluorides, monofluorination of phosphonate esters and 1,3-dicarbonyl compounds, and electrophilic fluorination of aromatics can be achieved in remarkably high yields, often with high regioselectivity (Figure 1).[42] Pyridine, quinolone, and quinoxaline derivatives react with elemental fluorine in the presence of iodine to give the corresponding α-fluorination (i.e., fluorine vicinal to the nitrogen) products. This reaction proceeds by an addition–elimination mechanism, initiated by iodine monofluoride that is formed in situ. However, direct fluorination is largely limited to only a few specialized laboratories that can handle the otherwise extremely hazardous elemental fluorine gas.

2.2 Fluorination of Alkenes

The major products of the electrophilic addition of F_2 to alkenes (fluorination being performed using dilute fluorine solutions in nitrogen at low temperatures) are the corresponding *syn*-difluoroalkanes.[46,47] In these reactions, the initially formed carbocation exists as the contact ion pair (i.e., the leaving group has not gone too far away from the originally attached carbon, and therefore the carbocation is not freely solvated), so that the fluoride ion capture effectively competes with the possible conformational interconversion through the C–C bond rotations (Figure 2).

Relatively more stable, easily handled, transportable, and storable electrophilic fluorinating agents, such as Selectfluor and xenon difluoride (XeF_2) are generally synthesized from fluorine gas on an industrial scale (Figure 3). These compounds may be substituted for the elemental fluorine for carrying out electrophilic additions to alkenes, electrophilic fluorination of enolsilyl ethers and aromatics, and desulfurative fluorinations of thioethers and dithiolanes (vide infra). Synthesis of XeF_2 can be achieved in multigram scale from the reaction of anhydrous fluorine with xenon in anhydrous HF (as the solvent) at ambient conditions in near-quantitative yields.[48] The ^{18}F-labeled Selectfluor or XeF_2 can also be synthesized using the ^{18}F-F_2, and used in the synthesis of the ^{18}F-radiolabeled PET radiopharmaceuticals (Please see Chapter 7 for the detailed aspects of the ^{18}F-radiolabeled compounds.).[10,49]

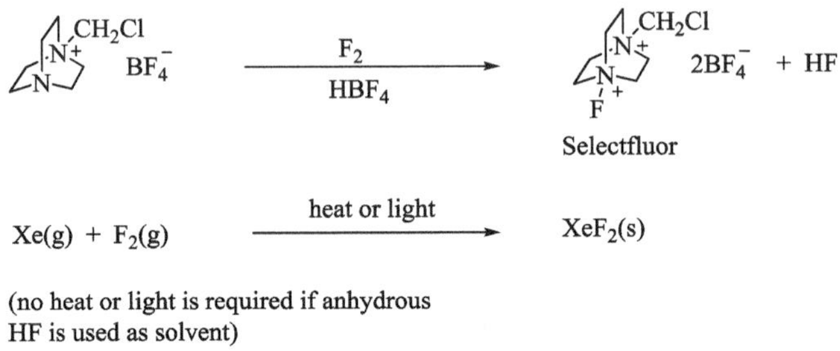

FIGURE 3 Elemental fluorine in the synthesis of Selectfluor and XeF_2.

3. HYDROGEN FLUORIDE AND ITS AMINE COMPLEXES

Neat hydrogen fluoride undergoes electrophilic additions to alkenes, alkynes, and nucleophilic substitutions with alkyl halides (or aryl sulfonates) to give the corresponding alkyl fluorides. However, these reactions require high-pressure equipment due to the low boiling point of anhydrous HF (bp=19.6 °C). In order to extend the liquid range of HF to relatively higher temperatures that are typically required for these reactions (about 50 °C), there have been many attempts to develop relatively nonvolatile reagents. These reagents are formed upon mixing of HF in oxygenated and amine solvents, such as tetrahydrofuran (THF), diethyl ether, alcohols, pyridine, and triethylamine, and upon mixing of HF with trialkylphosphines (Figure 4). Among several such solvent–HF combinations, PPHF (30:70 wt% of py and HF, respectively) emerged as the most convenient electrophilic fluorinating agent, through the efforts of Olah and coworkers in the early 1970s. This reagent is now generally referred to as Olah's reagent. PPHF retains liquid range up to 55 °C, and is commercially available.[11] Even though the commercially available PPHF has a pyridine to HF ratio of approximately 1:9, varying degrees of mol ratios, for example, pyridine to HF ratios ranging from 1:1 to 1:8 are used for various applications that require relatively less vigorous conditions.[12] The acidity of PPHF and other amine–HF complexes approaches to as high as that of concentrated sulfuric acid, with a Hammett acidity constant (H_o) of −11 (cf. H_o of anhydrous HF = −15.1).[50,51]

Olah and coworkers have also developed dimethyl ether-HF complexes (DMEPHF; mol ratio of ether:HF = 1:5), that are relatively mild nucleophilic fluorinating agents as compared to PPHF.[52] The *N*,*N*-dialkylimidazolium–HF based complexes (with a weight ratio of dialkylimidazolium:HF = 1:2.3) are somewhat less acidic, relatively more stable in air and moisture, and can be used for the halofluorination of alkenes and ring-opening fluorination of epoxides.[53] KHF_2 or $KF{\cdot}2H_2O$ has been used for ring-opening nucleophilic fluorination of aziridines to give the corresponding β-fluoroalkylamines.[54]

3.1 Hydrofluorination of Alkenes and Alkynes, and Deoxyfluorination of Alcohols

PPHF, as demonstrated by Olah and coworkers, is an effective fluorinating agent for hydrofluorination, halofluorinations, and nitrofluorination of

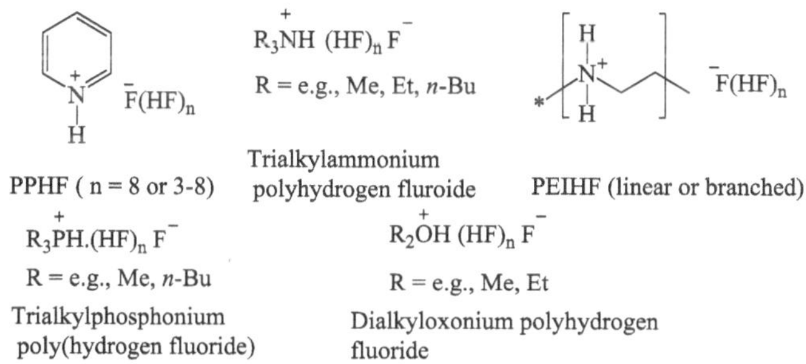

FIGURE 4 Structures of selected amine–$(HF)_n$, (pyridinium poly(hydrogen fluoride) (PPHF), poly(ethyleniminium) poly(hydrogen fluoride) (PEIHF) phosphine–$(HF)_n$, and alcohol–$(HF)_n$ reagents.

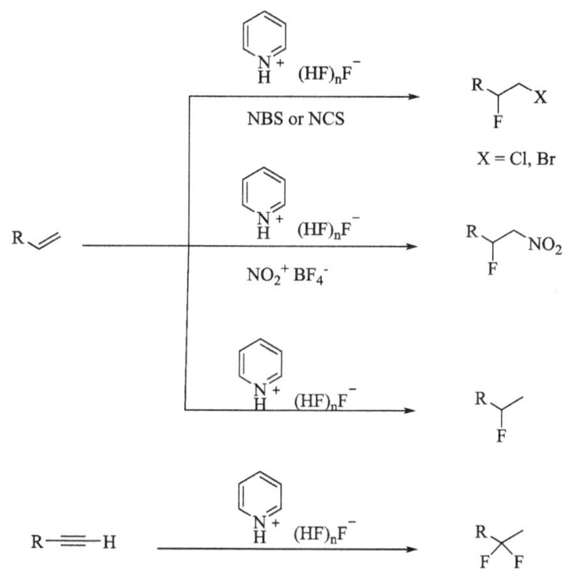

NBS = N-bromosuccinimide; NCS = N-chlorosuccinimide

FIGURE 5 Hydrofluorination reactions of alkenes and alkynes using pyridinium polyhydrogen fluoride (PPHF).

alkenes or alkynes.[11,12] Secondary, tertiary, and benzyl alcohols, react with PPHF to give the corresponding alkyl fluorides in moderate to high yields (Figure 5). As an added advantage, these fluorination reactions, using the PPHF, could be carried out in polyethylene bottles under normal atmospheric pressure at temperatures up to 50 °C. Triethylamine trihydrofluoride ($Et_3N{\cdot}3HF$), although much less reactive than the PPHF, could be used in certain fluorination reactions, and these reactions could be achieved even in the Pyrex glass flasks; however, caution should be exercised as the glass surface

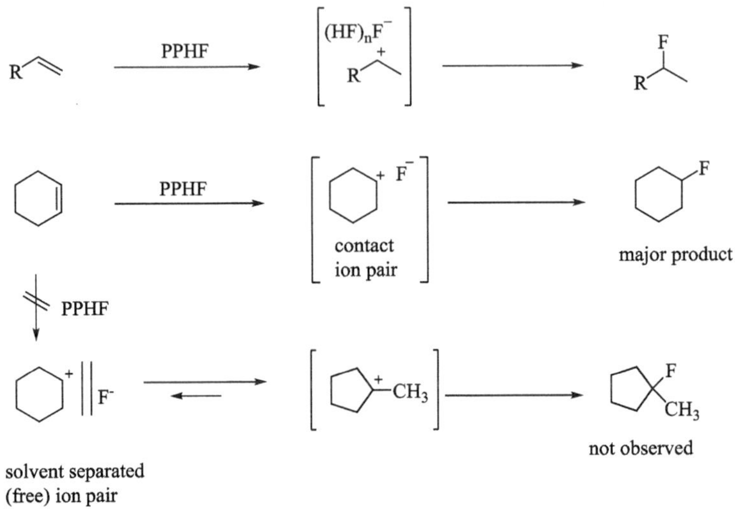

FIGURE 6 A generally accepted mechanism of the electrophilic hydrofluorination of alkenes.

gets etched by this reagent after relatively longer reaction times. Dichloromethane or THF are typically used as solvents for these reactions. Appropriate safety precautions[55] (as in the case of the HF), should be used when handling these amine–HF complex agents, as these reagents are in equilibrium with traces of HF.

The orientation of hydrofluorination of alkenes by PPHF is in accordance with Markovnikov's rule, and usually does not result in the formation of rearranged products, indicating that the initially formed carbocation intermediate in these reactions is a contact ion pair rather than a free carbenium ion. Hydrofluorination of cyclohexene also gives fluorocyclohexane as the predominant product using PPHF in THF at 0 °C, which further supports this hypothesis because the free cyclohexyl cation would be expected to readily rearrange to the thermodynamically more stable tertiary 1-methylcyclopentyl cation, which, if formed, would give the 1-fluoro-1-methylcyclopentane as the major product (Figure 6). Similarly, hydrofluorination of 2-norbornene and 1-methylcyclohexene gives 2-fluoronorbornane and 1-fluoro-1-methylcyclohexane as the major products. Secondary and tertiary alcohols, upon reaction with PPHF give the corresponding alkyl fluorides in good yields.

PPHF, Et_3N-3HF, Deoxo-Fluor, and DAST (vide infra) are among the most commonly used reagents for the preparation of glycosyl fluorides from the corresponding carbohydrate derivatives. Thus, reaction of the *O*-acetyl or *O*-benzyl α-D-glucopyranose with PPHF gives the corresponding α-D-glucosyl fluorides, stereoselectively, in high yields.[56,57] β-*N*-glycosyl triazole derivatives, on the other hand, give the corresponding β-glycosyl fluorides as the major products using PPHF.[58] The latter triazoles, in turn, were synthesized from the corresponding glycosyl azides through 1,3-dipolar cycloaddition reaction with di-*tert*-butyl acetylene dicarboxylate (Figures 7 and 8).

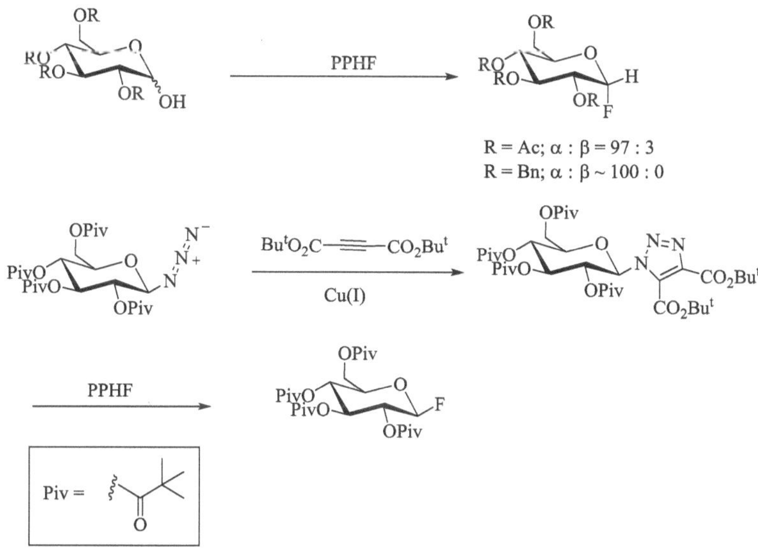

FIGURE 7 Preparation of glycosyl fluorides using pyridinium polyhydrogen fluoride (PPHF).

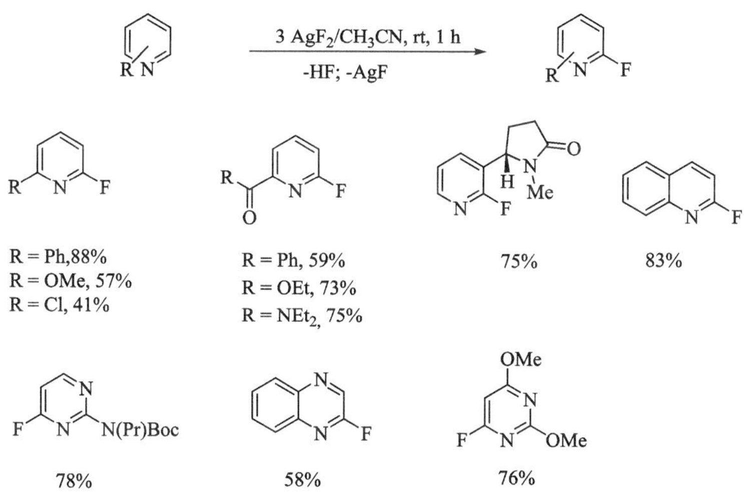

FIGURE 8 *ortho*-Fluorination of various pyridine derivatives using AgF_2.

4. SELECTIVE *ortho*-FLUORINATION OF PYRIDINES BY AgF_2

Hartwig and coworkers achieved the selective *ortho*-C–H fluorination of pyridines by reaction with AgF_2 (which is commercially available in bulk quantities and can be prepared by the reaction of elemental F_2 with Ag).[59] The use of three mol equivalents of AgF_2 (although only two mol equivalents of AgF_2 are needed by stoichiometry) gives high yields of the *ortho*-fluorinated pyridines, and the reaction tolerates a broad variety of functional groups, including carbonyl, ether, halogen, and amide functionalities.

The suggested mechanism for the reaction involves coordination of the AgF_2 to the pyridine, followed by the nucleophilic addition of the fluoride ion from Ag-F to the *ortho* position, and the *ortho*-hydrogen atom abstraction by a second molecule of AgF_2 (Eqn (2)).

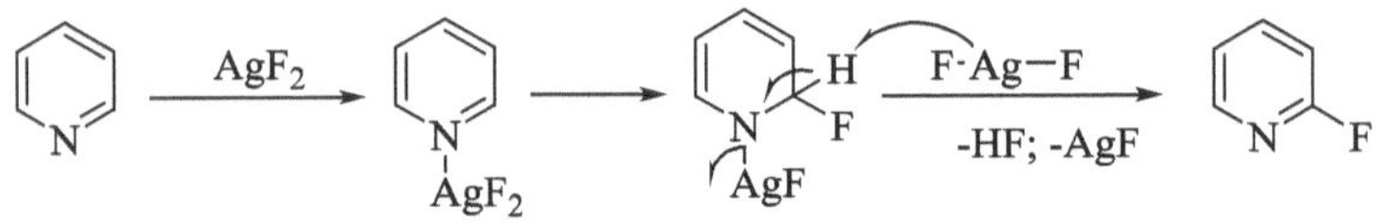

(2)

The latter selective *ortho*-CH fluorination could be used to synthesize the fluorinated analogs of the pharmaceutically interesting aza-aromatics. *O*-acyl derivative of the tropicamide (Mydriacyl, an anticholinergic drug), a calcitonin gene-related peptide (CGRP) antagonist, and roflumilast (Daliresp, a drug used to treat chronic obstructive pulmonary disease) are readily α-fluorinated to give the corresponding analogs (Figure 9).[59] Use of the ^{18}F-labeled AgF_2 in these reactions would provide access to the corresponding ^{18}F-labeled analogs, which are potentially useful as PET tracers. Further, the α-fluorine in these compounds is amenable to nucleophilic substitution reactions with amines and alcohols to give their corresponding amino and ether derivatives.

The fluoropyridine may also be used as a readily available synthon to synthesize other drug candidates or their derivatives; thus, the α-fluorinated version

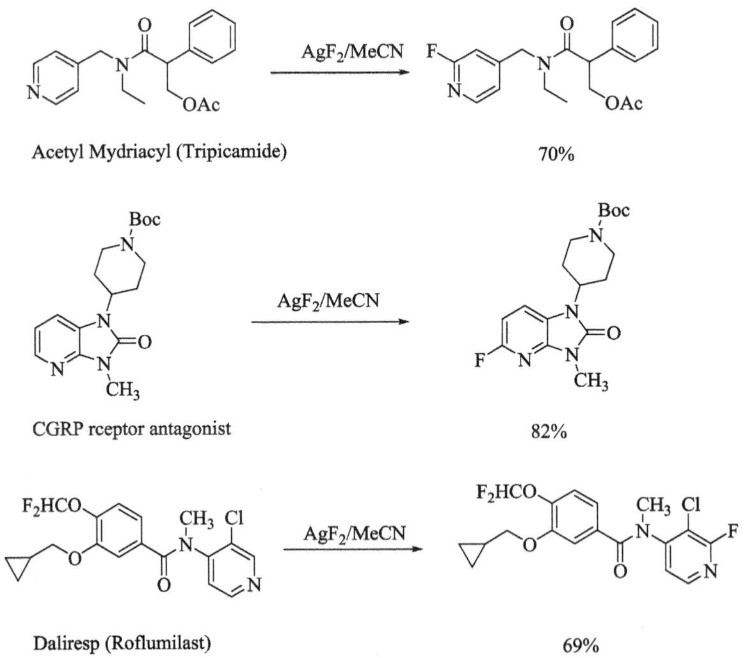

FIGURE 9 *ortho*-Fluorination of some pharmaceutically interesting heterocyclic compounds.

of the pioglitazone (Actos), a prescription drug used to treat diabetes, was prepared in a single step through the reaction of a α-fluoropyridine derivative with the corresponding thiazolidinedione (Eqn (3)).

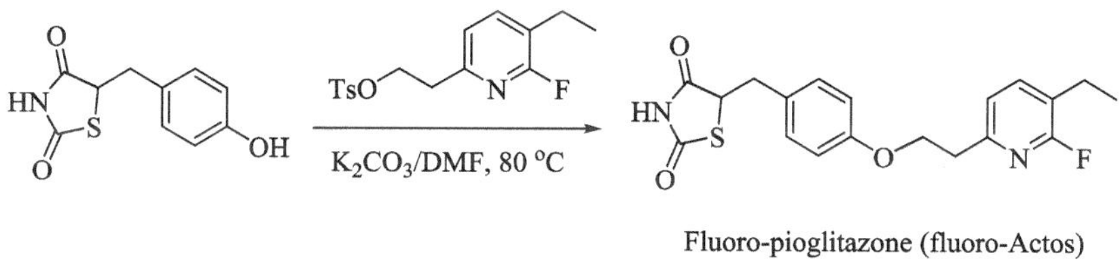

(3)

5. DEOXYFLUORINATIONS BY DAST AND RELATED REAGENTS

Reactions of secondary or tertiary alcohols with PPHF give the corresponding deoxyfluorination products (Eqn (4)). However, in these reactions the initially formed carbocation rearranges to the more stable carbocation and therefore unexpected products sometimes dominate over the expected products.

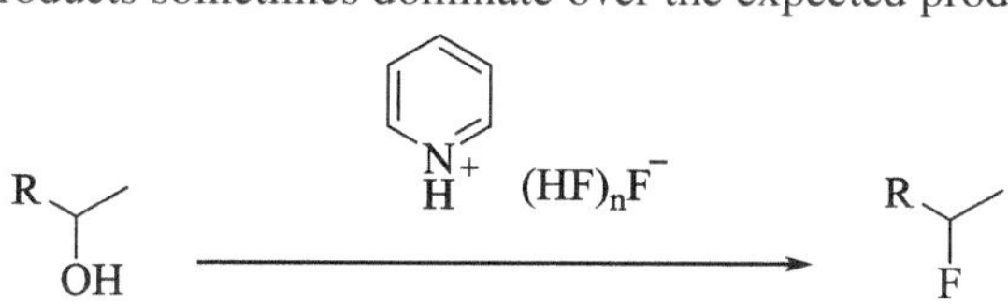

(4)

Various alternative reagents have been introduced to circumvent this problem. The most widely used reagent for this purpose is DAST (Et_2NSF_3), originally introduced by Middleton at DuPont.[60,61] DAST is commercially available and can be readily obtained through the reaction of SF_4 with one equivalent of diethylamine. The DAST-mediated deoxyfluorinations usually proceed with inversion of configuration. DAST (as well as SF_4/HF) also is useful for the conversion of carbonyl compounds to the *gem*-difluoro compounds. Use of DAST for *gem*-difluorination allows for a short synthesis of the gemcitabine, an anticancer drug. 1,3,5-tri-*O*-benzoyl-2-oxoribofuranose upon fluorination with DAST in dichloromethane (DCM), at low temperature, gives the corresponding *gem*-difluorofuranose derivative, which is an intermediate for the synthesis of the gemcitabine (Eqn (5)). Several other synthetic strategies for the synthesis of gemcitabine also rely on the *gem*-difluorination of the carbonyl group.[62]

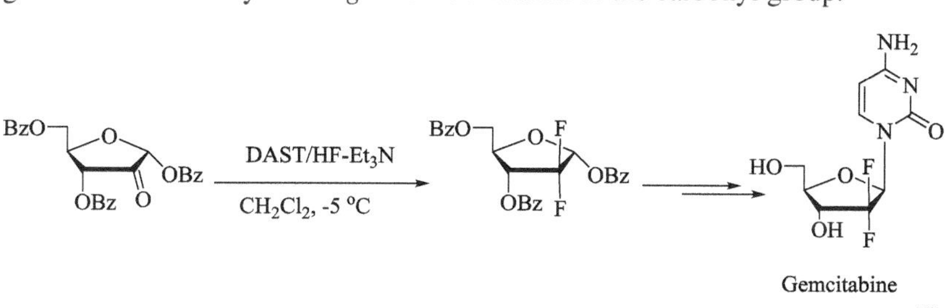

(5)

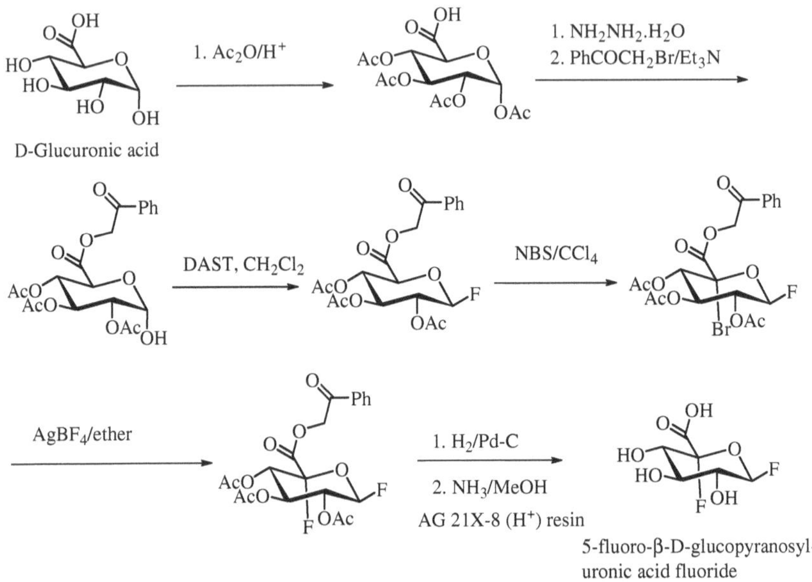

FIGURE 10 Synthesis of 5-fluoro-β-D-gluocopyranosyluronic acid fluoride.

DAST is used widely for the preparation of fluorinated carbohydrates, especially the anomeric fluorides, which are used as substrates for the glycosidases.[63] Synthesis of 5-fluoro-β-D-glucopyranosyluronic acid fluoride from D-glucuronic acid involves the key steps of the DAST-mediated conversion of the anomeric hydroxyl into the β-fluoride, and free radical bromination at C_5 followed by $AgBF_4$-mediated S_N1 reaction (Figure 10).[64] This compound has been used as a mechanistic probe for determining the mechanism of β-glucuronidase enzyme-catalysis.

DAST, although a widely used reagent, is relatively unstable at high temperatures.[65] Lal and coworkers have developed a related reagent, bis(methoxyethyl) aminosulfur trifluoride (Deoxo-Fluor), which is more thermally stable than DAST, and is the preferred reagent for reactions that proceed at relatively high temperatures.[66] Deoxo-Fluor is stable at temperatures up to 150 °C, whereas DAST starts to decompose violently at about 90 °C. The relatively high stability of the Deoxo-Fluor is rationalized as due to the coordination of the ether oxygen with vacant d-orbitals of sulfur, thereby decreasing the propensity of the intermolecular reactions. This would lead to the attenuated disproportionation and decomposition of the reagent at high temperatures. (Figure 11).[66]

Deoxo-Fluor also effects conversion of aldehydes and ketones to the corresponding *gem*-difluorides and carboxylic acids to the trifluoromethyl derivatives (Figure 12).

Morpholino-DAST (MOST) is relatively more stable than the DAST, and its use is preferred in place of DAST for the fluorination of alcohols.[65] Flow reactors may be used for these reagents for large-scale synthesis, so that the thermal degradation

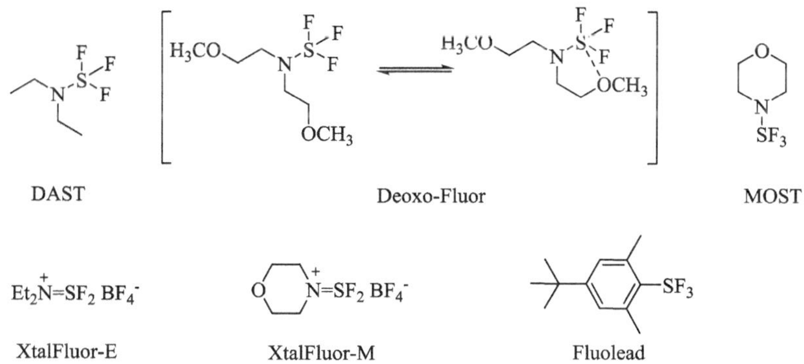

FIGURE 11 Widely used deoxyfluorination reagents.

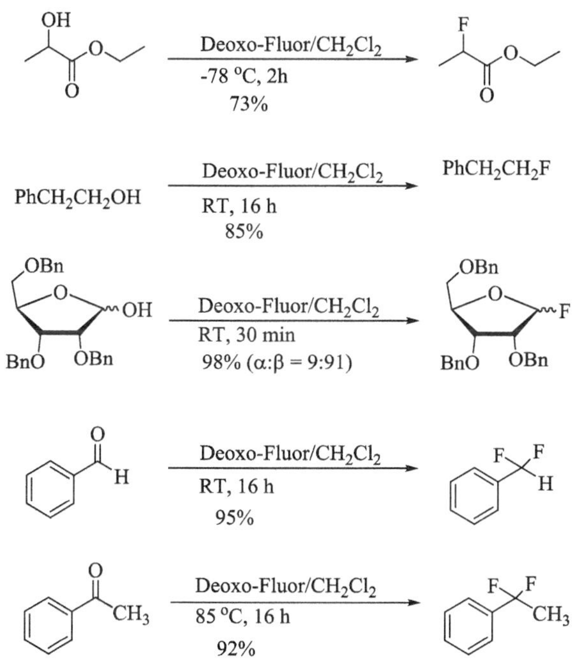

FIGURE 12 Selected Deoxo-Fluor-mediated conversions.

of the reagents is minimized. Typically, with these reagents deoxyfluorination of alcohols takes place at about −78 °C to room temperature. The reactivity order for the deoxyfluorination of alcohols is: benzylic >2°>3°>1°. The DAST-mediated reactions are relatively slow for the sterically crowded alcohols. The deoxyfluorination of anomeric hydroxyls of carbohydrates using DAST proceeds under relatively mild conditions.[67] XtalFluor-E, derived from DAST is a relatively mild deoxyfluorinating agent, and in the presence of Et_3N-HF, it has reactivity similar to that of DAST.[68] XtalFluor-E and its morpholino analog XtalFluor-M are significantly more stable than DAST and related reagents and can be used for the conversion of alcohols to alkyl fluorides, and carbonyls to the *gem*-difluoro compounds.[68] Umemoto's reagent, Fluolead (4-*tert*-butyl-2,6-dimethylphenylsulfur trifluoride), a crystalline and thermally stable reagent, provides another operationally simple reagent for the conversion of alcohols to alkyl fluorides, aldehydes and ketones to the corresponding *gem*-difluoro compounds, and carboxylic acids to their corresponding trifluoromethyl compounds (Figure 13).[28,69]

Deoxyfluorination of β-aminoalcohols using DAST or Deoxo-Fluor is usually accompanied by rearrangements due to the neighboring amino group participation, involving the formation of reactive aziridine intermediates. This rearrangement reaction was used in the synthesis of the erythro-(2*S*,3*S*)-3-fluoro-*N*-methyl-D-aspartic acid (β-fluoro-NMDA), an agonist of the NMDA ionotropic glutamate receptor, starting from the dimethyl 2,3-epoxyfumarate.[70] The epoxide ring opening of the latter compound using *N*-methyl-*N*-benzylamine, followed by reaction with Deoxo-Fluor results in the formation of an intermediate aziridinium ion, which upon ring-opening nucleophilic substitution by the fluoride anion (derived from the Deoxo-Fluor) gives the diastereomerically pure

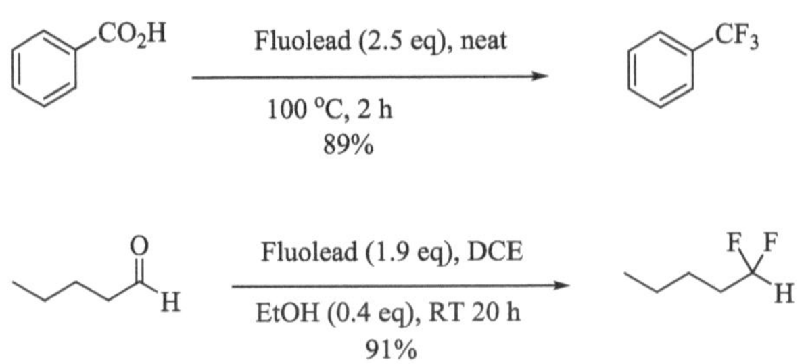

FIGURE 13 Selected conversions using Fluolead as the fluorinating reagent.

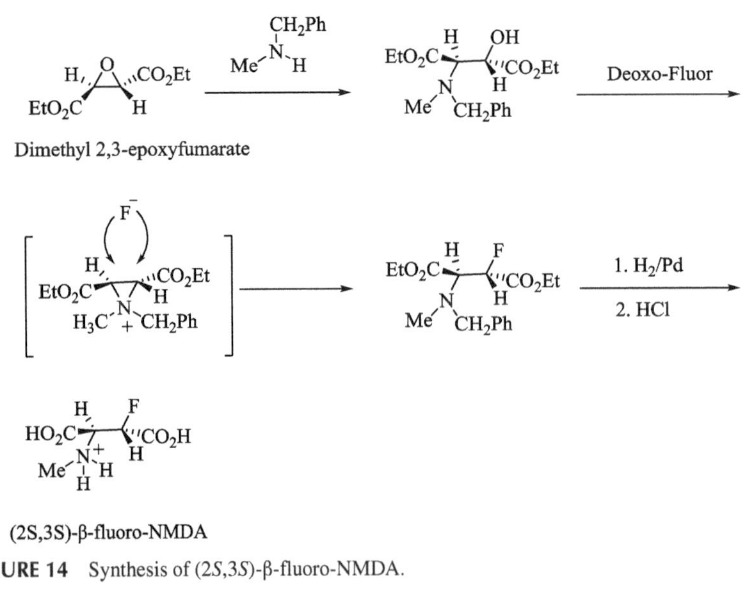

FIGURE 14 Synthesis of (2*S*,3*S*)-β-fluoro-NMDA.

diethyl 2-fluoro-3-(*N*-methylbenzylamino)-D-aspartate. Deprotection of the benzyl group by catalytic hydrogenation, followed by hydrolysis of the ester group, gives the (2*S*, 3*S*)-β-fluoro-NMDA (Figure 14). This compound adopts a gauche conformation, in which the C–F and C–N bonds are gauche to each other, at the receptor binding site, which results in electrostatic stabilization; however, the corresponding (2*S*, 3*R*)-diastereomer preferentially adopts the anti-conformer at the binding site, thereby resulting in its energeticallly unfavorable binding.[70]

The *R*- and *S*-enantiomers of 3-fluoro-γ-aminobutyric acid were also similarly synthesized from the corresponding β-amino alcohols; these fluorinated derivatives act as agonists of cloned γ-aminobutyric acid A receptors.[33,34]

6. DEDIAZONIATION–FLUORINATION OF AMINES

The Balz–Schiemann reaction, which involves diazotization of the primary aromatic amines followed by fluorinative dediazoniation, could be achieved in preparative scale by carrying the reaction in PPHF, in which case the isolation of the otherwise explosive intermediary diazonium salts is not required (Figure 15). This reaction has been achieved on an industrial scale in a continuous flow reactor, using HBF_4 instead of PPHF as the HF source. HBF_4 is a relatively safer reagent due to its ease of operation and operational safety. Using this method, large quantities of *ortho*-difluorobenzene could be prepared from the reaction of *ortho*-fluoroaniline and sodium nitrite, in 90% yield; the reaction time for the diazotization ranged from 4 to 15 s at 10–25 °C, whereas the fluorodediazoniation of the isolated diazonium salts proceeds in 1–7 min at 140–200 °C.[71]

The fluorodediazoniation reaction was also achieved in high yields in ionic liquid solvents.[72] Further, diazotization followed by fluorinative dediazoniation of α-amino acids gives α-fluorocarboxylic acids in high yields and with high stereoselectivities (retention of configuration) (Figure 16).[11,73,74] The corresponding 2-halo carboxylic acids are obtained when the diazotization is performed in the presence of the more nucleophilic alkaline metal halides (chlorides, bromides, and iodides).[30] The retention of configuration in these reactions can be rationalized as due to the intermediacy of an α-lactone intermediate.

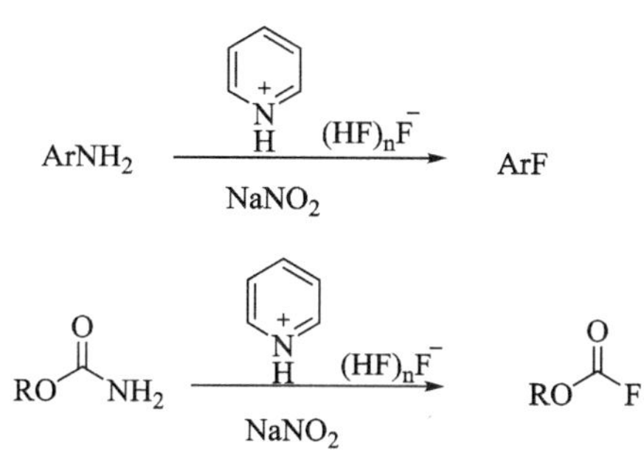

FIGURE 15 Balz–Schiemann fluorination reaction.

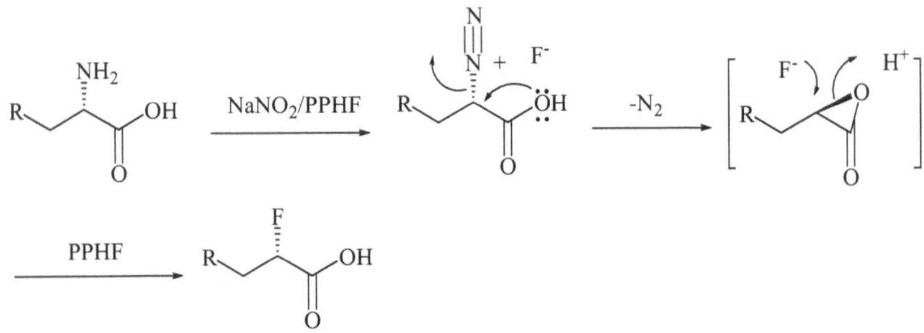

FIGURE 16 Fluorodediazoniation of α-amino acids.

7. *gem*-DIFLUORINATIONS

Carbonyl compounds react with DAST, Deoxo-Fluor, MOST, XtalFluor, or Fluolead (vide supra) to give the corresponding *gem*-difluoro compounds in good to excellent yields. Dithiolane derivatives of aldehydes and ketones can be transformed into their corresponding *gem*-difluoro compounds by desulfurative fluorination using PPHF in the presence of N-bromosuccinimide (NBS),[75] Selectfluor,[76a] sulfuryl chloride fluoride[76b], or nitrosonium tetrafluoroborate,[77] or by anodic desulfurization in the presence of $Et_3N{\cdot}3HF$.[78,79] Dithioketals also have been transformed into the *gem*-difluoro compounds using *p*-iodotoluene difluoride or BrF_3 (Figure 17).[80,81]

The *gem*-difluoro compounds have been synthesized through the reaction of aldehyde or ketone hydrazones with in situ generated BrF from the reaction of NBS with PPHF (Figure 18).[82] This obviates the use of the BrF and BrF_3 reagents, which are synthesized using the elemental fluorine.[46,47] The *gem*-difluoro compounds have significant applications in biological and medicinal chemistry due to their unique role as isosteres of ethereal oxygens and carbonyl compounds. Several pharmaceutical candidates such as gemcitabine owe their activity to the *gem*-difluoromethylene moiety.[83,84]

The nucleophilic difluoromethylation of *N-tert*-butanesulfinimines using the corresponding (2-pyridyl)sulfonyldifluoromethanes and sodium hexamethyldisilazide proceeds in good yields and with high diastereoselectivity to give the corresponding β-amino-α,α-difluorosulfones. These sulfone products could be transformed to the corresponding β-amino-α,α-difluorosulfonic acids, or α,α-difluorosulfinic acids, under mild reaction conditions. The α,α-difluorosulfones could also be transformed into the *gem*-difluoropeptidosulfonamides, which are potential sulfonylamidopeptide mimetics, and therefore may have broad-ranging applications in the medicinal chemistry (Figure 19).[85]

Ruppert–Prakash reagent ($TMSCF_3$; CF_3SiMe_3) forms the difluorocarbene (F_2C:) intermediate in the presence of various anion sources such as NaI, tetrabutylammonium triphenyldifluorosilicate (Bu_4N^+ $[Ph_3SiF_2]^-$; TBAT), tetramethylammonium fluoride (Me_4N^+ F^-; TMAF), and CaI_2. The carbene thus generated in situ reacts spontaneously with alkenes and alkynes to give their

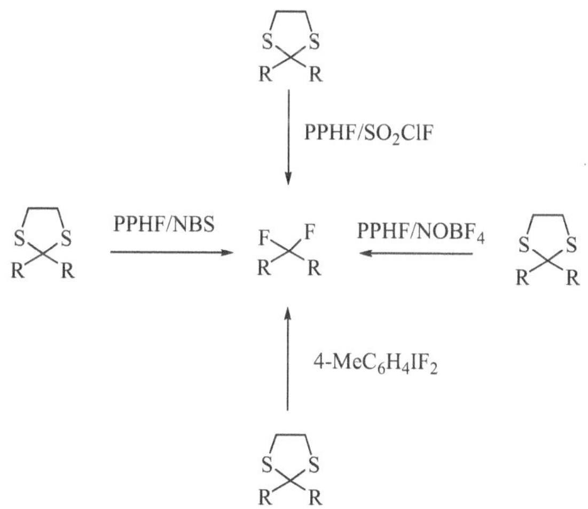

FIGURE 17 Desulfurative *gem*-difluorination of dithiolanes (NBS = N-bromosuccinimide; PPHF = pyridinium polyhydrogenfluoride).

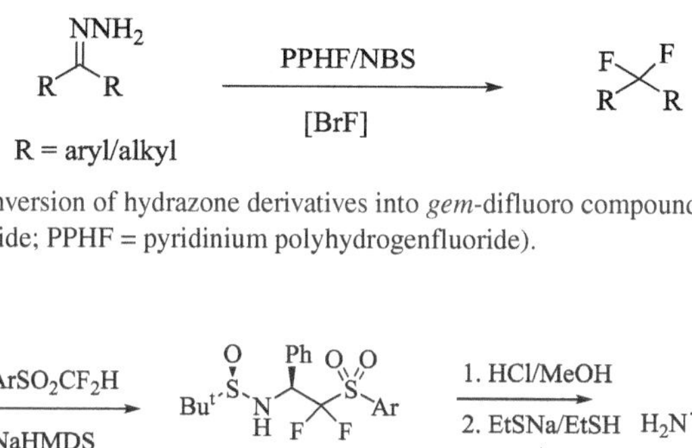

FIGURE 18 Conversion of hydrazone derivatives into *gem*-difluoro compounds (NBS = N-bromosuccinimide; PPHF = pyridinium polyhydrogenfluoride).

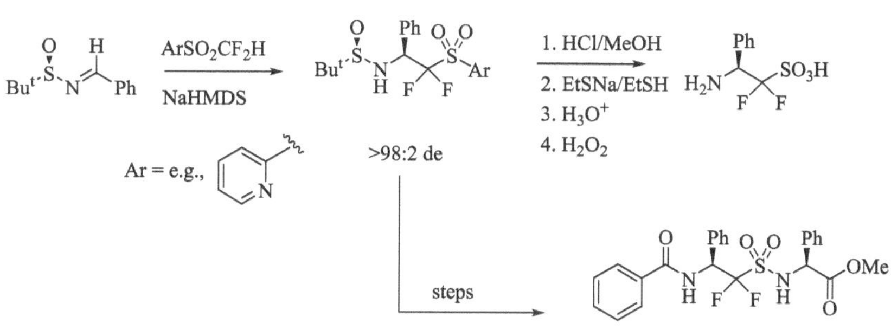

FIGURE 19 Nucleophilic difluoromethylation of *N-tert*-butanesulfinimines for the preparation of *gem*-difluorosulfonic acids and peptidosulfonamides (NaHMDS = hexamethyldisilazide).

corresponding *gem*-difluorocycloprapane or *gem*-difluorocyclopropene derivatives, respectively (Figure 20).[86]

The difluorocarbene (F_2C:), generated from the CF_3SiMe_3 and lithium iodide, can also be used to transform a variety of imidazoles, benzimidazoles,

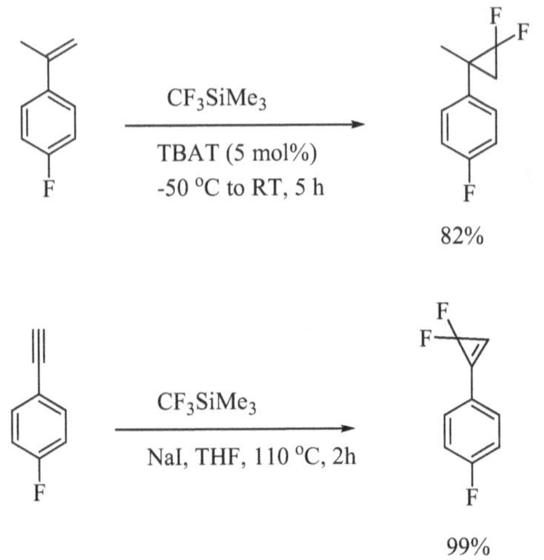

FIGURE 20 *gem*-Difluoromethylenation of alkenes and alkynes by the in situ formed difluorocarbene (THF = tetrahydrofuran; TBAT = tetrabutylammonium triphenyldifluorosilicate).

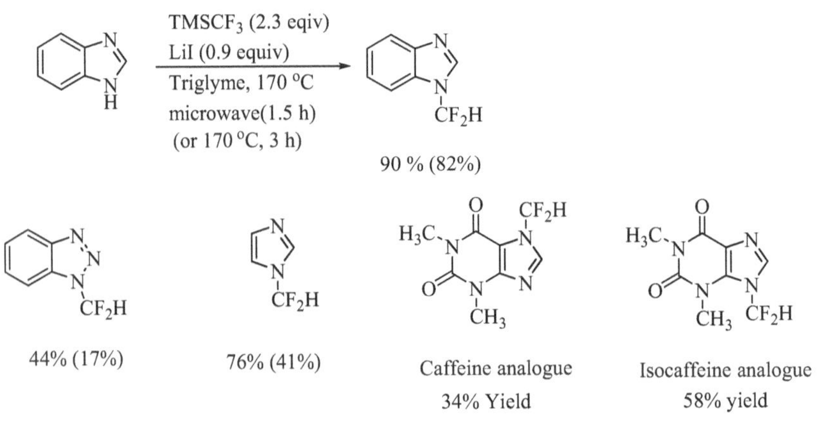

FIGURE 21 *N-gem*-difluoromethylation of azaheterocyclic compounds.

and related compounds into their *N*-difluoromethylated compounds in good to excellent yields under microwave irradiation or through conventional heating. Using this method, theophylline was converted into the *N-gem*-difluoromethylated analogs of caffeine and isocaffeine in good yields (Figure 21).[87] Interestingly, the X-ray crystal structure (as well as the ^{19}F- and ^{1}H NMR data) of the isocaffeine analog indicates the presence of weak intramolecular hydrogen bonding of the fluorines with the adjacent *N*-methyl hydrogens (Figure 21).[87] Although such hydrogen bonding interactions of the fluorine are not as common, there are many instances where fluorine can form stabilizing, weak hydrogen bonds with the neighboring hydrogens, as evident through their X-ray crystal structures.[88–91]

8. TRIFLUOROMETHYLATIONS

Trifluoromethyl substituent is one of the most frequently used groups in the drug development. The CF_3 group helps to improve metabolic stability of the drugs due to the high C–F bond strength and due to its increased oxidative stability. The CF_3 moiety often shows relatively increased binding selectivity at the enzyme active site and is frequently buried in the hydrophobic pocket of the enzymes. As it is relatively highly lipophilic, the trifluoromethylated compounds are generally permeable through the cell membranes, and thus trifluoromethylated compounds have improved bioavailability. A majority of the trifluoromethyl-containing pharmaceuticals have the CF_3 moiety incorporated into the aryl rings, while some others contain a trifluoromethyl ketone pharmacophore. For example, a norfloxacin analog in which the *N*-ethyl group has been replaced by the *N*-CF_3 moiety has similar antibacterial activity in the in vitro studies. Oxidative desulfuration–fluorination of the dithiocarbamate moiety using NBS and PPHF provided a key intermediate for the preparation of the norfloxacin (Eqn (6)).[92]

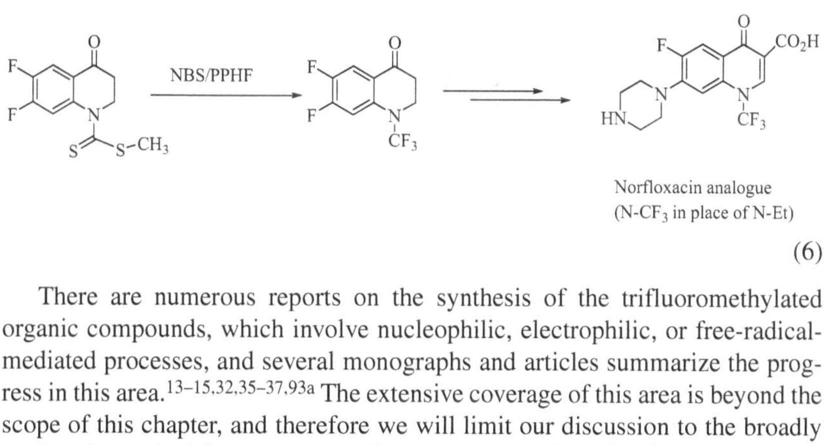

(6)

There are numerous reports on the synthesis of the trifluoromethylated organic compounds, which involve nucleophilic, electrophilic, or free-radical-mediated processes, and several monographs and articles summarize the progress in this area.[13–15,32,35–37,93a] The extensive coverage of this area is beyond the scope of this chapter, and therefore we will limit our discussion to the broadly applicable methodologies, as related to their potential applicability for the synthesis of pharmaceuticals.

8.1 Nucleophilic Trifluoromethylation

Ruppert–Prakash reagent, trifluoromethyltrimethylsilane (CF_3SiMe_3), is the reagent of choice for the trifluoromethylation of carbonyl compounds, since it is relatively easy to handle; is commercially available; and because these trifluoromethylations proceed under mild conditions that tolerate a variety of other functional groups. The trifluoromethyl anion, unlike other perfluoroalkyl anions, is rather unstable, even at low temperatures, and therefore the nucleophilic

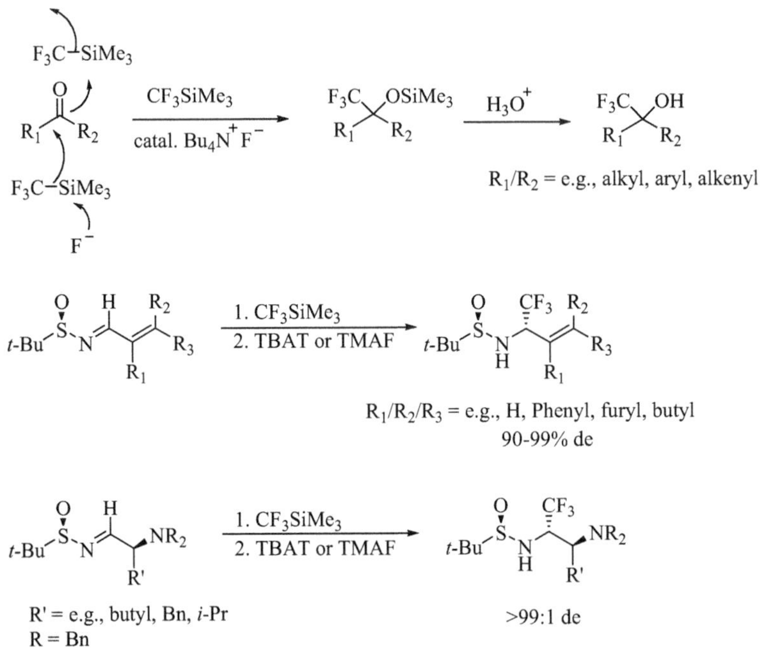

FIGURE 22 Trifluoromethylation of the carbonyl compounds and the chiral *N*-sulfinylimines (KHMDS = potassium hexamethyldisilazane; Bn = benzyl; *i*-Pr = isopropyl; TBAT = tetrabutylammonium triphenyldifluorosilicate; TMAF = tetramethylammonium fluoride).

trifluoromethyl anion is generated in situ from the CF_3SiMe_3 in the presence of tetrabutylammonium fluoride ($Bu_4N^+ F^-$) or other nucleophilic catalysts, including potassium *tert*-butoxide (*t*-BuOK) and trimethylamine oxide.[94] Only a catalytic amount of the fluoride source is sufficient to bring about the nucleophilic reaction, because the alkoxide species formed at each step of the reaction further propagates the reaction until all the starting material has reacted. These CF_3^- transfer reactions proceed through a pentavalent trigonal pyramidal silyl anion intermediate.[95]

Several diastereo- and enantioselective trifluoromethylation reactions of aldehydes and ketones using the CF_3SiMe_3 as the nucleophilic reagent have been reported.[37] Prakash and coworkers obtained high diastereoselectivities (90–99%) in the Michael addition reactions of the chiral α,β-unsaturated *tert*-butanesulfinimines with CF_3SiMe_3.[96] Similarly, the chiral *tert*-butanesulfinylimines of the α-amino aldehydes undergo nucleophilic trifluoromethylation with CF_3SiMe_3 in the presence of a catalytic quantity of TMAF to give the *N*-*tert*-butanesulfinyl-α-trifluoromethyl ethylenediamines in high yields and with high diastereoselectivity (>99:1 diastereomeric ratio) (Figure 22).[97] The chiral sulfinimines derived from various aromatic or aliphatic aldehydes were diastereoselectively trifluoromethylated using CF_3SiMe_3 in the presence

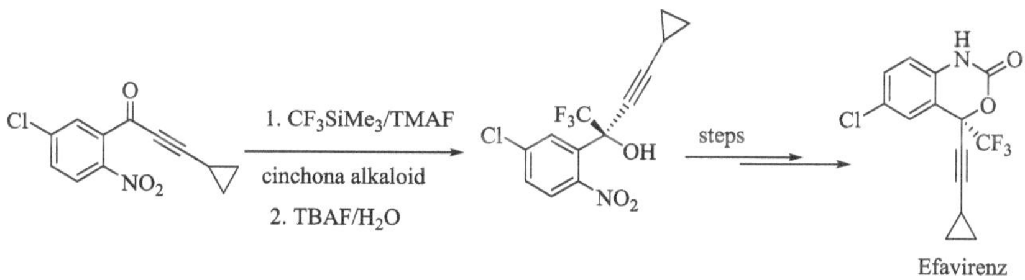

FIGURE 23 Nucleophilic trifluoromethylation towards the synthesis of efavirenz (TBAF = tetrabutylammonium fluoride).

of TBAT to give their corresponding α-trifluoromethyl amines in high yields and with excellent diastereoselectivity.[14,98] Dolbier and coworkers have developed an elegant and relatively convenient method for perfluoroalkylations, trifluoromethylations, and difluoromethylations of a series of electrophiles (e.g., aldehydes, ketones, α-keto esters, and acyl chlorides) using the corresponding perfluoroalkyl (and trifluoromethyl- and difluoromehyl-) halides and tetrakis(diethylamino) ethylene (TDAE) as the reductant.[99]

Whereas there are numerous reports of diastereoselective trifluoromethylation reactions, the enantioselective trifluoromethylation reactions are limited to only limited cases.[19,20,26,100–104] Relatively low enantioselectivities, up to 50% enantiomeric excess, were observed for the trifluoromethylation of carbonyl compounds using the cinchona alkaloids (e.g., *N*-[4(trifluoromethyl)benzyl]cinchonium fluoride).[105] The trifluoromethylation of aromatic aldehydes (e.g., naphthaldehyde) using cinchona-alkaloid-derived phase transfer catalyst in combination with TMAF gave the corresponding trifluoromethylated alcohols in high yields and with moderate to good enantioselectivities (50–70% ee).[20] Using this approach enantioselective synthesis of efavirenz, a nonnucleoside, reverse transcriptase inhibitor of human immunodeficiency virus-1 has been achieved, even though the key step of the trifluoromethylation gives only about 50% enantioselectivity (Figure 23).[106]

Several groups have reported trifluoromethylations using the otherwise greenhouse gas fluoroform (CHF_3) in the presence of electrogenerated bases[107,108] or other strong bases, such as dimsyl potassium (CH_3SOCH_2K) and *t*-BuOK, in dimethylformamide (DMF) solvent.[107,109–111] In these trifluoromethylations, DMF was presumed to stabilize the CF_3^- anion, precluding its dissociation into the difluoromethylcarbene ($:CF_2$). Prakash, Olah, and coworkers achieved direct trifluoromethylation of nonenolizable aromatic aldehydes and ketones, chalcones, formate esters, benzyl bromide, and methyl benzoate, using stoichiometric amounts of CHF_3 and potassium hexamethyldisilazane (Figure 24).[93a] This reaction proceeds in diethyl ether, THF, or toluene solvents at low temperatures, in moderate to good yields. Importantly, the use of DMF is not required to stabilize the intermediate trifluoromethyl anion (CF_3^-) (as the corresponding hemiaminolate), and the reaction workup conditions were greatly simplified. This method is also suitable for the preparation of CF_3SiMe_3 (requiring only stoichiometric amounts of fluoroform). The hitherto unobserved CF_3^- anion, associated with the [K(18-crown-6)] cation

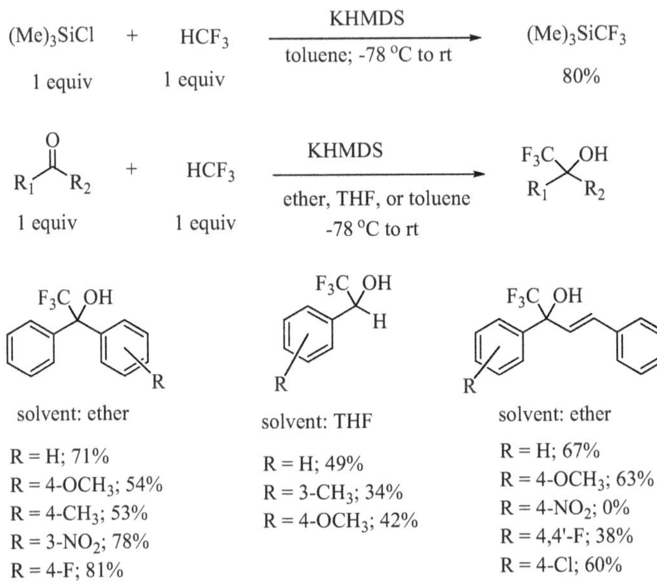

FIGURE 24 Direct trifluoromethylation of carbonyl compounds using fluoroform (KHMDS = potassium hexamethyldisilazide; THF = tetrahydrofuran).

in THF solution, has been unequivocally characterized by low-temperature ^{19}F and ^{13}C NMR spectroscopy, by Prakash, Olah, and coworkers.[93b]

Shibata and coworkers, using a sterically crowded "superbase" P4-*t*-Bu were able to generate the trifluoromethyl carbanion (CF_3^-) in THF, in the absence of the high-boiling DMF solvent, and carried out the trifluoromethylations of aromatic aldehydes and ketones to give their corresponding trifluoromethyl alcohols in good yields (Figure 25).[112] This method, however, requires use of the excess fluoroform.

8.2 Electrophilic Trifluoromethylation

Umemoto and coworkers have developed *S*-trifluoromethyl dibenzothiophenium salts and their oxonium, selenonium, and telluronium analogs as electrophilic trifluoromethylating agents. These compounds differ in their reactivity toward various nucleophiles and their reactivity also depends on the substituents on the aryl rings, with more electron-withdrawing groups (such as NO_2) on the aryl rings increasing their reactivity. The sulfonium-based reagents are relatively more reactive than the selenonium or telluronium analogs and are useful for the conversion of enolsilyl ethers to the α-trifluoromethyl ketones and trifluoromethylation of stabilized carbanions. The selenonium salts are useful for the conversion of highly reactive nucleophiles such as alkynyllithium and thiolates, while the oxonium salts are highly reactive, and can be used for the *O*- and *N*-trifluoromethylation of alcohols and amines (Figure 26).[25]

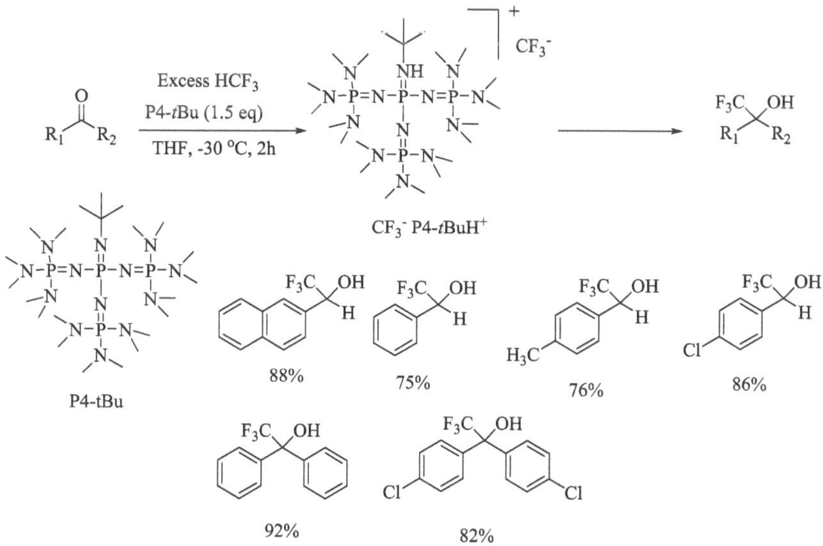

FIGURE 25 Trifluoromethylation of carbonyl compounds using fluoroform and a "superbase" (THF = tetrahydrofuran).

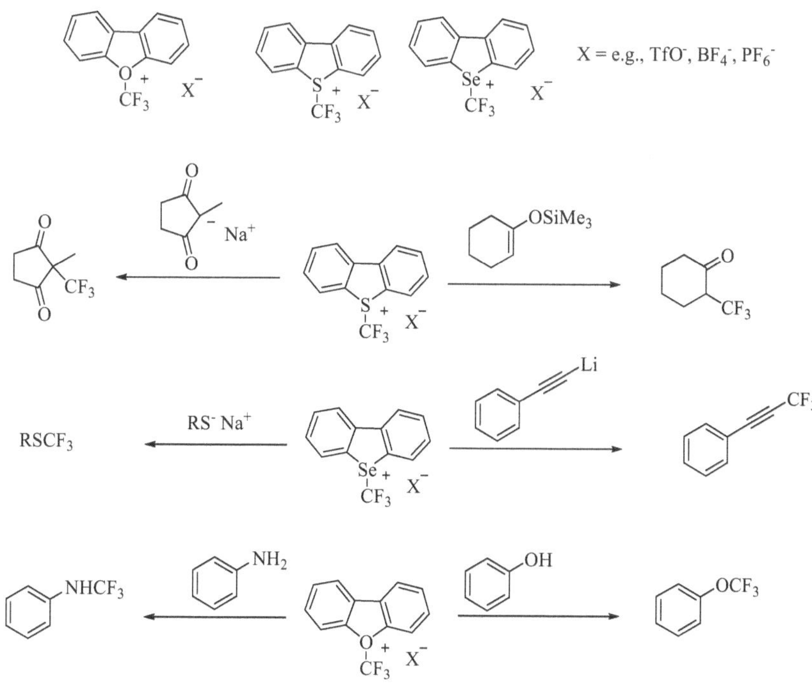

FIGURE 26 Electrophilic trifluoromethylation using sulfonium, selenonium, and oxonium salts (Umemoto's reagents).

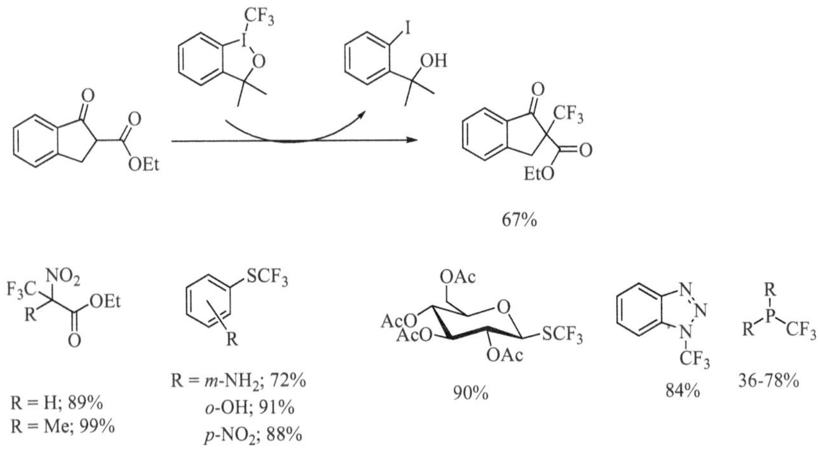

FIGURE 27 Electrophilic trifluoromethylation of carbon-, sulfur-, and phosphorus-centered nucleophiles using Togni's reagent.

Togni's reagent, a hypervalent iodine(III)-CF_3 compound, is a relatively cost-effective alternative to the Umemoto's electrophilic trifluoromethylating agents. It can be used for the α-trifluoromethylation of β-keto esters, α-nitro esters, *S*-trifluoromethylation of thiols, and *N*-trifluoromethylation of the triazoles and azoles (Figure 27).[21,23,24] The α-trifluoromethyl-α-nitro esters are useful precursors for the synthesis of α-trifluoromethyl amino acids.

8.3 Trifluoromethylation by Photoredox Catalysis

Redox photocatalysis mediated by iridium or ruthenium catalysts and a household fluorescent light bulb enables α-trifluoromethylation of aldehydes, using the commercially available trifluoromethyl iodide. Through this elegant photoredox catalysis, and using a chiral imidazolidinone catalyst, MacMillan and coworkers have obtained moderate to high yields of the α-trifluoromethyl aldehydes, with high enantioselectivity (ee 90–99% for 11 examples) (Figure 28).[113] These chiral aldehydes could be oxidized or reduced to the carboxylic acids or alcohols, respectively, or they could be transformed into other functional derivatives, such as α- or β-trifluoromethylamines, with high enantioselectivity.

Photoredox-catalysis-mediated by a ruthenium catalyst as the single-electron transfer agent is useful for the 1,2-difunctionalization of alkenes.[114] Terminal and internal alkenes react with Umemoto's reagent (*S*-trifluoromethyldibenzothiophenium triflate), in the presence of nitriles and catalytic amounts of a photoredox catalyst, to give the corresponding β-trifluoromethylamides (Figure 29).[114] Thus, in the presence of a redox catalyst, electrophilic trifluoromethylating reagents react through a radical mechanism.

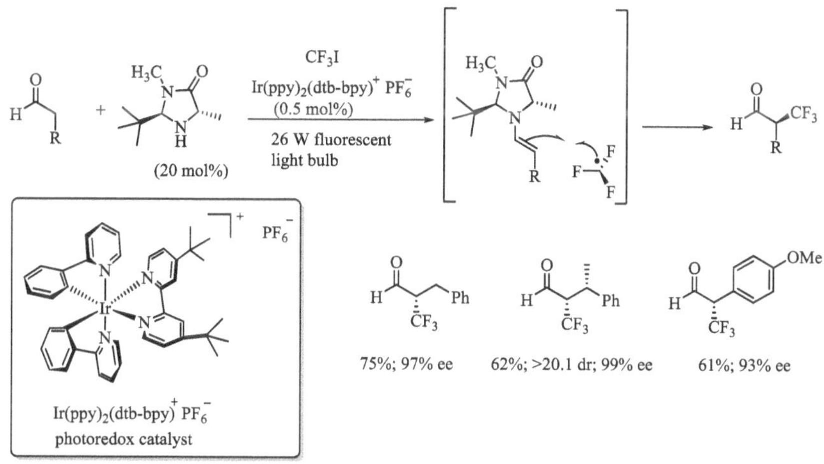

FIGURE 28 Photoredox catalysis for the enantioselective α-trifluoromethylation of aldehydes.

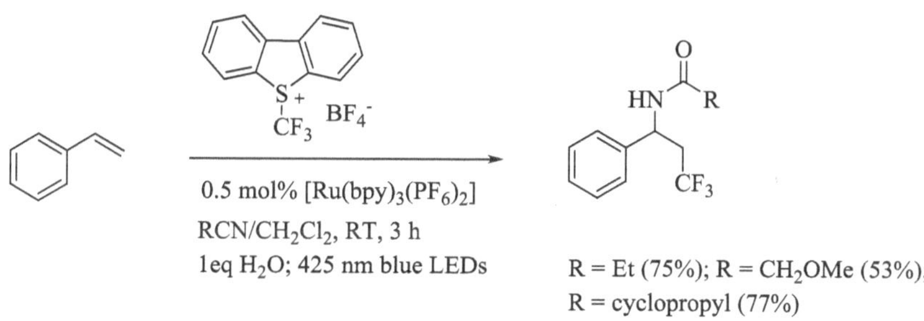

FIGURE 29 Photoredox catalysis for the preparation of β-trifluoromethylamides.

8.4 Pd-Catalyzed Aryl and Vinyl Trifluoromethylations

Buchwald and coworkers have pioneered the palladium-catalyzed trifluoromethylations of aryl chlorides (and aryl bromides, iodides, or triflates).[115] This reaction requires the use of sterically crowded phosphine ligands BrettPhos or Ruphos, and proceeds under relatively mild conditions, at 130–140 °C. High yields of the trifluoromethylations can be achieved using trifluoromethyltriethylsilane (CF_3SiEt_3) and KF (as initiator for the generation of the trifluoromethyl anion), and 3 mol% of the palladium catalyst ($Pd_2(dba)_3$ (dba = dibenzylideneacetone) or $[(allyl)PdCl]_2$). Other reagent combinations, such as CF_3SiMe_3 and CsF, give relatively poor yields. A wide range of aryl chlorides having nitrile, ester, acetal, ether, amide, and tertiary amine functional groups, are trifluoromethylated under these conditions in high yields (Figure 30). Typically 2–5% of the reduction products (ArH) are obtained as byproducts.

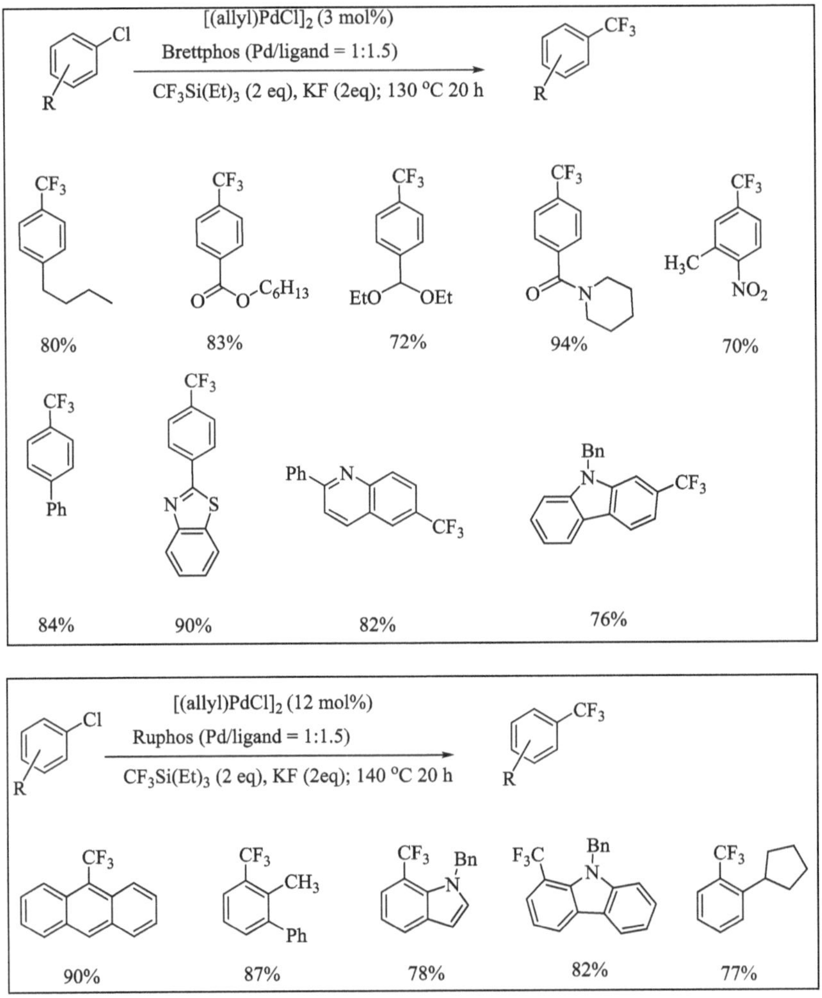

FIGURE 30 Pd(0)-catalyzed aryltrifluoromethylations.

The Pd(0)-catalyzed aryl trifluoromethylation reactions, as in other palladium-catalyzed transformations of aryl halides, involves three steps: (1) oxidative addition of aryl chlorides to Pd(0) to give $L_nPd(II)ArCl$, (2) transmetalation (exchange of the halide by trifluoromethyl anion), and finally (3) reductive elimination of the trifluoromethylarenes to regenerate the Pd(0) (Figure 31).[115] Relatively higher yields are obtained for these Pd(0)-catalyzed trifluoromethylation of aryl bromides in the presence of sodium dodecylsulfate.[116]

Vinyl triflates can be trifluoromethylated using palladium catalysis. However, these reactions do not proceed in the presence of BrettPhos ligand; analogous ligands that lack substituents at the *meta*-position of the upper aromatic ring, such as *tert*-BuXPhos, gave optimal yields of the products (Figure 32).[117]

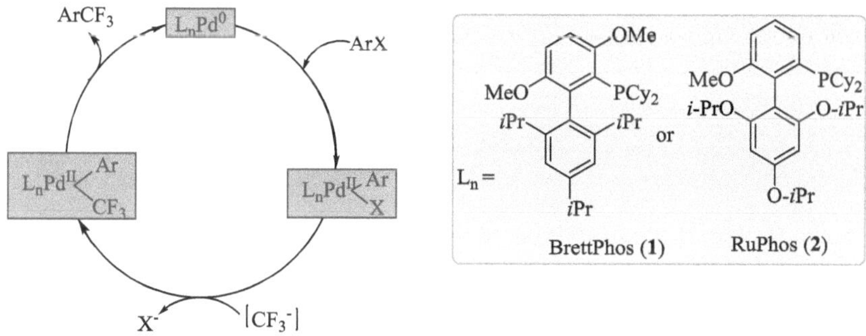

FIGURE 31 Schematic illustration of the Pd(0)-catalyzed trifluoromethylation of aromatic compounds (Cy = cyclohexyl); *i*Pr = isopropyl).[115]

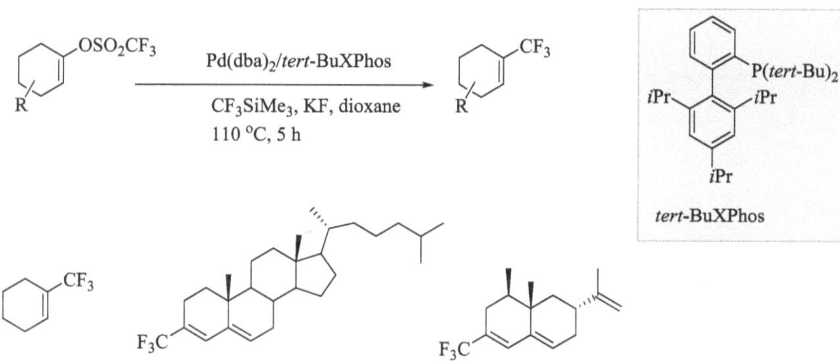

83% 82% 74%

FIGURE 32 Pd(0)-catalyzed trifluoromethylation of vinyl triflates.

9. PENTAFLUOROSULFANYL COMPOUNDS

The electronic properties of the pentafluorosulfanyl (SF_5) group are comparable with that of the CF_3 group (Hammett σ_p values: 0.68 (SF_5) and 0.54 (CF_3)[118]; electronegativities: 3.65 (SF_5) and 3.36 (CF_3)[119,120]). The SF_5 group imparts relatively more lipophilicity to the compounds as compared to that of the CF_3 group (Hansch hydrophobicity constants (π) are 1.51 and 1.09, respectively, for SF_5 and CF_3).[121] The SF_5 moiety has similar steric crowding as that for the *tert*-butyl group, whereas that for the CF_3 group is comparable to the isopropyl group.[122] Additionally, SF_5-derived compounds have relatively high thermal and chemical stabilities, and therefore there is growing interest in the incorporation of the SF_5 group in various pharmaceuticals, agrochemicals, and materials.[27] Synthetic challenges in incorporating the SF_5 group into the desired organic compounds are being continually addressed by several groups. Arylpentafluorosulfanyl compounds can be prepared on large scale either by Bowden's direct fluorination of diaryldisulfides (with electron-withdrawing substituents on the aryl ring, e.g., NO_2), or more conveniently,

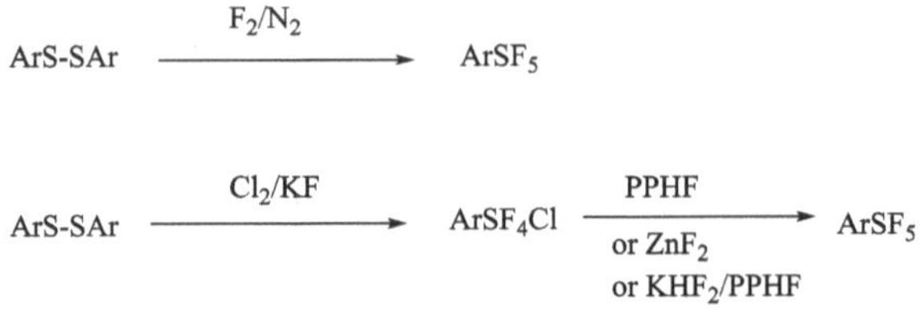

FIGURE 33 Preparation of pentafluorosulfanylarenes.

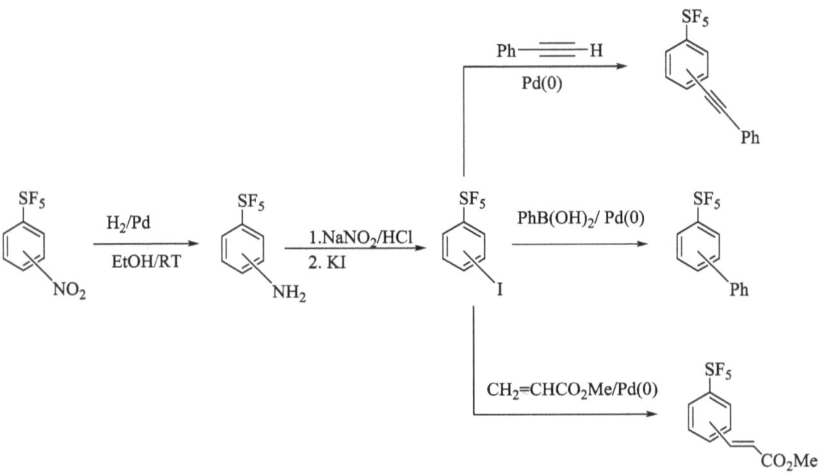

FIGURE 34 Conversion of the nitroaryl sulfur pentafluoride into the alkynyl-, phenyl-, and alkenyl-substituted pentafluorosulfanylarenes (EtOH = ethanol).

as reported by Umemoto, by first reacting diaryldisulfides (containing relatively broader variety of substituents) with Cl_2 in the presence of excess KF to give chlorotetrafluorosulfanylarenes, and then with PPHF or ZnF_2[27] (or using a mixture of KHF_2 and PPHF) (Figure 33).[123]

Bowden and coworkers have transformed nitroarylsulfur pentafluoride, which is readily synthesized from the corresponding diaryldisulfides, using the above procedure, into a variety of substituted pentafluorosulfanyl arenes. For example, catalytic hydrogenation of the nitro group followed by diazotization of the amine in the presence of KI gives the corresponding pentafluorosulfanyl iodoarenes. Heck, Sonogashira, or Suzuki reaction of the latter compound gives a broad range of alkenylated, alkynylated, and arylated derivatives of pentafluorosulfanyl arenes (Figure 34).[124]

Carreira and coworkers have reported the synthesis of the *meta*-SF_5-aryltrifluoroborate salts using Ir catalysis.[125] The Ir-catalyzed *meta*-aryl C-H activation of pentafluorosulfanyl arenes using bis(pinacolato)diboron followed by reaction with KHF_2 gives the corresponding SF_5-substituted potassium aryltrifluoroborates in high yields (72–94%), and these compounds serve as substrates for the Suzuki reactions of a series of aryl bromides (Figure 35).[125]

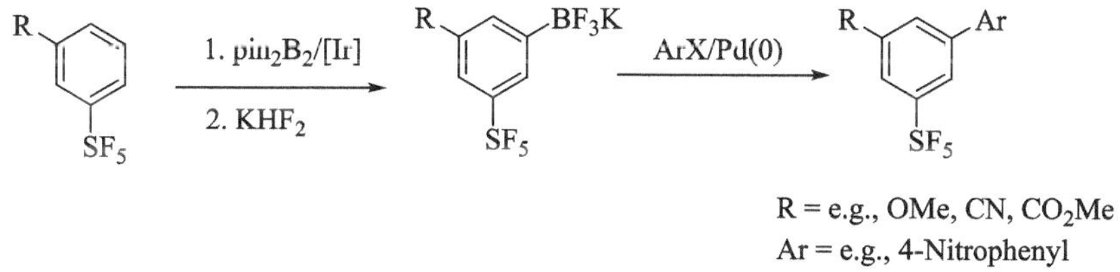

FIGURE 35 Regioselective arylation of pentafluorosulfanylarenes (pin_2B_2) = bis(pinacolato) diboron.

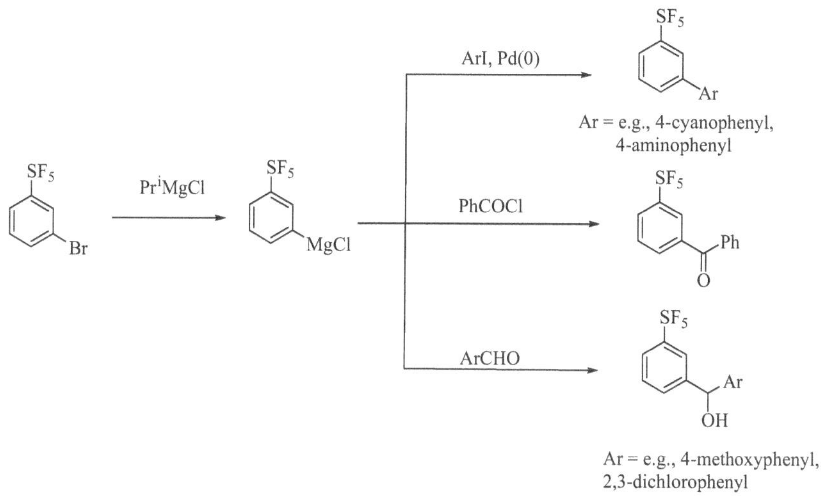

FIGURE 36 Selected Grignard reactions of *m*-bromopentafluorosulfanylbenzene.

Knochel and coworkers have developed synthetic methodologies for a range of polyfunctional SF_5-substituted indoles, benzothiazoles, and substituted aromatics, using zinc, copper, and magnesium intermediates.[126] Br/Mg exchange of 3-bromopentafluorosulfanylbenzene using isopropylmagnesium chloride followed by Pd(0)-mediated cross-coupling with various aryl halides gives 3-arylpentafluorosulfanylbenzene. The intermediate 3-pentafluorosulfanylmagnesium chloride also reacts with various carbonyl compounds to give their corresponding alcohols, or ketones (Figure 36).

Indole derivatives are particularly attractive as synthetic intermediates for many pharmaceuticals. Electrophilic bromination of 3-pentafluorosulfanylaniline using *N*-bromosuccinimide followed by Sonogashira reaction using trimethylsilylacetylene gives the SF_5-substituted 2-((trimethylsilyl) ethynyl)aniline. Treatment of this compound with potassium hydride in *N*-methylpyrrolidinone results in the intramolecular cyclization to give 6-pentafluorosulfanyl-1H-indole.[126] This SF_5-substituted indole derivative could provide valuable intermediate for the synthesis of a broad range of pharmaceuticals (Figure 37).

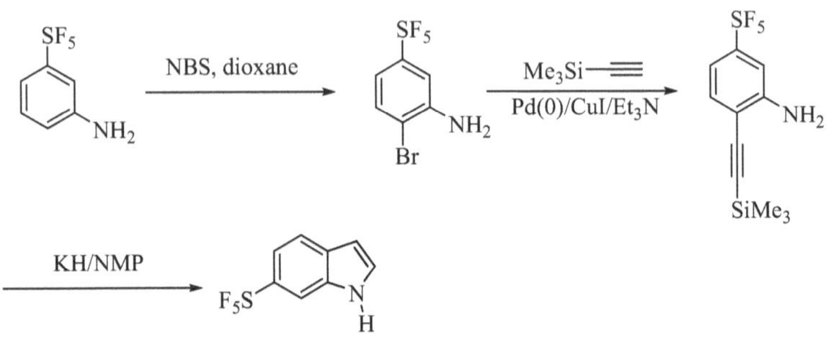

FIGURE 37 Synthesis of 6-pentafluorosulfanyl-1H-indole (NBS = N-bromosuccinimide; NMP = N-methyl-2-pyrrolidone).

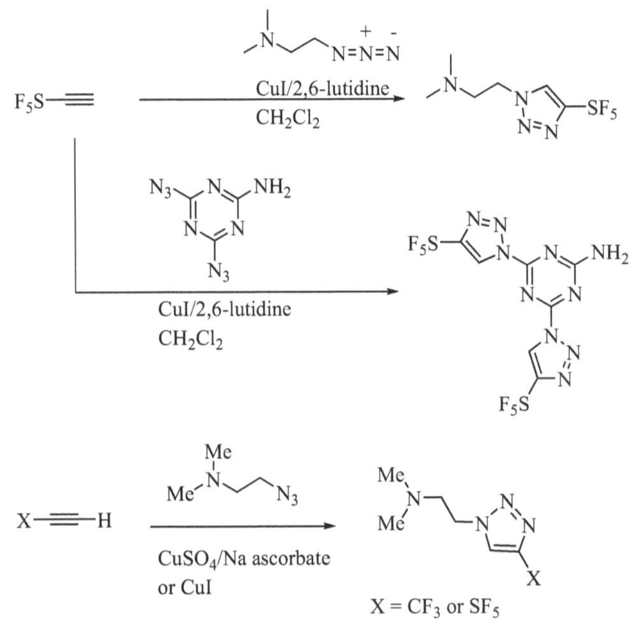

FIGURE 38 Click reactions of SF_5 acetylene.

9.1 Click Reactions of Pentafluorosulfanyl Acetylene

SF_5 acetylene undergoes the Cu(I)-catalyzed 1,3-dipolar cycloaddition (Sharpless Click reaction) to give a broad variety of the 1,4-disubstituted 1,2,3-triazoles with high regioselectivity. As compared to the corresponding trifluoromethyl-derived triazoles, these compounds have relatively more high energy densities.[127] Biologically useful SF_5-functionalized triazoles, analogous to those of the other fluorinated versions, are potentially amenable through this procedure. Similar to the copper(I)-catalyzed click reactions of the 3,3,3-trifluoropropyne, the SF_5 acetylene gives SF_5-substituted 1,2,3-triazoles, which are thermally stable up to 270 °C (Figure 38).[127,128]

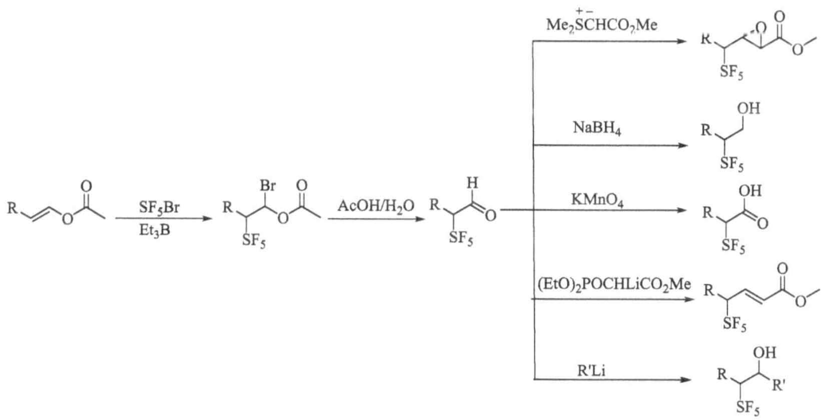

FIGURE 39 Synthesis and further transformations of α-pentafluorosulfanyl aldehydes.

Similarly, pentafluorosulfanyl-aryl diazonium salts, synthesized from the corresponding amines under anhydrous conditions using *tert*-butyl nitrite in the presence of boron trifluoride etherate, undergo Heck reactions with styrene derivatives, in situ, to give the corresponding pentafluorosulfanyl stilbenes in moderate yields.[129]

9.2 Pentafluorosulfanyl Carbonyl Compounds

α-Pentafluorosulfanyl aldehydes can be synthesized in high yields by the free-radical-catalyzed addition of bromopentafluorosulfane (SF_5Br/Et_3B) to the enol acetates, followed by hydrolysis in aqueous acetic acid.[122] The 2-pentafluorosulfanylalkanals thus obtained can be transformed into a variety of functionalized compounds, including the corresponding alcohols, carboxylic acids, glycidic esters, and α,β-unsaturated esters (Figure 39). Whereas the aryl SF_5 group is unstable to the alkyllithium reagents, the alkyl SF_5 moiety is robust to alkyllithium, including the Wittig–Horner reagent and α-silylalkyllithium. The SF_5 group is stable to strong oxidizing agents such as $KMnO_4$ and metal hydride reducing agents so that the α-SF_5 alkanals serve as the starting materials for the synthesis of variously functionalized derivatives.

9.3 SF_5-Containing Pharmaceuticals

9.3.1 SF_5-Mefloquine

Many pharmaceutical-analogs having a pentafluorosulfanylarene moiety have been synthesized for their potential medicinal applications. Wipf and coworkers have synthesized analogs of mefloquine in which 8-CF_3 group is replaced by 8-SF_5, 6-SF_5, and 7-SF_5 groups, towards the goal of reducing the blood–brain barrier and thereby attenuating the neurotoxicity (the central-nervous-system-mediated effects of seizures, depression, and hallucinations) associated with mefloquine (Figure 40).[30,130] In vitro studies of the latter

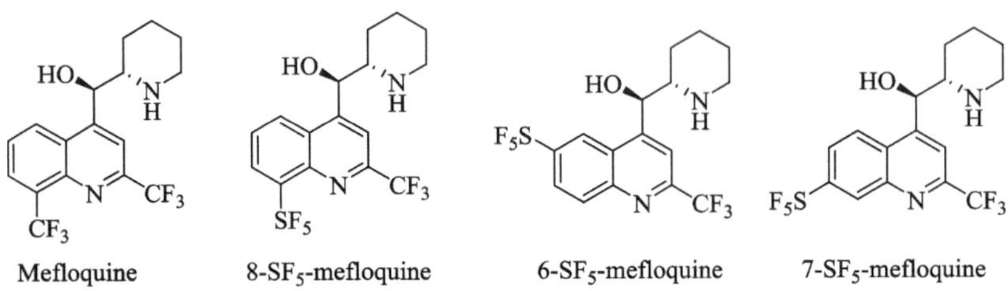

FIGURE 40 Structures of mefloquine and its SF_5 derivatives.

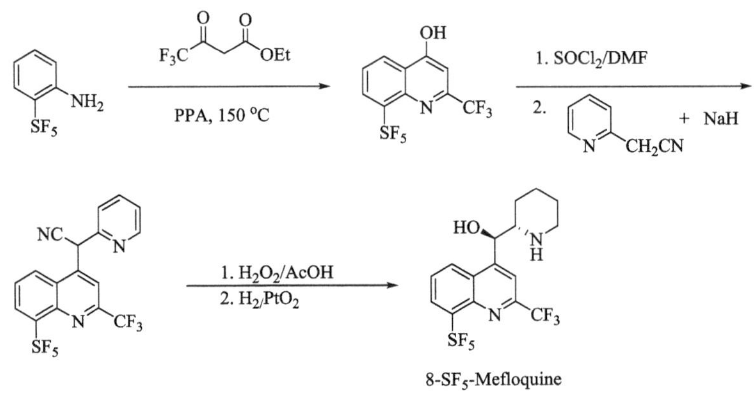

FIGURE 41 Synthesis of 8-SF_5 mefloquine (DMF = dimethylformamide; PPA = polyphosphoric acid).

compounds show that they exhibit equivalent or improved properties as compared to their CF_3 analogs.

The synthesis of the 8-SF_5 analog of mefloquine was achieved starting from the *ortho*-SF_5 aniline, which undergoes Conrad–Limpach reaction with ethyl 4,4,4-trifluoroacetoacetate in polyphosphoric acid (PPA) to give the 8-pentafluorosulfanylquinoline in high yields. Further synthetic transformations involving conversion of the hydroxyl group to chloride, nucleophilic aromatic substitution reaction using 2-pyridylacetonitrile in the presence of NaH, H_2O_2-mediated oxidation of the resulting C–CN bond to the ketone, and Pt-catalyzed reduction of the ketone and pyridine moieties of the latter compound gives the target molecule in moderate to good yields (Figure 41).[130]

9.3.2 *SF_5-Serotonin, Fluoxetine, and Fenfluramine*

Welch and coworkers have synthesized the SF_5 analogs of the selective serotonin (5-hydroxytryptamine) reuptake inhibitors—fluoxetine, fenfluramine, and norfenfluramine—in which the trifluoromethyl moiety is replaced by the pentafluorosulfanyl group (Figure 42).[29] These compounds have comparable or superior selectivity to those of the CF_3 analogs; for example, the SF_5

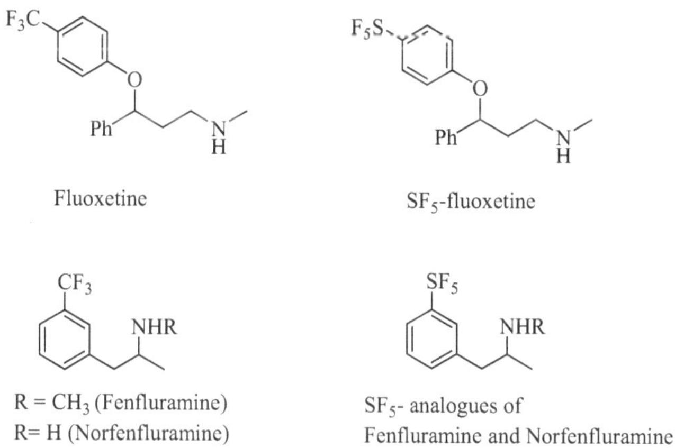

FIGURE 42 Structures of fluoxetine, fenfluramine, and their SF_5 derivatives.

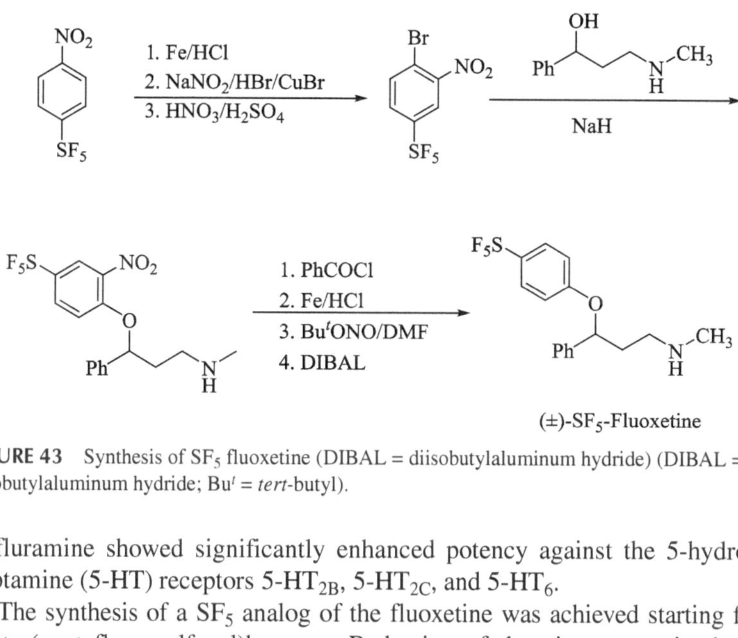

FIGURE 43 Synthesis of SF_5 fluoxetine (DIBAL = diisobutylaluminum hydride) (DIBAL = diisobutylaluminum hydride; Bu^t = *tert*-butyl).

fenfluramine showed significantly enhanced potency against the 5-hydroxytryptamine (5-HT) receptors 5-HT_{2B}, 5-HT_{2C}, and 5-HT_6.

The synthesis of a SF_5 analog of the fluoxetine was achieved starting from *p*-nitro(pentafluorosulfanyl)benzene. Reduction of the nitro group in the latter compound to the amino group, followed by diazotization, and dediazoniation–bromination, then nitration gives the 4-bromo-3-nitro (pentafluorosulfanyl) benzene. Activation of the bromine toward nucleophilic substitution reaction by the α-(*N*-methylethyl)benzyl alcohol is promoted by the *ortho*-nitro group. A subsequent series of reactions that involve protection of the secondary amino moiety, reduction of the nitro to the primary amine, diazotization, and reductive dediazoniation (accompanied by the simultaneous deprotection) gives the SF_5 analog of fluoxetine (Figure 43).[29] These synthetic transformations reveal that

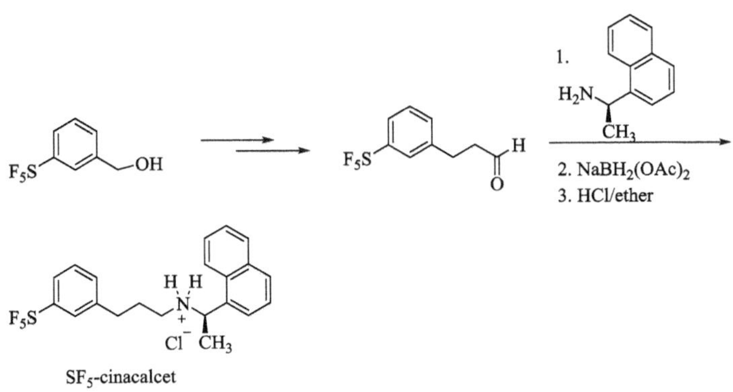

FIGURE 44 Synthesis of SF_5-cinacalcet.

the pentafluorosulfanyl group, as in the case of the trifluoromethyl group, tolerates a wide variety of reaction conditions, and is quite stable to acidic and basic workup.

9.3.3 *SF_5-Cinacalcet*

Cinacalcet (Sensipar, Amgen Pharmaceuticals) is an allosteric activator of the Ca^{2+} receptor, and is used for the treatment of hypercalcemia and hyperparathyroidism. A SF_5-substituted cinacalcet analog was prepared from the *m*-pentafluorosulfanylbenzyl alcohol, in several steps that involve chain homologation by using malonic ester synthesis, Dess–Martin periodinane oxidation of the primary alcohol to the aldehyde, and reductive amination using the (*R*)-1-(1-naphthyl)ethanamine (Figure 44).[31]

Assays using recombinant human calcium receptor showed that it has about 75% agonist activity as compared to that of the cinacalcet. Thus the SF_5 moiety mimics the CF_3 group in terms of its electronic and steric effects for selective binding to the proteins.[31]

10. SUMMARY AND OUTLOOK

Although elemental fluorine, in principle, can be used for the selective fluorination of organic compounds under controlled conditions, due to the technical difficulties involved in the handling of the fluorine gas this method is restricted to the specialized laboratories. Relatively stable electrophilic fluorinating reagents such as Selectfluor and XeF_2 can also be substituted for the elemental fluorine. Deoxyfluorination of alcohols and carbonyl compounds can be accomplished by the use of fluorinating agents, such as DAST, Deoxo-Fluor, and Fluolead. There is an emerging interest in the synthesis of trifluoromethylated pharmaceuticals, and accordingly much effort has been focused on the development of synthetic methodologies for the trifluoromethylations. The trifluoromethylations of carbonyl compounds and other reactions such as *S*-, *O*-, and

N-trifluoromethylations can be achieved through the use of the organometallic, electrophilic, nucleophilic, and radical fluorinating agents. Similarly synthetic methodologies have been developed that permit the synthesis of a variety of SF_5-containing pharmaceutical candidates.

REFERENCES

1. O'Hagan, D. Fluorine in Health Care: Organofluorine Containing Blockbuster Drugs. *J. Fluorine Chem.* **2010,** *131,* 1071–1081.
2. Lou, S.-J.; Xu, D.-Q.; Xia, A.-B.; Wang, Y.-F.; Liu, Y.-K.; Du, X.-H.; Xu, Z.-Y. $Pd(OAc)_2$-Catalyzed Regioselective Aromatic C–H Bond Fluorination. *Chem. Commun.* **2013,** *49,* 6218–6220.
3. Paull, D. H.; Scerba, M. T.; Alden-Danforth, E.; Widger, L. R.; Lectka, T. Catalytic, Asymmetric α-Fluorination of Acid Chlorides: Dual Metal-Ketene Enolate Activation. *J. Am. Chem. Soc.* **2008,** *130,* 17260–17261.
4. Watson, D. A.; Su, M.; Teverovskiy, G.; Zhang, Y.; Garcia-Fortanet, J.; Kinzel, T.; Buchwald, S. L. Formation of ArF from LPdAr(F): Catalytic Conversion of Aryl Triflates to Aryl Fluorides. *Science* **2009,** *325,* 1661–1664.
5. Fier, P. S.; Luo, J.; Hartwig, J. F. Copper-Mediated Fluorination of Arylboronate Esters. Identification of a Copper(III) Fluoride Complex. *J. Am. Chem. Soc.* **2013,** *135,* 2552–2559.
6. Ichiishi, N.; Canty, A. J.; Yates, B. F.; Sanford, M. S. Cu-Catalyzed Fluorination of Diaryliodonium Salts with KF. *Org. Lett.* **2013,** *15,* 5134–5137.
7. Ye, Y.; Schimler, S. D.; Hanley, P. S.; Sanford, M. S. $Cu(OTf)_2$-Mediated Fluorination of Aryltrifluoroborates with Potassium Fluoride. *J. Am. Chem. Soc.* **2013,** *135,* 16292–16295.
8. Chambers, R. D.; Sandford, G. Durham Microreactors for Direct Fluorination. *Chim. Oggi* **2004,** *22,* 6–8.
9. Chambers, R. D.; Spink, R. C. H. Microreactors for Elemental Fluorine. *Chem. Commun.* **1999,** 883–884.
10. Stenhagen, I. S. R.; Kirjavainen, A. K.; Forsback, S. J.; Jorgensen, C. G.; Robins, E. G.; Luthra, S. K.; Solin, O.; Gouverneur, V. [^{18}F]Fluorination of an Arylboronic Ester using [^{18}F] Selectfluor bis(triflate): Application to 6-[^{18}F]fluoro-l-DOPA. *Chem. Commun.* **2013,** *49,* 1386–1388.
11. Olah, G. A.; Welch, J. T.; Vankar, Y. D.; Nojima, M.; Kerekes, I.; Olah, J. A. Synthetic Methods and Reactions. 63. Pyridinium Poly(Hydrogen Fluoride) (30% Pyridine-70% Hydrogen Fluoride): A Convenient Reagent for Organic Fluorination Reactions. *J. Org. Chem.* **1979,** *44,* 3872–3881.
12. Reddy, V. P.; Prakash, G. K. S.; Olah, G. A. Onium-Poly Hydrogen Fluorides as Acid Catalysts, Ionic Liquids and Fluorinating Agents in Organic Reactions. *Adv. Org. Synth.* **2006,** *2,* 183–211.
13. Prakash, G. K. S.; Hu, J. Selective Fluoroalkylations with Fluorinated Sulfones, Sulfoxides, and Sulfides. *Acc. Chem. Res.* **2007,** *40,* 921–930.
14. Prakash, G. K. S.; Mandal, M. Nucleophilic Trifluoromethylation Tamed. *J. Fluorine Chem.* **2001,** *112,* 123–131.
15. Prakash, G. K. S.; Yudin, A. K. Perfluoroalkylation with Organosilicon Reagents. *Chem. Rev.* **1997,** *97,* 757–786.
16. Wiedemann, J.; Heiner, T.; Mloston, G.; Prakash, G. K. S.; Olah, G. A. Synthetic Methods and Reactions. Part 201. Direct Preparation of Trifluoromethyl Ketones from Carboxylic Esters: Trifluoromethylation with (Trifluoromethyl)trimethylsilane. *Angew. Chem. Int. Ed.* **1998,** *37,* 820–821.

17. Bastos, R. S. (Trifluoromethyl)trimethylsilane ($TMSCF_3$) - Ruppert's Reagent: An Excellent Trifluoromethylation Agent. *Synlett* **2008,** 1425–1426.
18. Gouverneur, V.; Lozano, O. Asymmetric Fluorination, Monofluoromethylation, Difluoromethylation, and Trifluoromethylation Reactions, Science of Synthesis. *Stereosel. Synth.* **2011,** *3,* 851–930.
19. Kawai, H.; Tachi, K.; Tokunaga, E.; Shiro, M.; Shibata, N. Cinchona Alkaloid-Catalyzed Asymmetric Trifluoromethylation of Alkynyl Ketones with Trimethylsilyl Trifluoromethane. *Org. Lett.* **2010,** *12,* 5104–5107.
20. Kawai, H.; Mizuta, S.; Tokunaga, E.; Shibata, N. Cinchona Alkaloid/TMAF Combination: Enantioselective Trifluoromethylation of Aryl Aldehydes. *J. Fluorine Chem.* **2013,** *152,* 46–50.
21. Eisenberger, P.; Kieltsch, I.; Armanino, N.; Togni, A. Mild Electrophilic Trifluoromethylation of Secondary and Primary Aryl- and Alkylphosphines Using Hypervalent Iodine(III)-CF_3 Reagents. *Chem. Commun.* **2008,** 1575–1577.
22. Fang, Z.; Ning, Y.; Mi, P.; Liao, P.; Bi, X. Catalytic C–H α-Trifluoromethylation of α,β-Unsaturated Carbonyl Compounds. *Org. Lett.* **2014,** *16,* 1522–1525.
23. Kieltsch, I.; Eisenberger, P.; Togni, A. Mild Electrophilic Trifluoromethylation of Carbon- and Sulfur-centered Nucleophiles by a Hypervalent Iodine(III)-CF_3 Reagent. *Angew. Chem. Int. Ed.* **2007,** *46,* 754–757.
24. Niedermann, K.; Frueh, N.; Senn, R.; Czarniecki, B.; Verel, R.; Togni, A. Direct Electrophilic *N*-Trifluoromethylation of Azoles by a Hypervalent Iodine Reagent. *Angew. Chem. Int. Ed.* **2012,** *51,* 6511–6515.
25. Umemoto, T. Recent Advances in Perfluoroalkylation Methodology. *ACS Symp. Ser.* **2005,** *911,* 2–15.
26. Umemoto, T.; Adachi, K. New Method for Trifluoromethylation of Enolate Anions and Applications to Regio-, Diastereo- and Enantioselective Trifluoromethylation. *J. Org. Chem.* **1994,** *59,* 5692–5699.
27. Umemoto, T.; Garrick, L. M.; Saito, N. Discovery of Practical Production Processes for Arylsulfur Pentafluorides and Their Higher Homologs, Bis- and Tris-(Sulfur Pentafluorides): Beginning of a New Era of Super-Trifluoromethyl Arene Chemistry and its Industry. *Beilstein J. Org. Chem.* **2012,** *8,* 461–471. No 453.
28. Umemoto, T.; Singh, R. P.; Xu, Y.; Saito, N. Discovery of 4-tert-butyl-2,6-dimethylphenylsulfur Trifluoride as a Deoxofluorinating Agent with High Thermal Stability as Well as Unusual Resistance to Aqueous Hydrolysis, and Its Diverse Fluorination Capabilities Including Deoxofluoro-Arylsulfinylation with High Stereoselectivity. *J. Am. Chem. Soc.* **2010,** *132,* 18199–18205.
29. Welch, J. T.; Lim, D. S. The Synthesis and Biological Activity of Pentafluorosulfanyl Analogs of Fluoxetine, Fenfluramine, and Norfenfluramine. *Bioorg. Med. Chem.* **2007,** *15,* 6659–6666.
30. Wipf, P.; Mo, T.; Geib, S. J.; Caridha, D.; Dow, G. S.; Gerena, L.; Roncal, N.; Milner, E. E. Synthesis and Biological Evaluation of the First Pentafluorosulfanyl Analogs of Mefloquine. *Org. Biomol. Chem.* **2009,** *7,* 4163–4165.
31. Chia, P. W.; Brennan, S. C.; Slawin, A. M. Z.; Riccardi, D.; O'Hagan, D. Allosteric Agonists of the Calcium Receptor (CaR): Fluorine and SF_5 Analogues of Cinacalcet. *Org. Biomol. Chem.* **2012,** *10,* 7922–7927.
32. Kirsch, P. *Modern Fluoroorganic Chemistry: Synthesis, Reactivity, Applications*, 2nd ed.; Wiley-VCH, 2013.
33. Uneyama, K. Recent Advances in the Syntheses of Fluorinated Amino Acids. In *Fluorine in Medicinal Chemistry and Chemical Biology;* Ojima, Iwao, Ed.; John Wiley: Chichster, UK, 2009; pp 213–256.
34. Uneyama, K. *Organofluorine Chemistry;* Blackwell Publishing Ltd.: Oxford, UK, 2006.

35. Ma, J.-A.; Cahard, D. Strategies for Nucleophilic, Electrophilic, and Radical Trifluoromethylations. *J. Fluorine Chem.* **2007,** *128,* 975–996.
36. Liu, X.; Xu, C.; Wang, M.; Liu, Q. Trifluoromethyltrimethylsilane: Nucleophilic Trifluoromethylation and beyond. *Chem. Rev.* **2014**. http://dx.doi.org/10.1021/cr400473a. Ahead of Print, *Chem. Abstr.* 2014:670890.
37. Ma, J.-A.; Cahard, D. Update 1 of: Asymmetric Fluorination, Trifluoromethylation, and Perfluoroalkylation Reactions. *Chem. Rev.* **2008,** *108,* PR1–PR43. Web Only Content.
38. Wang, J.; Sanchez-Rosello, M.; Acena, J. L.; del Pozo, C.; Sorochinsky, A. E.; Fustero, S.; Soloshonok, V. A.; Liu, H. Fluorine in Pharmaceutical Industry: Fluorine-Containing Drugs Introduced to the Market in the Last Decade (2001–2011). *Chem. Rev.* **2014,** *114,* 2432–2506.
39. Chambers, R. D. Footsteps of a Fluorine Chemist. *J. Fluorine Chem.* **2010,** *131,* 665–675.
40. Adcock, J. L.; Cherry, M. L. Aerosol Direct Fluorination: A Developing Synthesis Technology and an Entry Level Mechanistic Tool. A Short Review. *Ind. Eng. Chem. Res.* **1987,** *26,* 208–215.
41. Chambers, R. D.; Fox, M. A.; Sandford, G.; Trmcic, J.; Goeta, A. Elemental Fluorine. Part 20. Direct Fluorination of Deactivated Aromatic Systems Using Microreactor Techniques. *J. Fluorine Chem.* **2006,** *128,* 29–33.
42. Sandford, G. Elemental Fluorine in Organic Chemistry (1997–2006). *J. Fluorine Chem.* **2007,** *128,* 90–104.
43. Adcock, J. L.; Cherry, M. L. Aerosol Direct Fluorination: Direct Synthesis of Perfluorinated Glyme Ethers. *J. Fluorine Chem.* **1985,** *30,* 343–350.
44. Chambers, R. D.; Grievson, B.; Drakesmith, F. G.; Powell, R. L. Free Radical Chemistry. Part 5. A New Approach to the Synthesis of Perfluorinated Ethers. *J. Fluorine Chem.* **1985,** *29,* 323–339.
45. Turnbull, M. D.; Carter, N. B.; Dennison, S.; Deacon, J.; Holley, R. Microfluidics-Based Reactors for Safe Fluorinations Using Elemental Fluorine. *Chimia* **2004,** *58,* 159–162.
46. Rozen, S. Direct Fluorination of Organic Compounds with Elemental Fluorine. In *Efficient Preparation of Fluorine Compounds;* Roesky, H., Ed.; Wiley: New York, 2013; pp 146–153.
47. Rozen, S. Electrophilic Fluorination with Elemental Fluorine and Reagents Derived from It. *Adv. Org. Synth.* **2006,** *2,* 3–41.
48. Breddemann, U.; Debackere, J. R.; Schrobilgen, G. J. A Room-Temperature Non-Irradiative Synthesis of XeF_2. *Eff. Prep. Fluorine Compd.* **2013,** 11–15.
49. Vasdev, N.; Pointner, B. E.; Chirakal, R.; Schrobilgen, G. J. On the Preparation of Fluorine-18 Labelled XeF_2 and Chemical Exchange between Fluoride Ion and XeF_2. *J. Am. Chem. Soc.* **2002,** *124,* 12863–12868.
50. Gillespie, R. J.; Liang, J. Superacid Solutions in Hydrogen Fluoride. *J. Am. Chem. Soc.* **1988,** *110,* 6053–6057.
51. Olah, G. A.; Surya Prakash, G. K.; Molnar, A.; Sommer, J. *Superacid Chemistry,* 2nd ed.; John Wiley & Sons, Inc.: Hoboken, New Jersey, 2009.
52. Bucsi, I.; Toeroek, B.; Marco, A. I.; Rasul, G.; Prakash, G. K. S.; Olah, G. A. Stable Dialkyl Ether/Poly(Hydrogen Fluoride) Complexes: Dimethyl Ether/Poly(Hydrogen Fluoride), a New, Convenient, and Effective Fluorinating Agent. *J. Am. Chem. Soc.* **2002,** *124,* 7728–7736.
53. Yoshino, H.; Matsumoto, K.; Hagiwara, R.; Ito, Y.; Oshima, K.; Matsubara, S. Fluorination with Ionic Liquid $EMIMF(HF)_{2.3}$ as Mild HF Source. *J. Fluorine Chem.* **2006,** *127,* 29–35.
54. Noritake, S.; Shibata, N.; Kawai, H.; Pandy, M. K.; Nakamura, S.; Toru, T. Ionic Liquids as Media for Nucleophilic Ring Opening Fluorination of Aziridines. *Heterocycl. Commun.* **2009,** *15,* 105–113.

55. Segal, E. B. First Aid for a Unique Acid, HF: A Sequel. *Chem. Health Saf.* **2000,** *7,* 18–23.
56. Hayashi, M.; Hashimoto, S.; Noyori, R. Simple Synthesis of Glycosyl Fluorides. *Chem. Lett.* **1984,** 1747–1750.
57. Szarek, W. A.; Grynkiewicz, G.; Doboszewski, B.; Hay, G. W. The Synthesis of Glycosyl Fluorides Using Pyridinium Poly(Hydrogen Fluoride). *Chem. Lett.* **1984,** 1751–1754.
58. Broder, W.; Kunz, H. A New Method of Anomeric Protection and Activation Based on the Conversion of Glycosyl Azides into Glycosyl Fluorides. *Carbohydr. Res.* **1993,** *249,* 221–241.
59. Fier, P. S.; Hartwig, J. F. Selective C–H Fluorination of Pyridines and Diazines Inspired by a Classic Amination Reaction. *Science* **2013,** *342,* 956–960.
60. Middleton, W. J. New Fluorinating Reagents. Dialkylaminosulfur Fluorides. *J. Org. Chem.* **1975,** *40,* 574–578.
61. Banks, R. E., Smart, B. E., Tatlow, J. C., Eds. *Organofluorine Chemistry: Principles and Commercial Applications;* Plenum: New York, 1994.
62. Brown, K.; Dixey, M.; Weymouth-Wilson, A.; Linclau, B. The Synthesis of Gemcitabine. *Carbohydr. Res.* **2014,** *387,* 59–73.
63. Albert, M.; Repetschnigg, W.; Ortner, J.; Gomes, J.; Paul, B. J.; Illaszewicz, C.; Weber, H.; Steiner, W.; Dax, K. Simultaneous Detection of Different Glycosidase Activities by ^{19}F NMR Spectroscopy. *Carbohydr. Res.* **2000,** *327,* 395–400.
64. Wong, A. W.; He, S.; Withers, S. G. Synthesis of 5-fluoro-β-D-glucopyranosyluronic Acid Fluoride and its Evaluation as a Mechanistic Probe of *Escherichia coli* β-glucuronidase. *Can. J. Chem.* **2001,** *79,* 510–518.
65. Messina, P. A.; Mange, K. C.; Middleton, W. J. Aminosulfur Trifluorides: Relative Thermal Stability. *J. Fluorine Chem.* **1989,** *42,* 137–143.
66. Lal, G. S.; Pez, G. P.; Pesaresi, R. J.; Prozonic, F. M.; Cheng, H. Bis(2-methoxyethyl)Aminosulfur Trifluoride: A New Broad-spectrum Deoxofluorinating Agent with Enhanced Thermal Stability. *J. Org. Chem.* **1999,** *64,* 7048–7054.
67. Posner, G. H.; Haines, S. R. A Convenient, One-step, High-yield Replacement of an Anomeric Hydroxyl Group by a Fluorine Atom Using DAST. Preparation of Glycosyl Fluorides. *Tetrahedron Lett.* **1985,** *26,* 5–8.
68. L'Heureux, A.; Beaulieu, F.; Bennett, C.; Bill, D. R.; Clayton, S.; La Flamme, F.; Mirmehrabi, M.; Tadayon, S.; Tovell, D.; Couturier, M. Aminodifluorosulfinium Salts: Selective Fluorination Reagents with Enhanced Thermal Stability and Ease of Handling. *J. Org. Chem.* **2010,** *75,* 3401–3411.
69. Xu, W.; Martinez, H.; Dolbier, W. R., Jr. Arylsulfur Trifluorides: Improved Method of Synthesis and Use as In situ Deoxofluorination Reagents. *J. Fluorine Chem.* **2011,** *132,* 482–488.
70. Chia, P. W.; Livesey, M. R.; Slawin, A. M. Z.; Mourik, T. v.; Wyllie, D. J. A.; O'Hagan, D. 3-Fluoro-*N*-methyl-L-aspartic Acid (3*F*-NMDA) Stereoisomers as Conformational Probes for Exploring Agonist Binding at NMDA Receptors. *Chem. Eur. J.* **2012,** *18,* 8813–8819. S8813/8811-S8813/8816.
71. Yu, Z.; Lv, Y.; Yu, C. A Continuous Kilogram-Scale Process for the Manufacture of *O*-Difluorobenzene. *Org. Process Res. Dev.* **2012,** *16,* 1669–1672.
72. Laali, K. K.; Gettwert, V. J. Fluorodediazoniation in Ionic Liquid Solvents: New Life for the Balz–Schiemann Reaction. *J. Fluorine Chem.* **2001,** *107,* 31–34.
73. Olah, G. A.; Shih, J. G.; Prakash, G. K. S. Fluorine-Containing Reagents in Organic Synthesis. *J. Fluorine Chem.* **1986,** *33,* 377–395.
74. Qing, F.-L.; Zheng, F. Synthesis of Trifluoromethylated and *gem*-difluoromethylenated Biologically Interesting Compounds from Fluorine-containing Synthons. *Synlett* **2011,** 1052–1072.

75. Sondej, S. C.; Katzenellenbogen, J. A. *gem*-Difluoro Compounds: A Convenient Preparation from Ketones and Aldehydes by Halogen Fluoride Treatment of 1,3-dithiolanes. *J. Org. Chem.* **1986,** *51,* 3508–3513.
76. (a) Reddy, V. P.; Alleti, R.; Perambuduru, M. K.; Welz-Biermann, U.; Buchholz, H.; Prakash, G. K. S. *gem*-Difluorination of 2,2-diaryl-1,3-dithiolanes by Selectfluor and Pyridinium Polyhydrogen Fluoride. *Chem. Commun.* **2005,** 654–656; (b) Prakash, G. K. S.; Hoole, D.; Reddy, V. P.; Olah, G. A. Synthetic methods and reactions. 187. Simplified preparation of α,α-difluorodiphenylmethanes from benzophenone 1,3-dithiolanes with sulfuryl chloride and pyridinium polyhydrogen fluoride. *Synlett* **1993,** 691–693.
77. York, C.; Prakash, G. K. S.; Olah, G. A. Synthetic Methods and Reactions. 196. Desulfuration Fluorination Using Nitrosonium Tetrafluoroborate and Pyridinium Poly(Hydrogen Fluoride). *Tetrahedron* **1996,** *52,* 9–14.
78. Fuchigami, T.; Mitomo, K.; Ishii, H.; Konno, A. Electrolytic Partial Fluorination of Organic Compounds Part 44. Anodic *gem*-difluorodesulfurization Using Triarylamine Mediators. *J. Electroanal. Chem.* **2001,** *507,* 30–33.
79. Yoshiyama, T.; Fuchigami, T. Electrolytic Partial Fluorination of Organic Compounds. 5. Anodic *gem*-difluorination of Dithioacetals. *Chem. Lett.* **1992,** 1995–1998.
80. Motherwell, W. B.; Wilkinson, J. A. Observations on the Reaction of Dithioketals with *para*-iodotoluene Difluoride: A Novel Route to *gem*-Difluoro Compounds. *Synlett* **1991,** 191–192.
81. Cohen, O.; Rozen, S. A General Route for the Synthesis of β,β-difluorocarboxylic Acids. *Tetrahedron* **2008,** *64,* 5362–5364.
82. Prakash, G. K. S.; Reddy, V. P.; Li, X. Y.; Olah, G. A. Synthetic Methods and Reactions. 156. Convenient Transformation of Carbonyl Compounds via Their Hydrazones into *gem*-Difluoro Compounds with *N*-bromosuccinimide/pyridinium Poly(Hydrogen Fluoride) or Poly-4-vinylpyridinium Poly(Hydrogen Fluoride). *Synlett* **1990,** 594–596.
83. Reddy, V. P.; Perambuduru, M.; Alleti, R. Synthetic Approaches to *gem*-Difluoromethylene Compounds. *Adv. Org. Synth.* **2006,** *2,* 327–351.
84. Qiu, X.-L.; Yue, X.; Qing, F.-L. Fluorine-Containing Chiral Drugs. *Chiral Drugs* **2011,** 195–251.
85. Prakash, G. K. S.; Ni, C.; Wang, F.; Zhang, Z.; Haiges, R.; Olah, G. A. Difluoro(sulfinato) methylation of N-sulfinyl imines facilitated by 2-pyridyl sulfone: Stereoselective synthesis of difluorinated β-amino sulfonic acids and peptidosulfonamides. *Angew. Chem., Int. Ed.* **2013,** *52,* 10835–10839.
86. Wang, F.; Luo, T.; Hu, J.-B.; Wang, Y.; Krishnan, H. S.; Jog, P. V.; Ganesh, S. K.; Prakash, G. K. S.; Olah, G. A. Synthesis of *gem*-Difluorinated Cyclopropanes and Cyclopropenes: Trifluoromethyltrimethylsilane as a Difluorocarbene Source. *Angew. Chem. Int. Ed.* **2011,** *50,* 7153–7157.
87. Prakash, G. K. S.; Krishnamoorthy, S.; Ganesh, S. K.; Kulkarni, A.; Haiges, R.; Olah, G. A. *N*-Difluoromethylation of Imidazoles and Benzimidazoles Using the Ruppert-Prakash Reagent under Neutral Conditions. *Org. Lett.* **2014,** *16,* 54–57.
88. Schneider, H.-J. Hydrogen Bonds with Fluorine. Studies in Solution, in Gas Phase and by Computations, Conflicting Conclusions from Crystallographic Analyses. *Chem. Sci.* **2012,** *3,* 1381–1394.
89. Dalvit, C.; Vulpetti, A. Intermolecular and Intramolecular Hydrogen Bonds Involving Fluorine Atoms: Implications for Recognition, Selectivity, and Chemical Properties. *Chem. Med. Chem.* **2012,** *7,* 262–272.
90. Howard, J. A. K.; Hoy, V. J.; O'Hagan, D.; Smith, G. T. How Good Is Fluorine as a Hydrogen Bond Acceptor? *Tetrahedron* **1996,** *52,* 12613–12622.
91. Dunitz, J. D.; Taylor, R. Organic Fluorine Hardly Ever Accepts Hydrogen Bonds. *Chem. Eur. J.* **1997,** *3,* 89–98.

92. Asahina, Y.; Araya, I.; Iwase, K.; Iinuma, F.; Hosaka, M.; Ishizaki, T. Synthesis and Antibacterial Activity of the 4-quinolone-3-carboxylic Acid Derivatives Having a Trifluoromethyl Group as a Novel *N*-1 Substituent. *J. Med. Chem.* **2005,** *48,* 3443–3446.
93. (a) Prakash, G. K. S.; Jog, P. V.; Batamack, P. T. D.; Olah, G. A. Taming of Fluoroform: Direct Nucleophilic Trifluoromethylation of Si, B, S, and C Centers. *Science* **2012,** *338,* 1324–1327; (b) Prakash, G. K. S.; Wang, F.; Zhang, Z.; Haiges, R.; Rahm, M.; Christe, K. O.; Mathew, T.; Olah, G. A. Long-Lived Trifluoromethanide Anion: A Key Intermediate in Nucleophilic Trifluoromethylations. *Angew. Chem., Int. Ed.* **2014,** *53,* 1–5.
94. Prakash, G. K. S.; Mandal, M.; Panja, C.; Mathew, T.; Olah, G. A. Preparation of TMS Protected Trifluoromethylated Alcohols Using Trimethylamine *N*-oxide and Trifluoromethyltrimethylsilane ($TMSCF_3$). *J. Fluorine Chem.* **2003,** *123,* 61–63.
95. Prakash, G. K. S.; Panja, C.; Vaghoo, H.; Surampudi, V.; Kultyshev, R.; Mandal, M.; Rasul, G.; Mathew, T.; Olah, G. A. Facile Synthesis of TMS-Protected Trifluoromethylated Alcohols Using Trifluoromethyltrimethylsilane ($TMSCF_3$) and Various Nucleophilic Catalysts in DMF. *J. Org. Chem.* **2006,** *71,* 6806–6813.
96. Prakash, G. K. S.; Mandal, M.; Olah, G. A. Asymmetric Synthesis of Trifluoromethylated Allylic Amines Using α,β-Unsaturated *N-tert*-butanesulfinimines. *Org. Lett.* **2001,** *3,* 2847–2850.
97. Prakash, G. K. S.; Mandal, M. Stereoselective Synthesis of Trifluoromethylated Vicinal Ethylenediamines with α-Amino *N-tert*-Butanesulfinimines and $TMSCF_3$. *J. Am. Chem. Soc.* **2002,** *124,* 6538–6539.
98. Prakash, G. K. S.; Mandal, M.; Olah, G. A. Stereoselective Nucleophilic Trifluoromethylation of *N*-(*tert*-butylsulfinyl)imines by Using Trimethyl(trifluoromethyl)silane. *Angew. Chem. Int. Ed.* **2001,** *40,* 589–590.
99. Medebielle, M.; Dolbier, W. R., Jr. Nucleophilic Difluoromethylation and Trifluoromethylation Using Tetrakis(dimethylamino)ethylene (TDAE) Reagent. *J. Fluorine Chem.* **2008,** *129,* 930–942.
100. Shibata, N.; Mizuta, S.; Kawai, H. Recent Advances in Enantioselective Trifluoromethylation Reactions. *Tetrahedron: Asymmetry* **2008,** *19,* 2633–2644.
101. Wu, S.; Zeng, W.; Wang, Q.; Chen, F.-X. Asymmetric Trifluoromethylation of Aromatic Aldehydes by Cooperative Catalysis with (IPr)CuF and Quinidine-Derived Quaternary Ammonium Salt. *Org. Biomol. Chem.* **2012,** *10,* 9334–9337.
102. Wu, S.; Guo, J.; Sohail, M.; Cao, C.; Chen, F.-X. The Enantioselective Trifluoromethylation of Aromatic Aldehydes by Quaternary Ammonium Bromide and (IPr)CuF at Low Catalyst Loading. *J. Fluorine Chem.* **2013,** *148,* 19–29.
103. Okusu, S.; Kawai, H.; Xu, X.-H.; Tokunaga, E.; Shibata, N. *N*-2-Iodobenzylcinchoninium Bromide Is Effective for Catalytic Enantioselective Trifluoromethylation of Azomethine Imines in Solkane 365MFC. *J. Fluorine Chem.* **2012,** *143,* 216–219.
104. Enders, D.; Herriger, C. Asymmetric Synthesis of 2-trifluoromethyl-1,2,3-triols. *Eur. J. Org. Chem.* **2007,** 1085–1090.
105. Iseki, K.; Nagai, T.; Kobayashi, Y. Asymmetric Trifluoromethylation of Aldehydes and Ketones with Trifluoromethyltrimethylsilane Catalyzed by Chiral Quaternary Ammonium Fluorides. *Tetrahedron Lett.* **1994,** *35,* 3137–3138.
106. Kawai, H.; Kitayama, T.; Tokunaga, E.; Shibata, N. A New Synthetic Approach to Efavirenz through Enantioselective Trifluoromethylation by Using the Ruppert–Prakash Reagent. *Eur. J. Org. Chem.* **2011,** *2011,* 5959–5961. S5959/5951-S5959/5919.
107. Shono, T.; Ishifune, M.; Okada, T.; Kashimura, S. Electroorganic Chemistry. 130. A Novel Trifluoromethylation of Aldehydes and Ketones Promoted by an Electrogenerated Base. *J. Org. Chem.* **1991,** *56,* 2–4.

108. Barhdadi, R.; Troupel, M.; Perichon, J. Coupling of Fluoroform with Aldehydes Using an Electrogenerated Base. *Chem. Commun.* **1998,** 1251–1252.
109. Zhang, Y.; Fujiu, M.; Serizawa, H.; Mikami, K. Organocatalysis Approach to Trifluoromethylation with Fluoroform. *J. Fluorine Chem.* **2013,** *156,* 367–371.
110. Folleas, B.; Marek, I.; Normant, J.-F.; Saint, J. L. Fluoroform: An Efficient Precursor for the Trifluoromethylation of Aldehydes. *Tetrahedron Lett.* **1998,** *39,* 2973–2976.
111. Langlois, B. R.; Billard, T. Fluoroform, Fluoral, Trifluoroacetic, and Trifluoromethanesulfinic Acid Derivatives as New Reagents for the Introduction of Polyfluorinated Moieties. *ACS Symp. Ser.* **2005,** *911,* 57–86.
112. Kawai, H.; Yuan, Z.; Tokunaga, E.; Shibata, N. A Sterically Demanding Organo-Superbase Avoids Decomposition of a Naked Trifluoromethyl Carbanion Directly Generated from Fluoroform. *Org. Biomol. Chem.* **2013,** *11,* 1446–1450.
113. Nagib, D. A.; Scott, M. E.; MacMillan, D. W. C. Enantioselective α-Trifluoromethylation of Aldehydes via Photoredox Organocatalysis. *J. Am. Chem. Soc.* **2009,** *131,* 10875–10877.
114. Yasu, Y.; Koike, T.; Akita, M. Intermolecular Aminotrifluoromethylation of Alkenes by Visible-Light-Driven Photoredox Catalysis. *Org. Lett.* **2013,** *15,* 2136–2139.
115. Cho, E. J.; Senecal, T. D.; Kinzel, T.; Zhang, Y.; Watson, D. A.; Buchwald, S. L. The Palladium-Catalyzed Trifluoromethylation of Aryl Chlorides. *Science* **2010,** *328,* 1679–1681.
116. Samant, B. S.; Kabalka, G. W. A Novel Catalytic Process for Trifluoromethylation of Bromoaromatic Compounds. *Chem. Commun.* **2011,** *47,* 7236–7238.
117. Cho, E. J.; Buchwald, S. L. The Palladium-Catalyzed Trifluoromethylation of Vinyl Sulfonates. *Org. Lett.* **2011,** *13,* 6552–6555.
118. Sheppard, W. A. Electrical Effect of the Sulfur Pentafluoride Group. *J. Am. Chem. Soc.* **1962,** *84,* 3072–3076.
119. Sthre, L. J.; Berrah, N.; Bozek, J. D.; Borve, K. J.; Carroll, T. X.; Kukk, E.; Gard, G. L.; Winter, R.; Thomas, T. D. Chemical Insights from High-Resolution X-ray Photoelectron Spectroscopy and ab Initio Theory: Propyne, Trifluoropropyne, and Ethynylsulfur Pentafluoride. *J. Am. Chem. Soc.* **2001,** *123,* 10729–10737.
120. Bergmann, D.; Hinze, J. Electronegativity and Charge Distribution. *Struct. Bond. Berl.* **1987,** *66,* 145–190.
121. Hansch, C.; Muir, R. M.; Fujita, T.; Maloney, P. P.; Geiger, F.; Streich, M. The Correlation of Biological Activity of Plant Growth Regulators and Chloromycetin Derivatives with Hammett Constants and Partition Coefficients. *J. Am. Chem. Soc.* **1963,** *85,* 2817–2824.
122. Welch, J. T. Preparation of Pentafluorosulfanyl Carbonyl Compounds. *Eff. Prep. Fluorine Compd.* **2013,** 275–278.
123. Lummer, K.; Ponomarenko, M. V.; Roeschenthaler, G.-V.; Bremer, M.; Beier, P. An Improved Method for the Fluorination of Arylsulfur Chlorotetrafluorides to Arylsulfur Pentafluorides. *J. Fluorine Chem.* **2014,** *157,* 79–83.
124. Bowden, R. D.; Comina, P. J.; Greenhall, M. P.; Kariuki, B. M.; Loveday, A.; Philp, D. A New Method for the Synthesis of Aromatic Sulfur Pentafluorides and Studies of the Stability of the Sulfur Pentafluoride Group in Common Synthetic Transformations. *Tetrahedron* **2000,** *56,* 3399–3408.
125. Joliton, A.; Carreira, E. M. Ir-Catalyzed Preparation of SF_5-substituted Potassium Aryl Trifluoroborates via C–H Borylation and Their Application in the Suzuki–Miyaura Reaction. *Org. Lett.* **2013,** *15,* 5147–5149.
126. Frischmuth, A.; Unsinn, A.; Groll, K.; Stadtmueller, H.; Knochel, P. Preparations and Reactions of SF_5-substituted Aryl and Heteroaryl Derivatives via Mg and Zn Organometallics. *Chem. - Eur. J.* **2012,** *18,* 10234–10238.

127. Garg, S.; Shreeve, J. M. Trifluoromethyl- or Pentafluorosulfanyl-substituted Poly-1,2,3-triazole Compounds as Dense Stable Energetic Materials. *J. Mater. Chem.* **2011,** *21,* 4787–4795.
128. Huang, Y.; Gard, G. L.; Shreeve, J. M. One-Pot Syntheses of 1,2,3-triazoles Containing a Pentafluorosulfanylalkyl Group via Click Chemistry. *Tetrahedron Lett.* **2010,** *51,* 6951–6954.
129. Okazaki, T.; Laali, K. K.; Bunge, S. D.; Adas, S. K. 4-(Pentafluorosulfanyl)Benzenediazonium Tetrafluoroborate: A Versatile Launch Pad for the Synthesis of Aromatic SF_5 Compounds via Cross Coupling, Azo Coupling, Homocoupling, Dediazoniation, and Click Chemistry. *Eur. J. Org. Chem.* **2014,** *8,* 1630–1644.
130. Mo, T.; Mi, X.; Milner, E. E.; Dow, G. S.; Wipf, P. Synthesis of an 8-pentafluorosulfanyl Analog of the Antimalarial Agent Mefloquine. *Tetrahedron Lett.* **2010,** *51,* 5137–5140.

Chapter 4

Fluorinated Amino Acids, Peptides, and Proteins

Chapter Outline

1. INTRODUCTION

Incorporation of fluorinated amino acid residues into peptides and proteins is increasingly becoming an attractive strategy to probe the enzyme–substrate complexes and mechanisms of protein aggregation, and to modulate the chemical and thermal stabilities of proteins.[1] The protein function is well preserved for moderately fluorinated proteins, and is comparable with that of the native proteins. Further, through fluorine substitution, modulation of the protein functions, for example, improved stability and substrate selectivity, can be achieved.

Fluorine, because of its highest electronegativity of all the elements, significantly polarizes the adjacent bonds resulting in an increase in the acidity and

Organofluorine Compounds in Biology and Medicine. http://dx.doi.org/10.1016/B978-0-444-53748-5.00004-6

in a decrease in the basicity of the proximal substituents. Further, the van der Waals radius of fluorine (1.47 Å) is comparable with that of oxygen (1.57 Å) and is only slightly larger than that for hydrogen (1.2 Å) atom, and therefore substitution of a fluorine atom for a hydrogen or a hydroxyl group minimally alters the steric environment of the lightly fluorinated amino acids (i.e., in case of mono- or difluorinated compounds), but may contribute to an increase in their lipophilicity, cell permeability, and metabolic stability. On the other hand, fluorinated amino acids consisting of polyfluoroalkyl- or trifluoromethyl groups have relatively increased hydrophobicity as well as lipophobicity. That is, they have reduced solubility in both organic solvents and water and tend to self-aggregate. This effect, called as fluorous effect, favors protein folding for some of the fluorinated peptides or proteins, resulting in their increased thermal stability.

Because of the fluorous effect, fluorinated compounds (e.g., amino acids, peptides, or proteins) aggregate through a self-sorting process. This self-sorting results in the separation of fluorous compounds from their mixtures with nonfluorinated compounds. For example, global substitution of the hydrophobic amino acid residues of coiled coil proteins by their fluorinated analogs (e.g., substitution of leucine by hexafluoroleucine) results in their self-sorting (i.e., self-assembly) in the presence of their corresponding nonfluorinated analogs.[2] This self-sorting by fluorous effect, however, does not apply to fluorinated aromatics, because the nonfluorinated aryl rings have face-to-face attractive, polar π-interactions with the fluorinated rings, which favors their mutual aggregation (vide infra).[1,3]

Even a single fluorine substitution can have a dramatic effect on the conformational characteristics and proteolytic stabilities of the peptides. Thus, by selective incorporation of fluorine into peptides, peptide-based therapeutics may be designed.[4–7] Further, fluorinated derivatives of the naturally occurring (canonical) amino acids, especially those of leucine, valine, tryptophan, tyrosine, and phenylalanine, are increasingly used as probes in the study of protein folding and protein–ligand binding interactions. This topic has been extensively reviewed by Koksch, and Marsh and coworkers.[7,8]

Protein engineering based on both site-directed and random mutagenesis using unnatural (noncanonical) amino acids is useful in designing a diverse range of proteins that have unique chemical and biological properties. Substitution of the hydrophobic amino acid residues by fluorinated noncanonical amino acid analogs results in enhanced chemical, proteolytic, and thermal stability of both soluble and membrane-bound proteins[9–11]; yet, site-selective or global substitution of the amino acid residues by fluorinated amino acids generally retains the biological activity of the wild-type proteins (although in some cases the enzyme activities are reduced).[7] Peptide drug candidates derived from the canonical amino acids (naturally occurring amino acids) have low proteolytic stabilities, thereby limiting their therapeutic applications. In this context, side chain fluorinated peptides exhibit increased proteolytic stabilities in the human blood plasma.[6] Thus, the increased proteolytic stability of fluorinated peptide drugs may have significant impact on the development of peptide-based drugs.

Small fluorinated proteins can be synthesized by the conventional solid-phase peptide synthesis, but this approach cannot be used for the synthesis of large proteins. There currently exist various biosynthetic methods of protein synthesis by site-directed or random mutagenesis, and with continued improvements in these recombinant techniques it is expected that synthesis of large fluorinated proteins tailored to specific functions can be achieved.[7,8,12] Although ribosomal translation of monofluorinated amino acids into proteins can be achieved relatively easily using conventional recombinant techniques, biological incorporation of trifluoromethylated amino acids such as trifluoroleucine, trifluorovaline, hexafluorovaline, and hexafluoroleucine remains still a challenge.[12]

1.1 Fluorinated Amino Acids in Protein Engineering

The rapid advances in the synthetic methodologies for organofluorine compounds enable the synthesis of fluorinated amino acids of either stereochemical configuration in high enantiomeric excess; fluorinated versions of most of the canonical amino acids are now either commercially available or can be readily synthesized, and can be incorporated into peptides and proteins through solid-phase peptide synthesis or recombinant techniques.[5,13–19] Fluorinated amino acids have wide ranging applications in the study of protein–protein interactions.[7] Figure 1 shows the fluorinated amino acids that are commonly used in protein engineering for increasing protein stability, for probing the biological mechanisms, and in the design of novel therapeutics. A variety of fluorinated amino acids—such as hexafluoroleucine, trifluoroleucine, trifluorovaline, trifluoroisoleucine, pentafluoroleucine, and trifluoroethylglycine (TfeGly)—which are either commercially available or can be synthesized in the laboratory, are now routinely being used in the design of fluorinated peptides or proteins.[12]

1.2 Synthesis and Purification of Fluorinated Peptides and Proteins

Selective incorporation of the fluorinated amino acids into proteins can be achieved enzymatically by site-directed mutagenesis.[14,20–25] Alternatively, incorporation of fluorinated amino acids into oligopeptides can be accomplished by solid-phase peptide synthesis.[14,20,21] In solid-phase peptide synthesis, often the removal of the unreacted amino acids and intermediate peptides from the final full-length peptides is tedious and experimentally challenging. The peptide purification in these cases can be simplified through the fluorous capping of the terminal amino groups of the unreacted peptides, which involves, for example,

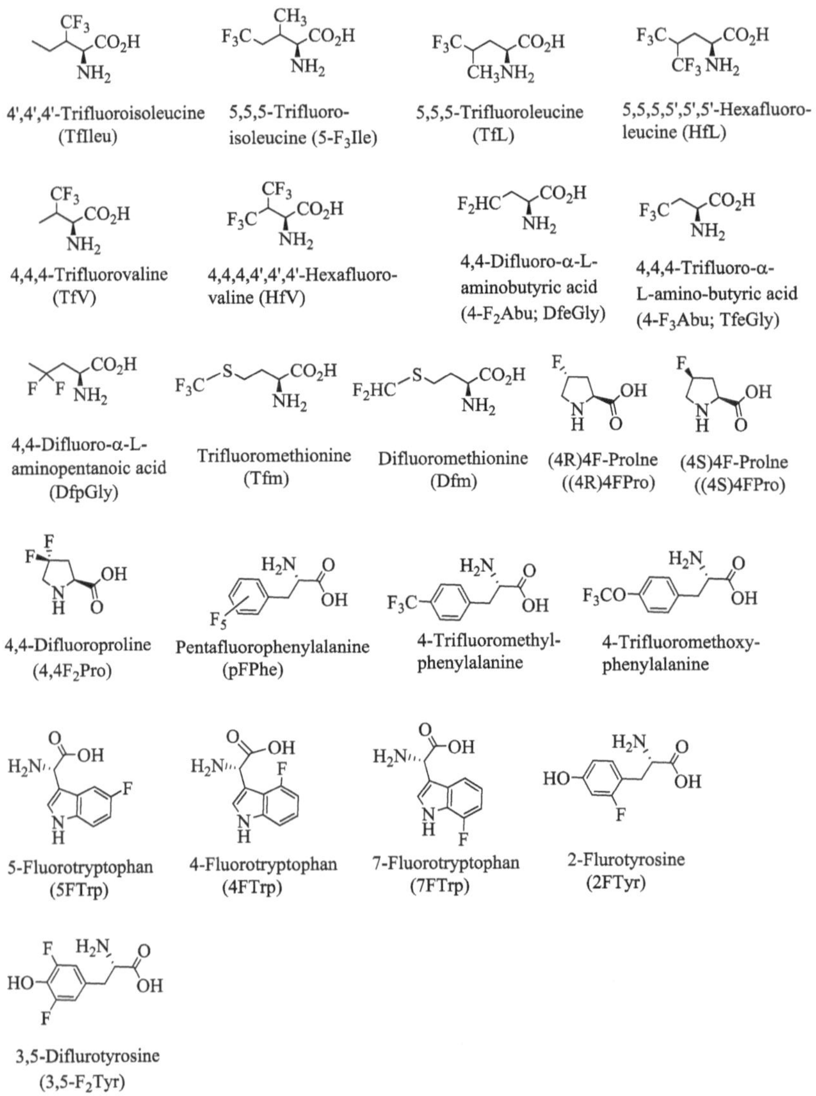

FIGURE 1 Structures of some of the commonly used fluorinated amino acids used in protein engineering.

trifluoroethylation using the corresponding trivalent iodonium salts as the alkylating agents. The *N*-trifluoroethylated amino groups are deactivated toward further peptide couplings, and thus do not interfere with the subsequent peptide coupling steps. At the end of the peptide synthesis, the fluorinated oligomeric impurities can be removed by fluorous silica gel chromatography, or in some cases, by simple centrifugation (Figure 2).[26–28]

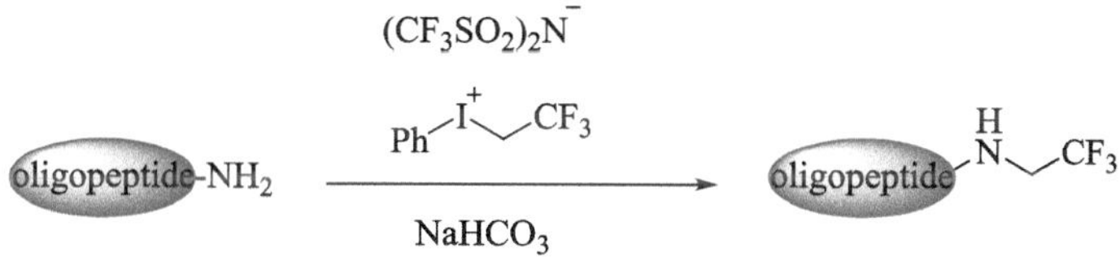

undesired interemediate peptide fragments, formed during solid phase peptide synthesis

removed by flash fluorous silica gel chromatography, or by centrifugation at the end of the peptide synthesis.

FIGURE 2 Fluorous capping of the terminal amino groups of the unreacted oligomeric peptides during solid-phase peptide synthesis; at the end of peptide synthesis these fluorous-tagged impurities can be separated from the final full-length peptide through fluorous silicagel chromatography or by simple centrifugation.

2. FLUORINATED LEUCINE

The highly fluorinated, hydrophobic amino acid 5,5,5,5′,5′,5′-hexafluoroleucine (HfLeu), when substituted for the hydrophobic amino acids (such as leucine or valine) in peptides or proteins affords them unique biological properties. In general, peptides containing fluorinated amino acid residues have relatively lower helix propensities, whereas they stabilize the β-sheet domains.[7] For example, a 19-residue peptide with a single hexafluoroleucine residue was synthesized by solid-phase peptide synthesis using fluorenylmethoxycarbonyl (Fmoc) protecting group strategy, and through circular dichroism spectroscopy it was determined that the α-helix propensity of the fluorinated peptide is significantly reduced as compared to that of the peptide that has nonfluorinated leucine (helix propensities are 0.128 and 1.06 for hexafluoroleucine and leucine, respectively). The helix propensity of the tetrafluoroleucine, 0.445, in this peptide is relatively higher than that for the hexafluoroleucine as expected from its relatively lower degree of fluorination.[29] The hydrophobic side chains of leucine shield the intrahelical hydrogen bonds from the exterior water so that the α-helix has increased stability. In case of the fluorinated amino acids such as HfLeu, on the other hand, the highly hydrophobic fluoroalkyl moiety may be partially or fully buried in the unfolded conformation, leading to their high energies of activation for folding into α-helix.

Marsh and coworkers, based on their studies on a model four-helix bundle protein, α4H, consisting of varying proportions of hexafluoroleucine at the hydrophobic core, demonstrated the enhanced stability of the fluorinated peptides to chemical and thermal denaturation, and enhanced stability against proteolytic degradation, as compared to that of the nonfluorinated amino acid analogs.[30] This increased stability of the fluorinated peptides was attributed to the fluorous interactions of the perfluoroalkyl side chains. Due to the increased chemical and thermal stability, the fluorinated proteins would retain their activity in the presence of organic solvents, and at relatively high temperatures. Thus, the fluorinated peptides or proteins may have practical applications including their use as catalysts in chemoenzymatic organic synthesis and in the design of advanced biomaterials.

2.1 Fluorinated Glucagon-Like Peptide-1

Fluorinated peptides, due to their enhanced hydrolytic stabilities, may have therapeutic potential. For example, the helical glucagon-like peptide GLP-1(7–36) is a potent antihyperglycemic gut hormone and helps maintain homeostasis of the plasma glucose levels.[31] It stimulates insulin secretion, increases pancreatic β-cell mass, and suppresses glucagon secretion, its activity depending on the rise in the blood glucose levels. Exogeneous GLP-1 results in the lowering of the blood glucose levels and therefore there is a great interest in developing therapeutics based on GLP-1. However, the plasma half-life of GLP-1 is only 2 min because GLP-1 is hydrolyzed rapidly by the serine protease, dipeptidyl peptidase-4, limiting its therapeutic value. A peptide that mimics GLP-1, exenatide ($t_{1/2}$=26 min), is currently in clinical use to treat type 2 diabetes.[4,32,33] Exenatide (Amylin Pharmaceuticals; Food and Drug Administration approved in 2012) acts as a GLP-1 agonist to lower the blood glucose level, although it is not stable enough for oral use.

Kumar and coworkers have investigated the possibility of enhancing the proteolytic stability of GLP-1 by a systematic substitution of hexafluoroleucine (HfLeu, represented as X in Figure 3) for the amino acid residues at the peptide cleaving sites, Ala7 and Glu9. Several mutants, including Ala8-HfLeu, Glu9-HfLeu, Gly10-HfLeu, and a double mutant, Ala8Glu9-HfLeu, were synthesized and tested for their proteolytic stability and affinity to the cognate human receptor (GLP-1R) in order to modulate their signal transduction efficiency.[4] All the tested mutants had significantly increased proteolytic stabilities. Especially, the Ala8-HfLeu mutant and the double mutant Ala8Glu9-HfLeu exhibited no detectable hydrolysis even after 24 h. Their binding affinity to the GLP-1 receptor (GLP-1R), however, is relatively diminished. EC_{50} values for Ala8-HfLeu, Glu9-HfLeu, Gly10-HfLeu, and the double mutant Ala8Glu9-HfLeu are 73.0, 2.0, 67.3, and 374 nM, respectively, as compared to the EC_{50} value of 1 nM for the GLP-1 (Figure 3). Interestingly, the P2′ site (Gly10) has a significant role in the proteolytic hydrolysis, as shown by a dramatic increase in the proteolytic stability of the Gly10-HfLeu mutant. The Glu9-HfLeu mutant has comparable binding affinity to the GLP-1R as that of GLP-1 despite the substitution of the polar glutamic acid residue by the nonpolar hexafluoroleucine, which may be due to the involvement of C–F multipolar interactions with the neighboring OH, NH_2, and C–H bonds of the protein side chains.[4]

	789	EC_{50}(nM)
GLP-1	NH_2-**HAEG**TFTSDVSSYLEGQAAKEFIAWLVKGR-$CONH_2$	1
Ala8-HfLeu (X)	NH_2-**HXEG**TFTSDVSSYLEGQAAKEFIAWLVKGR-$CONH_2$	73
Glu9-HfLeu (X)	NH_2-**HAXG**TFTSDVSSYLEGQAAKEFIAWLVKGR-$CONH_2$	2.0
Gly10-HfLeu (X)	NH_2-**HAEX**TFTSDVSSYLEGQAAKEFIAWLVKGR-$CONH_2$	67.3
Ala8Glu9-HfLeu (X)	NH_2-**HXXG**TFTSDVSSYLEGQAAKEFIAWLVKGR-$CONH_2$	374

FIGURE 3 Site-specific hexafluoroleucine mutants of GLP-1; all these mutants show dramatically increased proteolytic stability, and Glu9-HfLeu has similar binding affinity to the receptor as that of the GLP-1; X = hexafluoroleucine.

2.2 Fluorinated Antimicrobial Peptides

Incorporation of hexafluoroleucine (HfLeu) into the therapeutically useful antimicrobial peptides, for example, magainin2 amide and buforin II, results in their relatively enhanced or similar antimicrobial activity as that of the nonfluorinated peptides. These fluorinated peptides also showed increased stability toward trypsin-catalyzed hydrolysis.[34–36] Thus, appropriately fluorinated peptides, with improved proteolytic stability, would be potentially useful as therapeutics.[37]

5,5,5-trifluoroleucine (TFL) substitution of Leu9 and Leu18, or substitution of all four intrinsic leucine residues in a 26-residue membrane-binding protein melittin (GIGAV**L**KV**L**TTG**L**PA**L**ISWIKRKRQQ) results in its increased aggregation within the lipid bilayaer.[38] The extent of self-association and affinity for lipid bilayer depends on the position of substitution, and also on the stereochemistry of the trifluoroleucine side chain. Tirrell and Niemz have shown that the fluorinated peptides not only can control the tertiary and quaternary interactions in aqueous media but also can enhance protein–protein interactions within the lipid bilayers. The enhanced aggregation of the fluorinated peptide within the lipid membrane disrupts the cell membrane, forming discrete pores through which water and other small molecules may leak into the cells, eventually causing cell lysis.[38]

2.3 Fluorinated Chloramphenicol Acetyltransferase

Although fluorinated proteins, in general, show increased thermal stabilities, there are some instances where the fluorinated proteins have relatively lower thermal stabilities than their nonfluorinated analogs. For example, Tirrell and Montclare have shown that global replacement of leucine residues in chloramphenicol acetyltransferase (CAT) by TFL actually results in a 20-fold reduction in thermal stability.[11] CAT is a homotrimeric protein with each polypeptide chain consisting of 13 leucine residues. It provides bacterial resistance to chloramphenicol by catalyzing the acetylation of the chloramphenicol hydroxyl groups through acetyl coenzyme A. The half-life ($t_{1/2}$) of thermal inactivation at 60 °C for the fluorinated enzyme, CAT T (CAT expressed in media depleted of leucine and supplemented with TFL), was 5 min as compared with a $t_{1/2}$ of 101 min for the wild-type enzyme (CAT L; wild-type CAT), showing loss of the fluorinated enzyme activity at elevated temperature.[11] The specific activity of this fluorinated enzyme CAT T is also lower than that of the wild-type CAT $k_{cat}/K_m = 14.4\,\mu M/min$ and $8.4\,\mu M/min$, respectively, for the wild-type (CAT L) and its fluorinated version (CAT T), respectively. However, upon residue-specific mutations (K46M, S87N, and M142I) followed by a nonsense mutation that truncated the evolved polypeptide chain by one residue, the thermal stability of this globally TFL-substituted CAT enzyme (L2-A1 T; Figure 4) exceeded that of the wild-type enzyme; the $t_{1/2}$ at 60 °C for this mutant enzyme is 133 min, 30 min longer than that for the wild type. Interestingly, the K46M, S87N, and

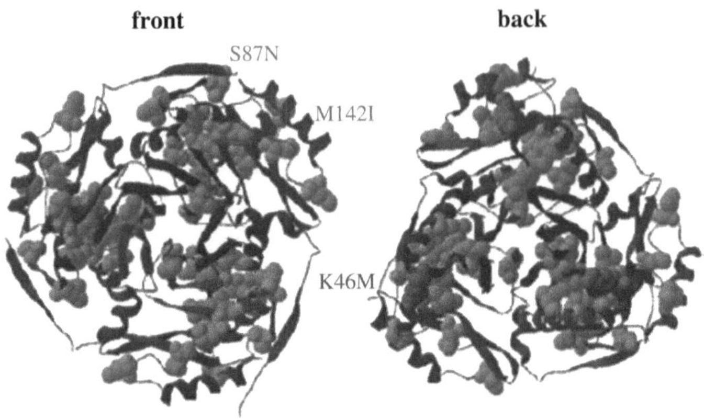

FIGURE 4 Structural model (front and back view) of the CAT (trimeric) showing three stabilizing mutations in L2-A1; mutations S87N and M142I are highlighted in red; K46M, in pink; leucine/TFL residues, in blue; and the substrate chloramphenicol, in red. (For interpretation of the references to color in this figure legend, the reader is referred to the online version of this book.) *Adapted with permission from Ref. 11. Copyright 2006, John Wiley and Sons.*

M142I mutations are more than 15 Å from the ligand chloramphenicol binding site and do not make direct contact with the Leu/TFL residues, yet there is a modest increase in the specific activity of the fluorinated enzyme L2-A1 T over that of the CAT T (k_{cat}/K_m = 14.4 μM/min, 12 μM/min and 8.4 μM/min, respectively, for wild-type CAT L, L2-A1 T, and CAT T, respectively). Thus, although global substitution of leucine residues by TFL attenuates the relative thermal stability and specific activity of the initially derived protein CAT T, directed evolution of this protein affords the protein L2A1 T with improved thermal stability (and with similar specific activity) as compared to the wild-type enzyme.

2.4 Fluorinated Coiled Coil Proteins

Tirrell and coworkers have shown that the incorporation of the highly hydrophobic trifluoroleucine into coiled coil proteins and leucine zipper peptides increases their thermal and chemical stability.[10] They have incorporated hexafluoroleucine into the coiled coil proteins, and demonstrated that these fluorinated proteins have relatively increased stability with respect to thermal and chemical denaturation compared to the wild-type protein. The hexafluoroleucine-containing protein is also relatively more stable than the protein mutant with trifluoroleucine[2]: the melting temperatures of the protein (HA1) modified by hexafluoroleucine and trifluoroleucine are 76 °C and 67 °C, respectively, as compared to the melting point of 54 °C for the wild-type protein.

Kumar and coworkers have synthesized a highly fluorinated peptide corresponding to the coiled coil region of the yeast transcription factor GCN4 by solid-phase peptide synthesis.[39] All four leucine residues and three valine residues of the peptide were substituted by trifluoroleucine and trifluorovaline,

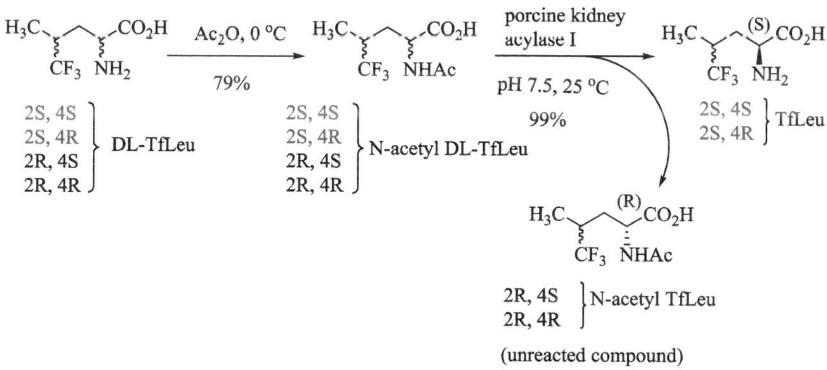

FIGURE 5 Resolution of the DL-TfLeu (racemic mixture) using porcine kidney acylase; the resolved trifluoroleucines (as a mixture of the two diastereomers) were used in the solid-phase peptide synthesis of the corresponding fluorinated peptides. Similar approach is used for the purification of the trifluorovaline diastereomers (Ac_2O = acetic anhydride).

respectively. The latter fluorinated peptide retains its α-helical structure at low micromolar concentrations and sediments as a dimeric species in the 5–30 μM range. This fluorinated peptide has a significant increase in thermal stability, with a melting temperature for the fluorinated peptide dimer being 15 °C higher than that for the corresponding nonfluorinated peptide. Guanidinium-hydrochloride-induced denaturation for this fluorinated protein is relatively slower than for the nonfluorinated version, that is, the free energy of unfolding for the fluorinated protein is relatively more positive by about 1 kcal/mol than for the corresponding hydrocarbon version of the peptide.

The diasteromeric mixture of (2S, 4S)- and (2S, 4R)-trifluoroleucines and (2S, 3S)- and (2S, 3R)-trifluorovalines, used for incorporation into the proteins, were obtained by porcine-kidney-acylase-mediated resolution of a diastereomeric mixture of *N*-acetyltrifluoroleucine and *N*-acetyltrifluorovaline, each containing all four possible diastereomers (Figure 5).[39]

2.5 Synthesis of L-5,5,5,5′,5′,5′-Hexafluoroleucine (HfLeu)

Cheng and Chiu have used combined chemoenzymatic synthesis for the preparation of the (2S)-5,5,5,5′,5′,5′-hexafluoroleucine (HfLeu) and the corresponding (2S)-5,5,5′,5′-tetrafluoroleucine in multigram scale.[29] Thus, Wittig reaction of the ylide **1** with hexafluoroacetone gave the enone **2**, which upon reductive hydrogenation under McMurry's conditions using aqueous $TiCl_3$ gave the α-keto ester **3** in high yields. Reductive amination of **3** under enzymatic catalysis in the presence of phenylalanyl dehydrogenase, ammonium acetate, and NADH afforded the HfLeu in high yields and with 98% enantiomeric excess (Figure 6). Using this short and efficient synthetic method, tetrafluorovaline was also synthesized starting from 1,1,3,3-tetrafluoroacetone and ylide **1**.

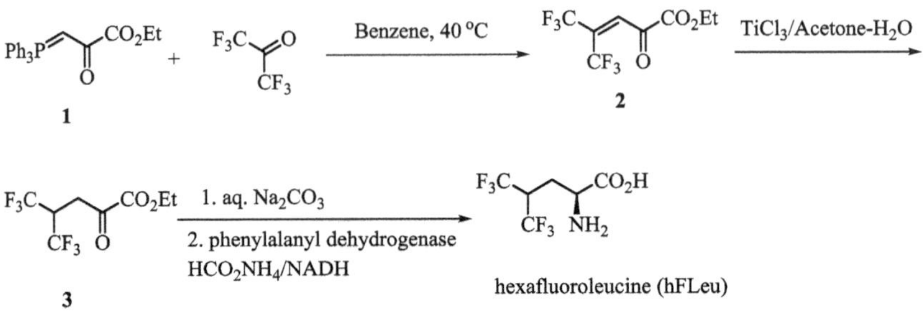

FIGURE 6 Synthesis of hexafluoroleucine (HfLeu) from the ylide **1**.

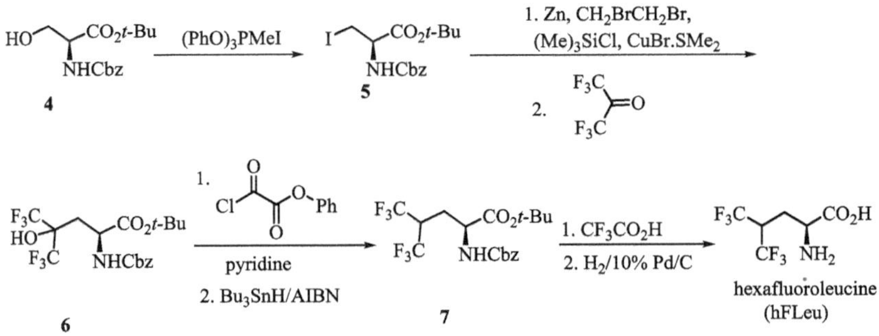

FIGURE 7 Synthesis of hexafluoroleucine (HfLeu) from the serine derivative **4** (AIBN = azobisisobutyronitrile).

Marsh and coworkers have developed synthesis of hexafluoroleucine, also in multigram scale, starting from *N*-Cbz-serine-*tert*-butyl ester (Cbz = benzyloxycarbonyl) (**4**).[40] The serine hydroxyl group was transformed into the iodide using methyltriphenoxyphosphonium iodide, and then into the corresponding zincate, which was reacted with hexafluoroacetone in the presence of Cu(I) catalyst to give **6**. Radical-mediated deoxygenation of **6**, followed by deprotection, gave the hexafluoroleucine in 50% overall yield, and with 99% enantiomeric excess (Figure 7).

3. FLUORINATED PROLINE AND ITS EFFECT ON COLLAGEN

Rational design of fluorinated peptides allows modulation of the thermal and chemical stabilities of proteins, and thus fluorinated amino acids play key role in protein engineering. In general, fluorinated proteins have relatively higher thermal stabilities than the nonfluorinated analogs, and in case of collagen, the stereochemical features of the fluorinated sites are also important in protein stabilization. Collagen protein has the repeating sequence (~300) of the tripeptide linkage, $(GlyProHyp)_n$ (Figure 8; Hyp = (2S,4R)-4-hydroxyproline), and three such polypeptide chains form an extended triple helix (Figure 9).

By substituting the hydroxyl group of the (2S,4R) 4-hydroxyproline (Hyp) by fluorine in a collagen-like polypeptide, its melting temperature (T_m) can be increased by about 20 °C, as demonstrated by Raines and coworkers; for example,

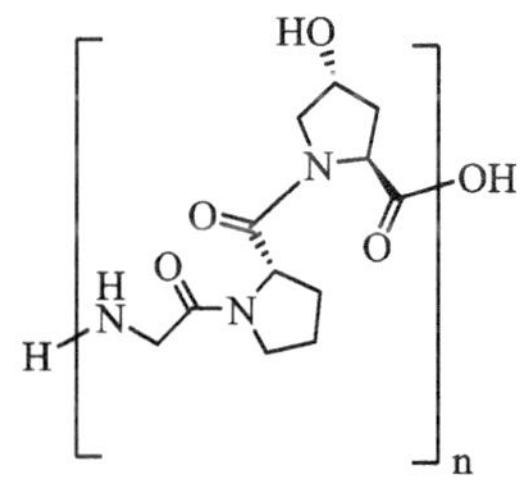

FIGURE 8 Structure of (GlyProHyp)$_n$ repeating sequence of the collagen peptide.

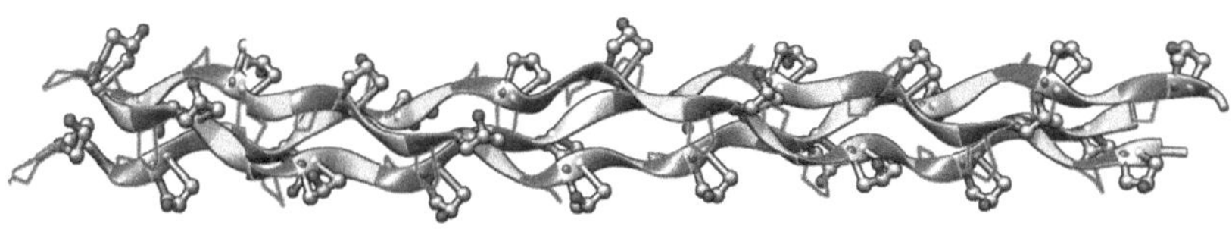

FIGURE 9 Ribbon diagram of a segment of collagen triple helix (GlyProHyp)$_n$; 4-hydroxyproline side chains are shown as ball and stick (oxygen = red and nitrogen = blue, respectively, and glycine and proline backbones are shown in cyan and purple, respectively) (the structure was created using UCSF Chimera software; PDB 1CGD). (For interpretation of the references to color in this figure legend, the reader is referred to the online version of this book.)

the T_m values of (Pro-Hyp-Gly)$_{10}$ is 69 °C, while that of (Pro-Flp-Gly)$_{10}$ (Flp = (2S,4R)-4-fluoroproline) is about 91 °C.[41,42] It is now accepted that the relatively higher thermal stabilities of fluorinated collagen as compared to the natural collagen is due to the stereoelectronic effects, and not solely through the hydrogen bonding interactions of the 4-hydroxyproline with bridging water molecules. The hydrogen bonding strength of carbon-bound fluorine is relatively insignificant as compared to hydroxyl groups, yet the substitution of the hydroxyl group by fluorine increases its thermal stability. Raines and coworkers have emphasized that if hydrogen bonding interactions of Hyp were the basis of the extra stability of the collagen, the fluorinated version would be expected to have lower T_m, contrary to their observations, and they have rationalized the relatively enhanced stability of the fluorinated version as due to the stereoelectronic effects exerted by the fluorine rather than due to the hydrogen bonding effects (vide infra).[42,43]

Fluorine is isosteric and isoelectronic with respect to OH group, and both these substituents favor the γ-exo conformation for the 4(R)-proline derivatives. In this conformation, the C_5–N bond is gauche to C–F so that there is maximum hyperconjugative delocalization of the adjacent anti-periplanar C-H bonds with the σ*-orbital of the C–F bond (Figure 10). The corresponding 4(S)-isomers exist predominantly in the γ-endo ring puckered conformation so that C_5–N and C–F bonds are gauche to each other (Gauche effect). The same conformational preference has been observed for the 3-fluoroproline: the γ-exo-conformer is favored for (3R)-3-fluoroproline, while γ-endo conformer is favored for the (3S)-3-fluoroproline. The stereochemistry of the fluorinated carbon at the γ-position is critical to the thermal stability of the collagen triple helix. Raines and coworkers have demonstrated that

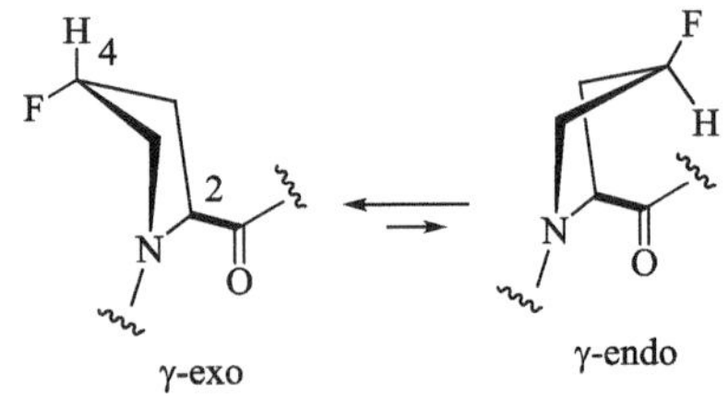

(2S,4R)-4-Fluoroproline (Flp)

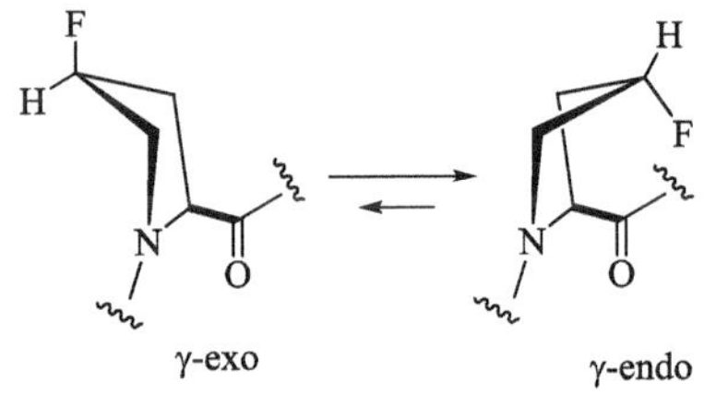

(2S,4S)-4-Fluoroproline (flp)

FIGURE 10 Gauche effect on the conformational stabilization of 4-fluoroproline; γ-exo conformer is favored for the 4R isomer (Flp), whereas γ-endo conformation is preferred for the 4S-conformer (flp).

replacement of the (2S,4R)-4-hydroxyproline (Hyp) in oligopeptide models of collagen by (2S,4S)-4-hydroxyproline (hyp) or (2S,4S)-4-fluoroproline (flp) actually destabilized the triple helix, showing that the stereoelectronic effects of the electronegative hydroxyl and fluoro substituents play a major stabilizing effect on the collagen triple helix structure.

The relative location of the Flp in the Pro-Hyp-Gly triad also plays a major role in the stabilization of the triple helix. For example, replacing the proline moiety in $(ProHypGly)_7$ by Flp or Hyp does not result in triple helix formation. On the other hand, the flp substitution for the Pro resulted in a stable triple helix with a T_m of 33 °C.[44]

4,4-Difluoroproline (DfP) lacks the ability to exert such Gauche effect and thus backbone torsion angles are not influenced by the fluorine. Using ^1H and ^{19}F nuclear magnetic resonance (NMR) spectroscopy it was shown that there is no conformational preference for the *N*-acetyl-4,4-difluoroproline methyl ester (Ac-DfP-OMe), and both γ-exo and γ-endo conformers were observed in water and chloroform solutions. Replacing Hyp of collagen-related peptides with 4,4-difluoroproline, as expected, does not show increased stability to the peptides. Thus, conformational preference due to the stereoelectronic factors rather than the increased hydrophobicity of the peptides due to the fluorine substitution seems to be the major factor in the stabilization of the triple helix of the Flp-derived collagen peptides.[42]

The ring conformation has subtle effect on the *cis–trans* equilibrium constants for the proline peptide bonds, which seems to control the relative

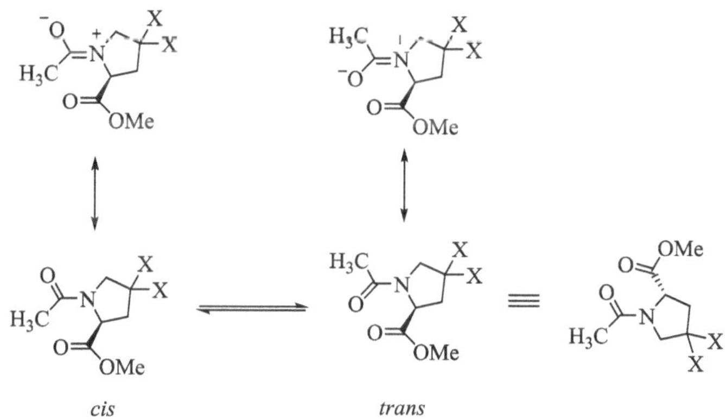

FIGURE 11 *cis-trans* Equilibration in Ac-Pro-OMe and its fluorinated derivatives; the fluorine substitution at C_4 results in relatively lower electron density on the nitrogen so that amide C–N bond has relatively more single bond character (i.e., the enolate contribution is diminished), and thereby has lower rotational activation barrier.

stability of the collagen homotrimeric triple helix. K_{trans}/K_{cis} values for Ac-DfP-OMe was estimated to be 3.6 in water, which is much smaller than that for N-acetyl-Flp methyl ester (Ac-Flp-OMe) (6.7) or Ac-Hyp-OMe (6.1).[42] On the other hand, the (4S)-isomers of these compounds, N-acetyl(2S,4S)-4-hydroxyproline methyl ester (Ac-hyp-OMe) and Ac-flp-OMe, show much reduced K_{trans}/K_{cis} values of 2.5, and 2.4, respectively, showing their inability to spontaneously self-assemble into triple helix. Interestingly, similar trend was found for the 3-fluorosubstituted N-acetylproline methyl ester (Ac-3-F-Pro-OMe) derivatives: Ac-(3R)-3F-Pro-OMe ($K_{trans}/K_{cis}=8.9$), Ac-(3S)-3F-Pro-OMe ($K_{trans}/K_{cis}=4.3$), and N-acetyl-3,3-difluoroproline methyl ester (Ac-3,3-DfPro-OMe) ($K_{trans}/K_{cis}=3.4$).[7] Thus, the *cis-trans* equilibrium constant is correlated to the ring conformation of the proline moiety and is an indicator of the stability of the collagen peptide triple helix.

Furthermore, the rotational barriers for the *cis-trans* isomerization are reduced due to the fluorine substitution. Thus, the rotational barrier for *cis-trans* isomerization for Ac-Pro-OMe, Ac-Flp-OMe, and Ac-DfP-OMe are 84.5, 82.6, and 80.8 kJ/mol, respectively.[7] This decreasing trend in the activation barriers is in accordance with pK_a lowering ability of the electron-withdrawing fluorine group. The lowering of the pK_a of the proline derivative results in the relatively decreased amide C–N bond order (i.e., resembling more of a single bond), and therefore results in its relatively lower rotational barrier (Figure 11).

3.1 Synthesis of 4-fluoroproline

Diethylaminosulfur trifluoride (DAST) or its analogs including Deoxo-Fluor and morpholinosulfur trifluoride can be used to displace a secondary hydroxyl

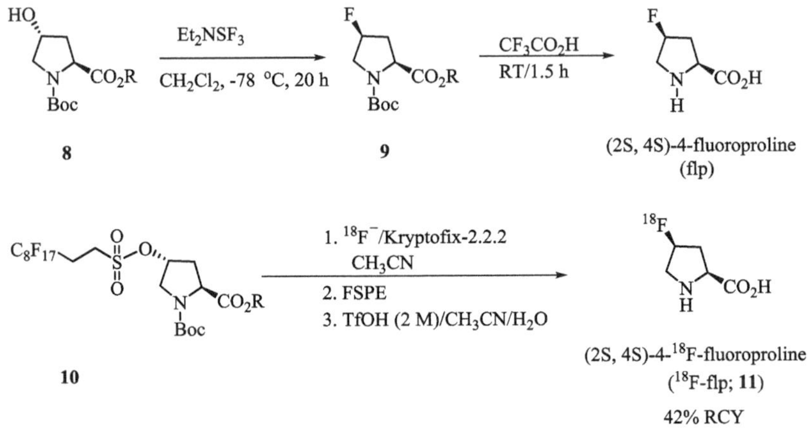

FIGURE 12 Synthesis of (2S, 4S)-^{18}F-fluoroproline (FSPE = fluorous solid phase extraction; TfOH = triflic acid; RCY = radiochemical yield).

group stereoselectively with inversion of configuration.[45,46] This method has been applied in the preparation of either Flp or flp starting from *N*-Boc-protected (2S, 4R) 4-hydroxy proline ester **8** or its (4S)-diastereomer, respectively.[46] The ^{18}F-labeled *N*-Boc-(2S, 4S)-4-^{18}F-fluoroproline ester **11** (^{18}F-flp) was synthesized through nucleophilic substitution (S_N2) reaction of the corresponding 4-sulfonate ester **10** using K^{18}F and the aza-crown ether Kryptofix-2.2.2 as the phase transfer catalyst, followed by hydrolysis (Figure 12).[46] Use of fluorous side chain in this sulfonate ester facilitated workup conditions as the unreacted fluorous precursor and the byproduct fluorous sulfonic acid ($C_8F_{17}CH_2CH_2SO_3H$) could be removed by fluorous solid-phase extraction (FSPE). After quantitative deprotection of the fluorous sulfonate moiety using triflic acid, followed by FSPE, the radiolabeled (2S, 4S)-4-^{18}F-fluoroproline (^{18}F-flp) was obtained in a radiochemical yield of 42%.[46] These radiolabeled amino acids are useful as tracers in positron emission tomography (Chapter 7 provides a detailed discussion of the biomedical applications of the various PET tracers).

3.2 Therapeutic Applications of Collagen Peptide Mimetics

The ideal collagen peptide mimetics that have potential therapeutic value should be incapable of self-assembly to the collagen triple helix, but should be able to anneal to natural collagen. The collagen mimetic peptides incorporating Flp residues bind tightly to mammalian collagen and mouse wounds ex vivo.[44] Raines and coworkers have synthesized a collagen peptide mimetic, Ac-(flp-Flp-Gly)$_7$-(Gly–Ser)$_3$-LysOH with lysine as the end group, the ε-amino group of which can be conjugated with various fluorophores or drug candidates through amide bond formation (Figure 13). Using fluorophore-conjugated peptide mimetics, Raines and coworkers demonstrated the strong binding affinity of this fluorinated peptide mimetic at the wound site.[44] This methodology opens up various possibilities for

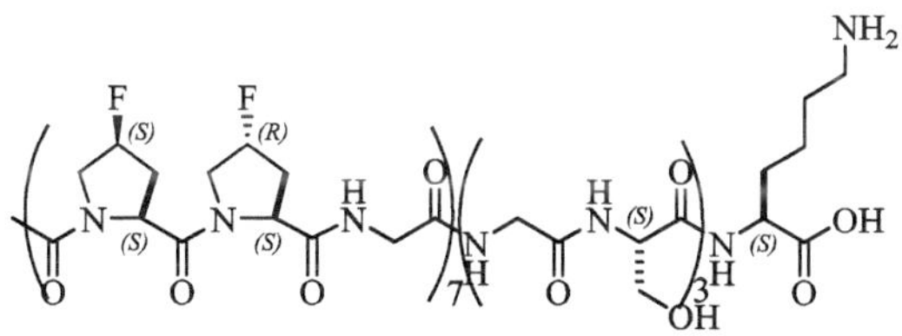

FIGURE 13 Structure of Ac-(flp-Flp-Gly)$_7$-(Gly-Ser)$_3$-LysOH, a collagen mimetic peptide for wound healing applications; the terminal lysine residue is useful for conjugating to fluorophores or drug candidates.

the drug delivery at the wound site, and facilitates the development of collagen-based biomaterials useful in skin grafting and restorative procedures.[44]

4. FLUORINATED METHIONINES

[S-Monofluoromethyl]methionine (MfMet) is relatively unstable under physiological conditions, thus limiting its biological applications, although protected versions of MfMet have been synthesized using XeF_2 or DAST reagents.[47,48] However, the corresponding [S-difluoromethyl]methionine (DfMet) and [S-trifluoromethyl]methionine (TfMet) (Figure 14) are relatively stable molecules and these fluorinated versions have been incorporated into peptides and proteins and their effect on the protein stabilities and protein folding have been studied.[49] Fluoroalkyl group, due to its high electronegativity reduces the electron density on sulfur and thereby enhances the hydrophobicity of the methionine residues, while at the same time the steric crowding of the molecule at the vicinity of the sulfur increases in the case of TfMet, because the trifluoromethyl moiety is relatively sterically crowded and has approximately equal van der Waals volume as that of the isopropyl group. Characterization of the proteins modified by fluoromethionines using enzyme or bioactivity assays revealed minimal perturbation of the activities of the proteins.[49] DfMet and TfMet, in preference over all other possible fluorinated methionines, are typically used for protein modifications.[27]

Honek and coworkers have substituted DfMet and TfMet for all the methionine residues in the bacteriophage lambda lysozyme (LaL), which contains three methionine residues at positions 1, 14, and 107, using methionine auxotrophic bacterial strains. DfMet was incorporated into the protein to the extent of 95%, while similar incorporation of TfMet could be achieved to only about 70%. The fluorination did not significantly affect the enzyme activity. ^{19}F NMR studies on the protein in which all three methionines are substituted by TfMet revealed that TfMet at positions 1 and 14 show a single signal as expected, but the TfMet at position 107, unexpectedly shows two signals due to its unusually different environment that restricts the free rotation around the S-CF_3 bond.[27] The M14L mutant, having TfMet, only at positions 1 and 107, on the other hand, showed a single signal per residue. Thus, in the M14L mutant the protein adapts a conformer in which the free rotation around the S-CF_3 has a relatively

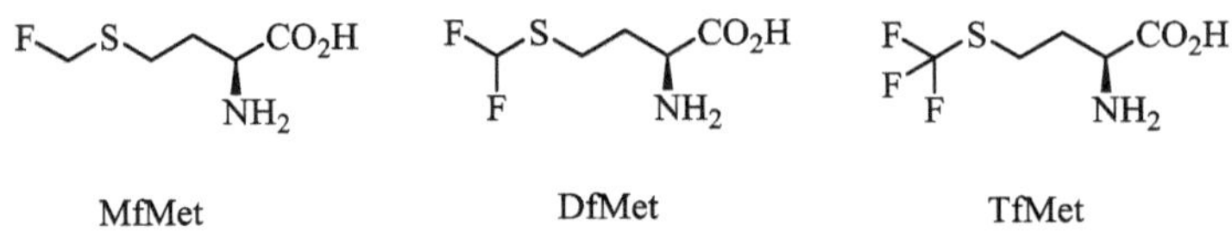

FIGURE 14 Structures of monofluoro-, difluoro-, and trifluoromethionines (MfMet, DfMet, and TfMet).

lower energy barrier.[27] Thus, incorporation of these crowded fluorinated amino acids into the proteins results in a subtle effect on their conformations.

The DfMet and TfMet, despite their weakened nucleophilicity at the sulfur, participate as axial ligands at Cu(II) in *Pseudomonas aeruginosa* azurin. The reduction potential of the azurin complex is increased by more than 227 mV when DfMet and TfMet are substituted for the conserved methionine in this protein, and the reduction potential correlates with the hydrophobicity of the methionine analogs; the higher the hydrophobicity of the methionine ligand, the greater the reduction potential.[26]

Budisa and coworkers have substituted the only two available methionine residues of a mutant version of enhanced green fluorescent protein (2M-EGFP; Met233, Met78, Met153, and Met88 in EGFP were mutated by Lys, Leu, Thr, and Leu, respectively) by the TfMet.[50] The latter protein, derived from the wild-type green fluorescent protein (GFP) from *Aequorea victoria* is predominantly a β-strand barrel surrounding the central helix anchoring the chromophore. Although the levels of incorporation of the TfMet were low due to the cellular toxicity of the TfMet, sufficient quantity of the TfMet-labeled protein could be obtained for its characterization by ^{19}F NMR. The ^{19}F NMR spectrum of this fluorinated protein shows two δ^{19}F signals for each of the TfMet residues. The solvent-exposed *N*-terminal TfMet shows a sharp signal, while the buried TfMet-218 shows a broadened peak (Figure 15). The ^{19}F NMR thus confirms the protein folding for this fluorinated version of the 2M-EGFP.

L-TfMet is effective against the toxicity of the human pathogens that express L-methionine γ-lyase (MGL-PLP; EC 4.4.1.11), a pyridoxal-phosphate(PLP)-containing enzyme. This latter enzyme converts L-methionine into α-oxobutyrate, ammonia, and methanethiol. Similar enzymatic reaction of L-TfMet results in the formation of trifluoromethanethiol (CF_3SH) as the byproduct, which spontaneously decomposes into the thiocarbonyl difluoride (SCF_2), and thereby is responsible for the cellular toxicity in the pathogens (Figure 16).[51]

4.1 Synthesis of Fluorinated Methionines

The TfMet and DfMet (S-trifluoromethylhomocysteine) could be readily synthesized from the inexpensive *N*-acetylcysteine thiolactone (**12**) using trifluromethyl iodide or Freon-22 ($CHClF_2$) under photochemical conditions (Figure 17).[47,52,53]

Alternatively, L-homocystine is reduced using sodium in liquid ammonia and the resulting homocysteine is treated with potassium *tert*-butoxide followed

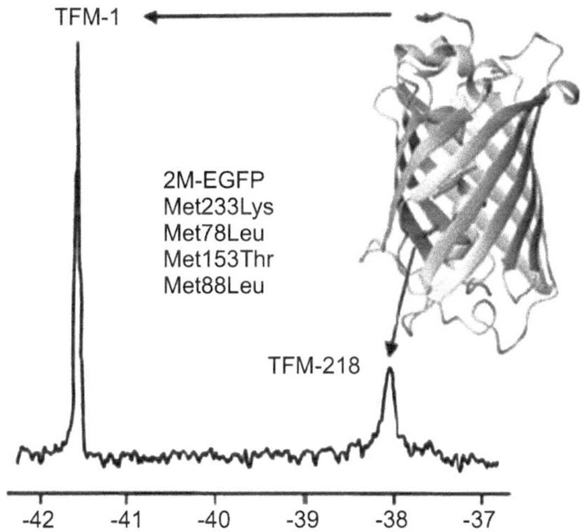

FIGURE 15 ^{19}F NMR of a TfMet-substituted mutant version of EGFP (2M-EGFP); the observation of two distinct signals with differing broadness indicates that the protein is in the folded state. *Adapted with permission from Ref. 50. Copyright 2004, Verlag Helvetica Chimca Acta AG, Zurich (John Wiley and Sons, Publishers).*

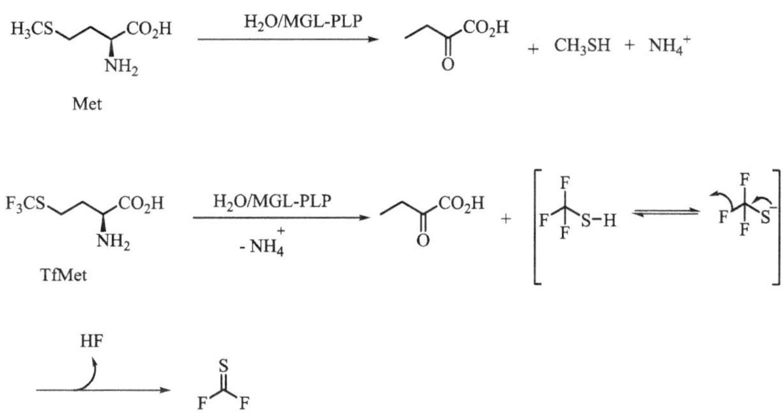

FIGURE 16 Cellular metabolism of Met and TfMet (MGL-PLP = pyridoxal 5′-phoshpate dependent methionine γ-lyase).

by iodotrifluoromethane or chlorodifluoromethane in ethanol under ultraviolet (UV) irradiation to give (S)-TfMet and (S)-DfMet, respectively (Figure 18).[51]

5. FLUORINATED TYROSINES

The fluorinated tyrosines, unlike the fluorinated phenylalanines, are relatively more hydrophilic rather than being hydrophobic. This effect is due to the strong electron-withdrawing inductive effect and consequent increase in the acidity of the hydroxyl group. The aryl-fluorinated tyrosine hydroxyl is therefore a better

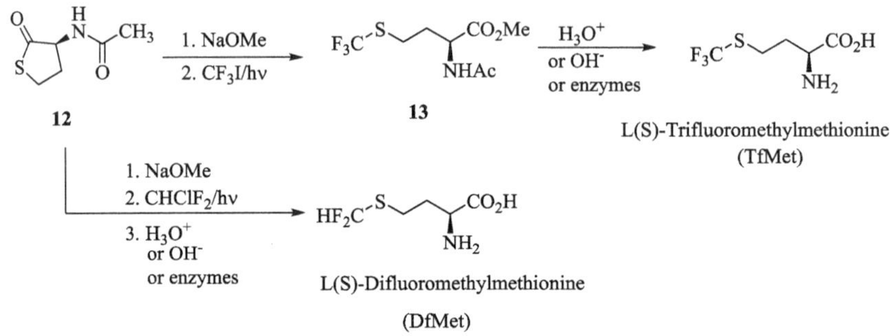

FIGURE 17 Synthesis of TfMet and DfMet.

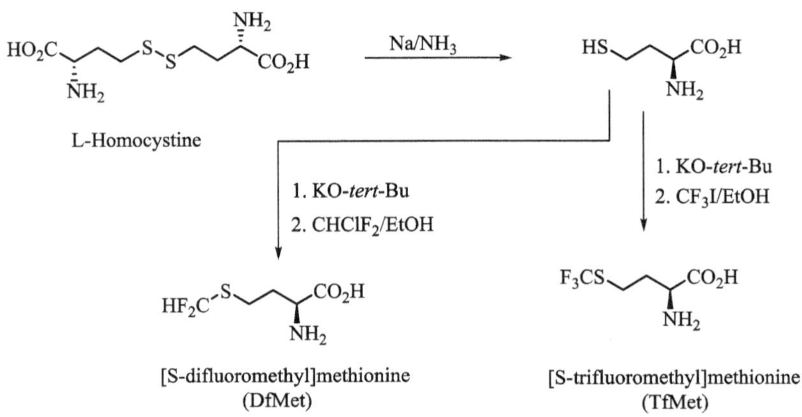

FIGURE 18 Synthesis of TfMet and DfMet from L-homocystine.

hydrogen bond donor than the tyrosine. The pK_a of the hydroxyl groups depends on the location and the number of fluorines in the aryl ring. The 3-fluorotyrosine (3FTyr) hydroxyl is slightly more acidic than 2FTyr ($\Delta pK_a=0.6$), and the 2,3,5,6-tetrafluorotyrosine (2,3,5,6-F_4Tyr) hydroxyl has dramatically higher acidity ($pK_a=5.2$) (Figure 19).[7] The higher acidity of the tyrosine hydroxyl groups due to fluorine substitution facilitates their kinase-catalyzed phosphorylation; for example, the semisynthetic proteins containing fluorinated tyrosines are relatively better substrates for kinases as compared to their nonfluorinated analogs.[54]

The 3FTyr is metabolically transformed in the liver into the toxic compound, fluoroacetate, which participates as an intermediate of the citric acid cycle, and thereby exhibits lethal effects in experimental models.[28] The aryl-fluorinated tyrosines are useful to probe the mechanistic details of the enzyme-catalyzed reactions, since the hydroxyl acidity can be fine-tuned by the number and position of the fluorine atoms. Through systematic substitution of the active site tyrosine moieties by fluorinated tyrosines in *Escherichia coli* riboncleotide

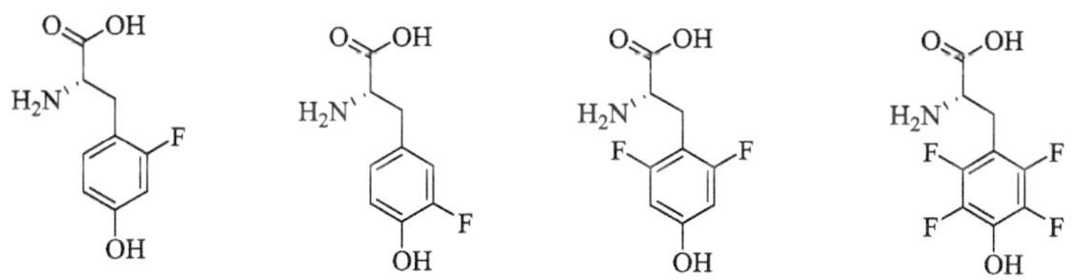

2FTyr 3FTyr $2,6F_2Tyr$ $2,3,5,6\text{-}F_4Tyr$

pK_a: 9.0 8.4 8.1 5.2

FIGURE 19 Structures of some of the fluorinated tyrosines and their hydroxyl pK_a values; due to the wide variations in the hydroxyl pK_a values, the fluorinated tyrosines are useful as biological probes to investigate the enzyme mechanisms.

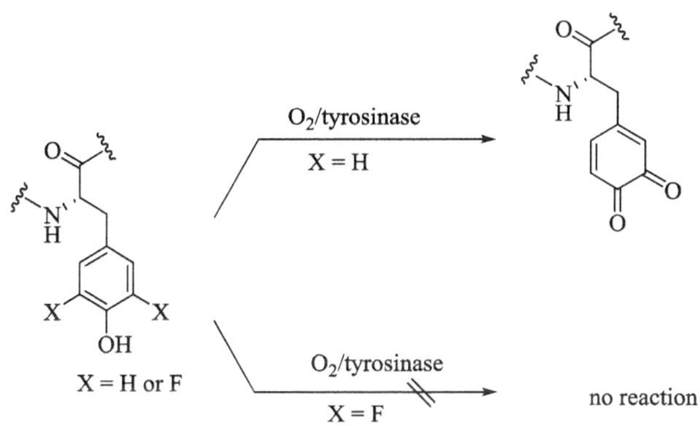

FIGURE 20 Tyrosinase-catalyzed oxidation of tyrosine residues to the corresponding *ortho*-quinones; this oxidation is disfavored in case of the corresponding fluorinated tyrosine.

reductase, it was shown that Tyr356 is involved as a redox-active amino acid along the radical propagation pathway.[55]

In order to achieve the site-selective incorporation of the fluorinated tyrosines into a GFP, the hydroxyl group of the fluorinated tyrosines was masked by a light-cleavable protecting group, *ortho*-nitrobenzyl moiety, and thus protected *O*-(*ortho*-nitrobenzyl)-2-fluorotyrosine, *O*-(*ortho*-nitrobenzyl)-3-florotyrosine, and *O*-(*ortho*-nitrobenzyl)-2,6-difluorotyrosine were genetically encoded in Escherichia coli.[56] This protecting group renders the amino acids unrecognizable by the endogeneous protein biosynthetic machinery, and thereby prevents the global incorporation of the fluorinated tyrosines into proteins. In the absence of the latter protecting group, the site selective incorporation of the fluorinated tyrosine residues into peptides could only be achieved using solid-phase peptide synthesis. Further, the 3,5-difuorotyrosine derivatives and their O-protected derivatives are stable towards tyrosinase catalyzed oxidation into the corresponding *ortho*-quinone derivatives (Figure 20).[56,57] Site-specific incorporation of 3FTyr at Tyr66 results in a red shift of

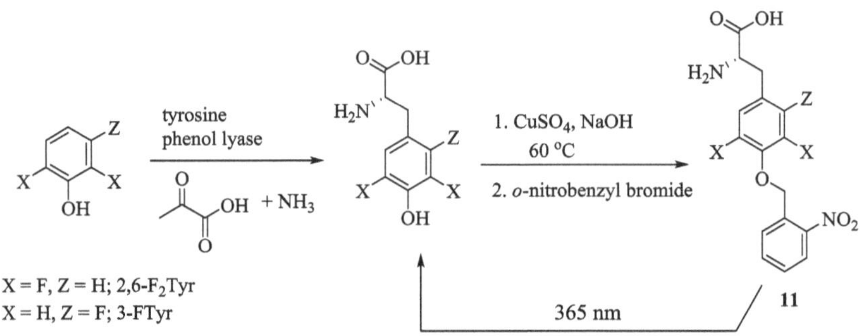

FIGURE 21 Enzyme-catalyzed synthesis of the fluorinated tyrosines, and protection of the hydroxyl group as the *ortho*-nitrobenzyl ether; the protecting group can be photochemically removed at the end of the peptide synthesis.

the fluorescence emission, whereas similar incorporation of 2,6-difluorotyrosine results in a blue shift. Thus, the site-specific incorporation of fluorinated amino acids modulates the fluorescent properties of the protein, and thus it is useful in the directed evolution of the photoswitching properties of the autofluorescent proteins (GFP).[56]

5.1 Synthesis of Fluorinated Tyrosines

In order to achieve site-specific incorporation of fluorotyrosines into proteins in *Escherichia coli*, and to prevent their celllular metabolism, it is necessary to temporarily mask the hydroxyl group, for example, using *ortho*-nitrobenzyl protecting group (*vide supra*). The latter protecting group can be photochemically cleaved to give the corresponding phenols simply by exposing the proteins to 365-nm UV radiation (Figure 21). The fluorinated tyrosines and their derivatives could be synthesized by tyrosine-phenol-lyase-catalyzed reaction of the fluorinated phenols with pyruvic acid in the presence of buffered ammonia. Complexation of thus obtained tyrosines with Cu^{2+}, followed by S_N2 alkylation using *ortho*-nitrobenzyl bromide gave the corresponding O-protected tyrosines. Using these temporarily masked substrates, it was possible for the site specific incorporation of various fluorinated tyrosines (2FTyr, 3FTyr, and 2,6F_2Tyr) into the *E.Coli* proteins.[56]

6. FLUORINATED PHENYLALANINE

Gao and coworkers have synthesized the 35-residue wild-type protein α_2D (an adrenergic receptor protein), as well as its double mutant in which the phenylalanines at positions 10 and 29 were substituted by pentafluorophenylalanine (Z) in order to determine whether quadrupole stacking between heterodimers exists and to probe the extent of the polar π-interactions in the protein stabilization.[58] They have achieved the synthesis of the double mutant (F10Z,F29Z) and the nonfluorinated analog through solid-phase peptide synthesis.

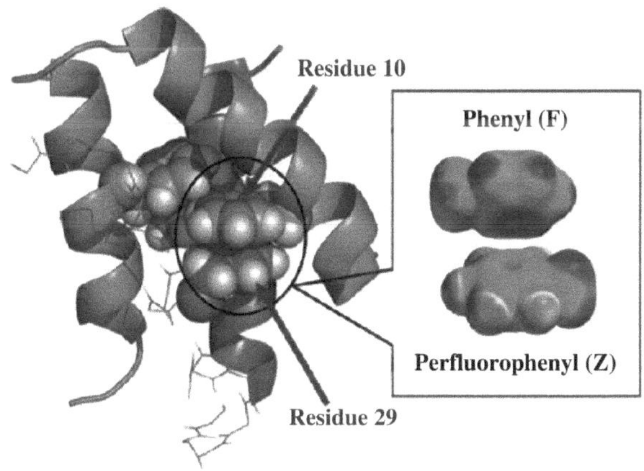

FIGURE 22 A truncated view of the structure of the dimeric structure of α_2D showing face-to-face π-stacking of the Phe10 from one chain (green) and the Phe29 from the second chain (blue). The inset shows the computed electrostatic potential surfaces for the phenyl and the pentafluorophenyl rings (blue color indicates positive and the red color indicates negative electrostatic potentials). (For interpretation of the references to color in this figure legend, the reader is referred to the online version of this book.) *Adapted with permission from Ref. 58. Copyright 2010, American Chemical Society.*

The wild-type protein α_2D exists as a homodimer, each of the monomeric peptides being in the helix-loop-helix conformation. The phenylalanine residues at positions 10 and 29 are stacked face to face with those from the second peptide chain through the polar π-interactions (Figure 22), and these dimers are in equilibrium with the monomeric peptide chains. The double-mutant α_2D (F10Z,F29Z; Z=pentafluorophenylalanine) also exists as a homodimer, and has dramatically increased thermal stability (T_m=79.8 °C) as compared to the wild-type protein (T_m=29.8 °C). This increased thermal stability of the fluorinated peptide was rationalized as due to the combined effect of hydrophobicity of the fluorinated aryl ring and the polar π-interactions between the aryl rings.[58] These aryl polar π-interactions are expected to be greater for the heterodimers in which the pentafluorophenyl ring stacks with the phenyl ring, since the quadrupole moments of the two aromatic rings of Z and F are similar in magnitude but opposite in sign. That is, the phenyl ring has a net negative charge at the center (quadrupole moment $=-29.9\times10^{-40}$ C/m^2), while the pentafluorophenyl ring has a net positive charge (quadrupole moment $=31.7\times10^{-40}$ C/m^2), so there is a relatively strong electrostatic interaction between the rings when they are stacked face to face (Figure 23).[1]

Upon mixing equal concentrations of the homodimers of the wild-type and the double-mutant version, they equilibrate rapidly to form the heterodimer as the predominant form, as evidenced by the ^{1}H- and ^{19}F NMR, and also by fluorescence resonance energy transfer experiments. The ^{19}F NMR for the homodimer shows four absorptions, one each for the *ortho-* and *meta*-fluorines and two absorptions for the *para*-fluorines. The ^{19}F NMR of the mixture (heterodimer) shows splitting of the *ortho-* and *meta*-fluorine signals due to the π-stacking interactions (Figure 23).

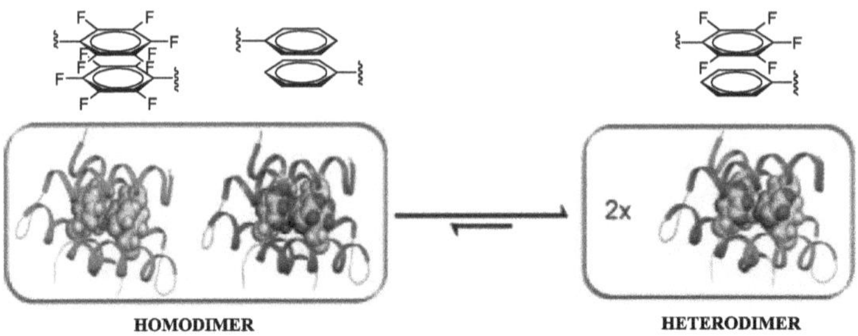

FIGURE 23 Cartoon diagram showing the reversible formation of the heterodimer F′Z′ from the equimolar mixture of homodimers F′F′ and Z′Z′ (Z′=F(10,29)Z-α_2D; F′=wt α_2D protein). The strong electrostatic π-interactions between the aryl and fluoroaryl rings contribute to the relatively greater stability of the heterodimer. *Adapted with permission from Ref. 59. Copyright 2010, Wiley-VCH Verlag GmbH & Co. KGaA, Weinheim.*

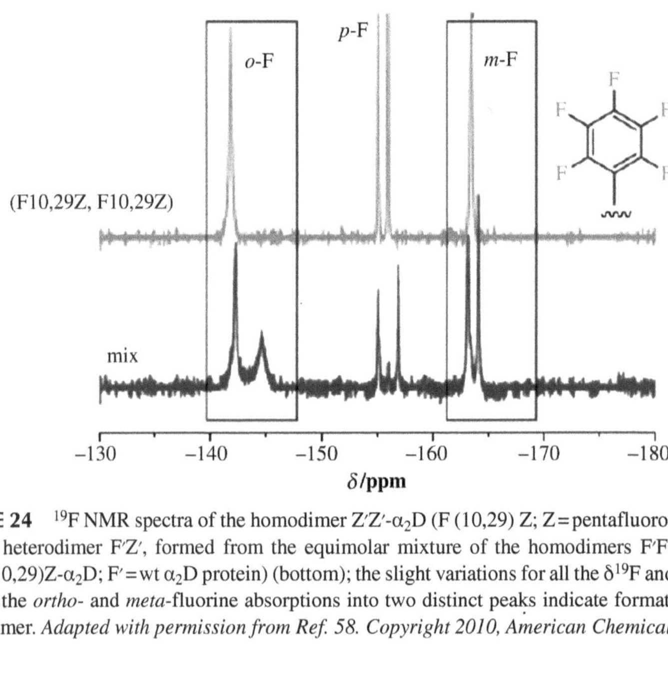

FIGURE 24 ^{19}F NMR spectra of the homodimer Z′Z′-α_2D (F (10,29) Z; Z=pentafluoroPhe) (top) and the heterodimer F′Z′, formed from the equimolar mixture of the homodimers F′F′ and Z′Z′ (Z′=F(10,29)Z-α_2D; F′=wt α_2D protein) (bottom); the slight variations for all the δ^{19}F and the splittings of the *ortho-* and *meta-*fluorine absorptions into two distinct peaks indicate formation of the heterodimer. *Adapted with permission from Ref. 58. Copyright 2010, American Chemical Society.*

The equilibration of the homodimers to heterodimers implies quadrupole interaction between the aryl and fluoroaryl rings. The electrostatic potential is negative at the center of the aryl rings, while it is positive at the center of the fluoroaryl ring, and therefore the heterodimeric proteins are relatively more stabilized as compared to either of the homodimers, the equilibrium favoring the heterodimer (Figure 24).

The aromatic π-stacking is an attractive strategy in the design of peptide-based materials and therapeutics. Synthetic collagen, for example, based on this noncovalent supramolecular aggregation is of therapeutic interest.

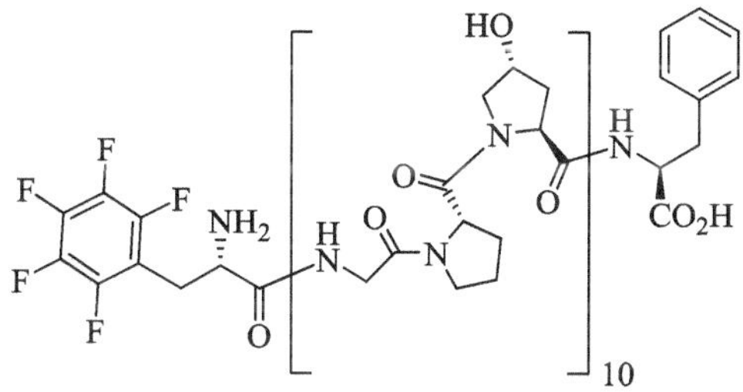

FIGURE 25 Structure of Z-(Gly.Pro.Hyp)$_{10}$-F (Z=pentafluorophenylalanine); the aryl rings of the terminal F and Z, through face-to-face polar π-interactions, self-assemble to form the thrombogenic collagen fibrils.

Conjugation of perfluorophenylalanine (Z) and phenylalanine (F) to the *N*- and *C*-terminal residues, respectively, of the collagen peptides (Gly.Pro. Hyp)$_{10}$ resulted in self-assembly of the material into collagen-like thrombogenic micrometer-scale fibrils.[3] Computational studies showed that the intermolecular Z–F aryl stacking that gives rise to the self-assembly is 6 kcal/mol more stable than that for the F–F stacking.[60]

The 32-mer 8-nm single-strand collagen peptide Z(Gly.Pro.Hyp)$_{10}$F (Figure 25) self-assembles into a collagen-like peptide, forming triple helical building blocks that further self-assemble into composite fibrils by noncovalent, π-stacking interactions between the terminal Z and F aryl rings. These collagen mimetics were able to induce platelet aggregation with a potency similar to the native type I collagen.[60]

Similarly, the aryl π-stacking interactions were exploited in the preparation of two-component fibrils and hydrogels based on the Fmoc-protected pentafluorophenylalanine (Fmoc-Z; **12**) and Fmoc-Phe (**13**).[61] Under these experimental conditions, Fmoc-Phe does not self-assemble to form the fibrils or hydrogels. Similar aggregation also resulted in the formation of two-component hydrogels for equimolar mixtures of Fmoc–phenylalanines with the corresponding monohalogenated Fmoc-Phe-X derivatives (X=e.g., Cl, Br). Although structural information is lacking, the formation of the two-component fibrils and hydrogels indicates the formation of self-assembled structures, presumably through the aryl π-stacking arrangement (Figure 26).

6.1 Synthesis of Fluorinated Phenylalanines

The proline-derived glycine Schiff base-Ni(II) complex **14** and its analogs are versatile precursors for the preparation of chiral amino acids.[62–64] The Schiff base **14** has been used as the chiral glycine equivalent in the preparation of the chirally pure pentafluorophenylalanine, and *ortho*-, *meta*-, and *para*-fluorophenylalanines in high yields (Figure 27).[65] Use of alkyl bromides as the alkylating agents in this synthesis provides access to the (S)-pentafluorophenylalanine as well as other fluorinated amino acids.

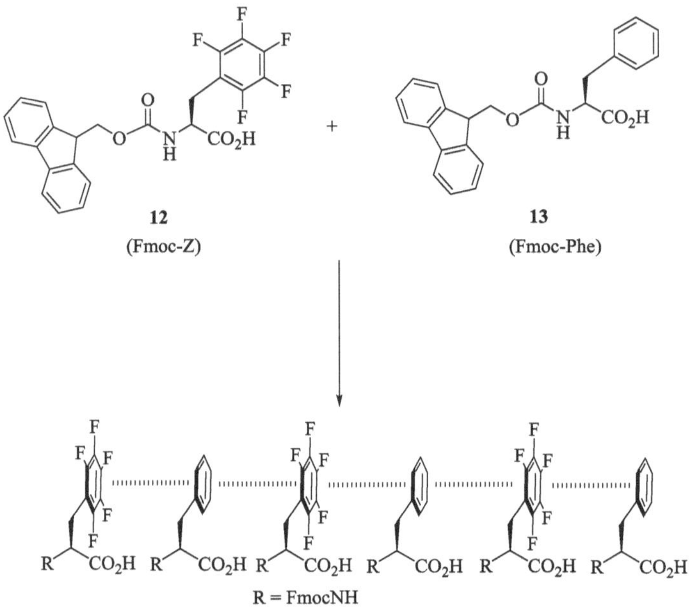

FIGURE 26 Formation of oligomeric π-stacked aggregates from equimolar mixture of the Fmoc-Z and Fmoc=Phe (Z=pentafluorophenylalanine).

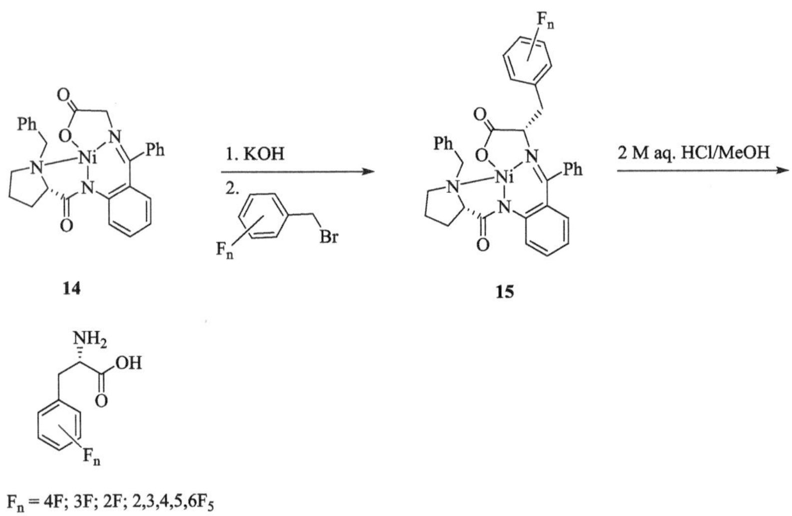

FIGURE 27 Asymmetric synthesis of fluorinated phenylalanines using a proline-derived Schiff base-Ni(II) complex.

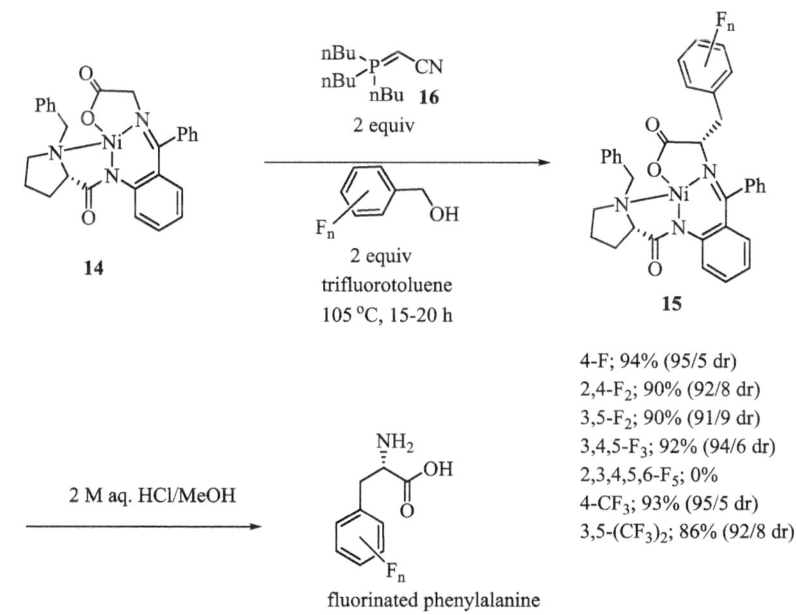

FIGURE 28 Synthesis of fluorinated phenylalanines using Mitsunobu–Tsunoda reaction.

Diastereoselective alkylation of the Schiff base-Ni(II) complex (**14**) using fluorinated benzyl alcohols and cyano(methylene)tributylphosphorane (**16**) followed by acid-catalyzed hydrolysis gives the corresponding α-alkylated products **15** in high yields and with high diastereoselectivities.[13] However, alkylation of the Schiff base using pentafluorobenzyl alcohol could not be achieved using this synthetic strategy (Figure 28).

Various other fluoroalkyl amino acids, such as TfeGly, could also be synthesized using this synthetic method. The mechanism of this reaction involves as a key step the reaction of the fluorinated alcohols with **16** to give the corresponding phosphonium salts **17**. Deprotonation of the Schiff base **14** by the acetonitrile anion (from **17**) gives the chiral carbanion **18**, which undergoes nucleophilic substitution (S_N2) reaction with **17** to give the α-alkylated product **15** (Figure 29).[13]

7. PEPTIDE MIMETICS IN DRUG DISCOVERY

The major drawbacks in the design of peptide-based drugs lie in their rapid hydrolysis by the intracellular peptidases. Peptide mimetics that are hydrolytically and proteolytically stable would circumvent this drawback. Of the many such peptide mimetics that are useful in the synthesis of peptide-based drugs, fluoroalkene (CH=CF), trifluoroethylamine, and CF_3-(*E*-alkene) isosteres have attracted much attention since these peptide isosteres have comparable charge distribution and dipole moments as those of the amide bonds.[66–71]

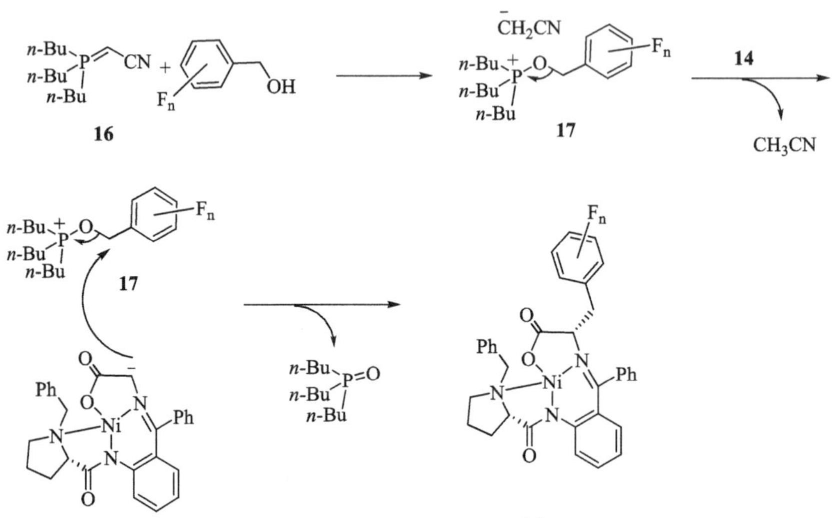

FIGURE 29 Mechanistic rationalization of the formation of **15** from the corresponding fluorinated benzyl alcohols.

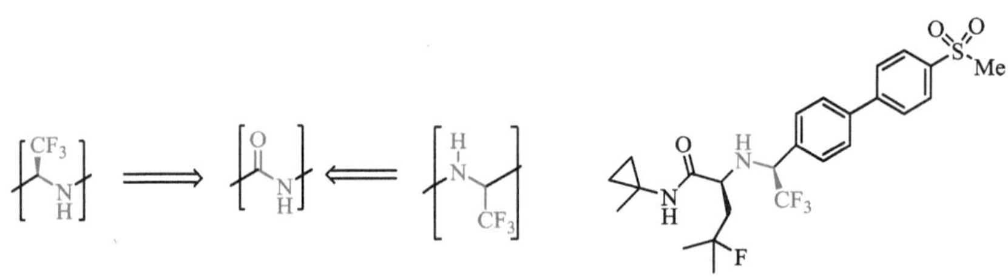

FIGURE 30 Trifluoroethylamine peptide isosteres.

The trifluoroethylamine moiety has been used as peptide mimetic (Figure 30) in the context of developing dipeptide analogs of cathepsin K inhibitors.[71,72] Odanacatib (Merck), a trifluoroethylamine isostere, is a selective cathepsin K inhibitor that is effective against osteoporosis in the experimental studies.[73]

Arg-Gly-Asp (RGD) tripeptide shows high recognition for the $\alpha_v\beta_3$ integrin receptors, and several analogs of RGD peptides were tested as effective recognition elements for these integrin receptors, in pursuit for developing cancer therapeutics. However, a trifluoroethylamine-based RGD peptide mimetic, (X[CH(CF_3)NH]Gly, showed a loss in $\alpha_v\beta_3$ integrin affinity, presumably due to the bulky CF_3 moiety not being well tolerated in the integrin-binding pocket.[74]

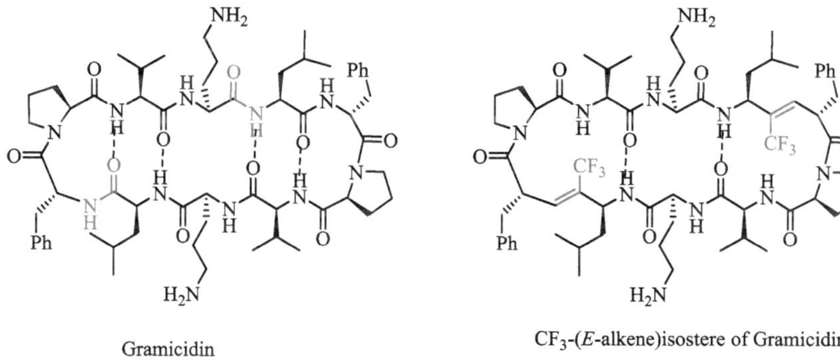

FIGURE 31 Structures of gramicidin and its CF_3-(*E*-alkene)peptide isostere.

Wipf and coworkers have synthesized a CF_3-(*E*-alkene) peptide isostere analog of the gramicidin, a cyclic peptide antibiotic (Figure 31).[68] The CF_3-(*E*-alkene) would allow mimicry of the electrostatic potential surface of the peptide bond with its relatively large dipole moment ($\mu = 2.3$ D; compare with $\mu = 3.6$ D for the peptide moiety).[69] Variable-temperature NMR and 2D NOESY NMR (two-dimensional nuclear Overhauser effect nuclear magnetic resonance spectroscopy) evidence indicate the occurrence of two interstrand hydrogen bonds for the NH(Leu) and C=O(Val) in the gramicidin peptide isostere. X-ray crystallography study of this peptide isostere also confirms the hydrogen bonding between these residues (NH···O=C distances of 1.96 and 2.00 Å). Thus two of the four hydrogen bonds in gramicidin are conserved in the CF_3-(*E*-alkene) peptide isostere. However, X-ray structure shows a twist of 70° for the C=C planes for these isosteres, due to the steric crowding of the trifluoromethyl moiety. This gramicidin trifluoromethyl alkene peptide isostere, despite the steric crowding at the alkene moiety, has similar potency as that for gramicidin against *Bacillus subtilis*.[68]

The key steps involved in the synthesis of the CF_3-(*E*-alkene) isostere involves hydrozirconation–iodination of the stannyl alkyne **19** to give the vinyl iodide **20**, which upon lithium–halogen exchange and nucleophilic addition to the *N*-Boc imine **21** afforded the mixture of vinyl stannane **22** and its diastereomer, which after separation, followed by iododestannylation, desilylation, and acetylation gave the vinyl iodide **23**. The Cu-mediated trifluoromethylation of **23** followed by deacetylation and a two-step oxidation gave the dipeptide isostere, *N*-Boc-Leu(ψ-(CF_3CH=CH)Phe; **24**).[68] Peptide coupling of **24** with H-Pro-Val-Orn(Cbz)-OMe (**25**; Cbz = benzyloxycarbonyl protecting group) using EDC (N-ethyl-N′-(3-dimethylaminopropyl) carbodiimide) as a coupling agent gave **26**, which upon *N*-Boc removal, stepwise coupling, followed by a final macrolactonization afforded the gramicidin peptide isostere (Figure 32).

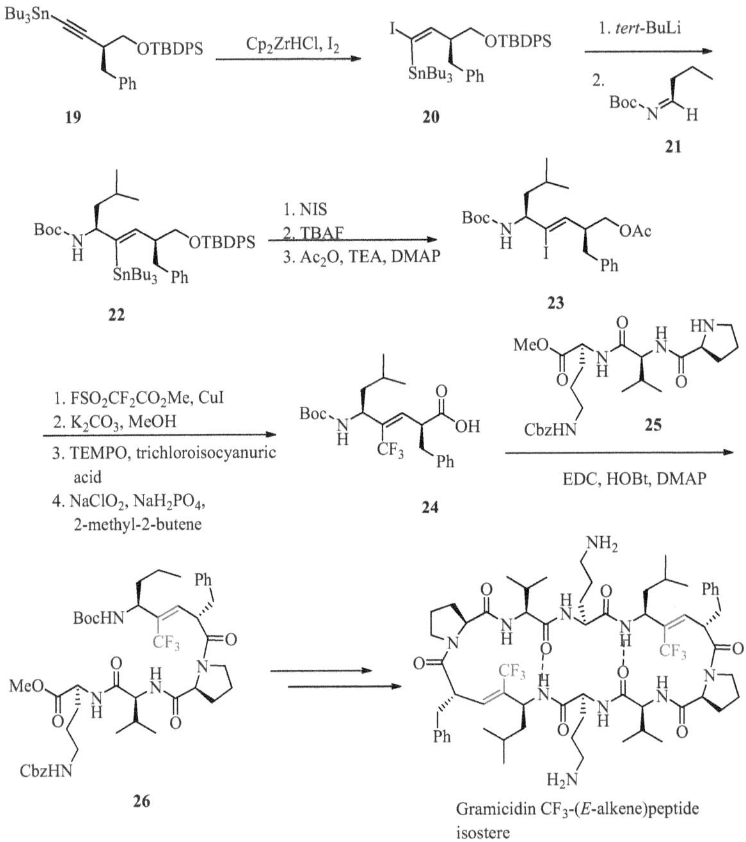

FIGURE 32 Synthesis of the Gramicidin CF_3-(*E*-alkene)peptide isostere (Ac_2O = acetic anhydride; DMAP = 4-dimethylaminopyridine; EDC = 1-ethyl-3-(3-dimethylaminopropyl)carbodiimide; HOBt = 1-hydroxybenzotriazole; NIS = N-iodosuccinimide; TEMPO = 2,2,6,6-tetramethyl-1-piperidinyloxy; TBAF = tetrabutylammonium fluoride).

8. SUMMARY AND OUTLOOK

A diverse range of proteins with unique chemical and biological properties can be designed through site-directed or random mutagenesis using fluorinated amino acids of either stereochemical configuration, whereas relatively small peptides can be synthesized through solid-phase peptide synthesis. Even though it is not completely predictable, fluorinated peptides and proteins often have improved proteolytic and thermal stabilities over the wild-type analogs. Fluorinated peptides, such as GLP-1 analogs and fluorinated collagens, because of their enhanced proteolytic and thermal stabilities, have potential therapeutic applications. The protein function is well preserved for moderately fluorinated proteins, and is comparable with that of the native proteins. Further, modulation of the protein functions, for example, improved stability and substrate selectivity, can be achieved using fluorine substitution. Peptide mimetics, such as those incorporating trifluoroethylamine and CF_3-(*E*-alkene) peptide isosteres, may have therapeutic applications, as in the case of the odanacatib, a cathepsin K inhibitor.

REFERENCES

1. Pace, C. J.; Gao, J. Exploring and Exploiting Polar-π Interactions with Fluorinated Aromatic Amino Acids. *Acc. Chem. Res.* **2013,** *46,* 907–915.
2. Bilgicer, B.; Xing, X.; Kumar, K. Programmed Self-sorting of Coiled Coils with Leucine and Hexafluoroleucine Cores. *J. Am. Chem. Soc.* **2001,** *123,* 11815–11816.
3. Cejas, M. A.; Kinney, W. A.; Chen, C.; Vinter, J. G.; Almond, H. R., Jr.; Balss, K. M.; Maryanoff, C. A.; Schmidt, U.; Breslav, M.; Mahan, A.; Lacy, E.; Maryanoff, B. E. Thrombogenic Collagen-mimetic Peptides: Self-assembly of Triple Helix-based Fibrils Driven by Hydrophobic Interactions. *Proc. Natl. Acad. Sci. U.S.A.* **2008,** *105,* 8513–8518.
4. Meng, H.; Krishnaji, S. T.; Beinborn, M.; Kumar, K. Influence of Selective Fluorination on the Biological Activity and Proteolytic Stability of Glucagon-like Peptide-1. *J. Med. Chem.* **2008,** *51,* 7303–7307.
5. Qiu, X. L.; Qing, F. L. Recent Advances in the Synthesis of Fluorinated Amino Acids. *Eur. J. Org. Chem.* **2011,** 3261–3278.
6. Asante, V.; Mortier, J.; Schlueter, H.; Koksch, B. Impact of Fluorination on Proteolytic Stability of Peptides in Human Blood Plasma. *Bioorg. Med. Chem.* **2013,** *21,* 3542–3546.
7. Salwiczek, M.; Nyakatura, E. K.; Gerling, U. I. M.; Ye, S.; Koksch, B. Fluorinated Amino Acids: Compatibility with Native Protein Structures and Effects on Protein–protein Interactions. *Chem. Soc. Rev.* **2012,** *41,* 2135–2171.
8. Buer, B. C.; Marsh, E. N. G. Fluorine: A New Element in Protein Design. *Protein Sci.* **2012,** *21,* 453–462.
9. Tang, Y.; Ghirlanda, G.; Petka, W. A.; Nakajima, T.; DeGrado, W. F.; Tirrell, D. A. Fluorinated Coiled-coil Proteins Prepared in Vivo Display Enhanced Thermal and Chemical Stability. *Angew. Chem. Int. Ed.* **2001,** *40,* 1494–1496.
10. Tang, Y.; Tirrell, D. A. Biosynthesis of a Highly Stable Coiled-coil Protein Containing Hexafluoroleucine in an Engineered Bacterial Host. *J. Am. Chem. Soc.* **2001,** *123,* 11089–11090.
11. Montclare, J. K.; Tirrell, D. A. Evolving Proteins of Novel Composition. *Angew. Chem. Int. Ed.* **2006,** *45,* 4518–4521.
12. Merkel, L.; Budisa, N. Organic Fluorine as a Polypeptide Building Element: in Vivo Expression of Fluorinated Peptides, Proteins and Proteomes. *Org. Biomol. Chem.* **2012,** *10,* 7241–7261.
13. Drouet, F.; Noisier, A. F. M.; Harris, C. S.; Furkert, D. P.; Brimble, M. A. A Convenient Method for the Asymmetric Synthesis of Fluorinated α-Amino Acids from Alcohols. *Eur. J. Org. Chem.* **2014,** *2014,* 1195–1201.
14. Hunter, L.; Butler, S.; Ludbrook, S. B. Solid Phase Synthesis of Peptides Containing Backbone-fluorinated Amino Acids. *Org. Biomol. Chem.* **2012,** *10,* 8911–8918.
15. Jubault, P.; Quirion, J.-C.; Lion, C.; Lemonnier, G. *Preparation of Fluorinated Cyclopropane Analogs of Glutamic Acid as MGluR4 Modulators Useful in the Treatment of Neurological Diseases* 2009–305702/2279997. Institut National des Sciences Appliquees de Rouen (INSA), Fr, 2011.
16. Tarui, A.; Sato, K.; Omote, M.; Kumadaki, I.; Ando, A. Stereoselective Synthesis of α-Fluorinated Amino Acid Derivatives. *Adv. Synth. Catal.* **2010,** *352,* 2733–2744.
17. Uneyama, K. Recent Advances in the Syntheses of Fluorinated Amino Acids. In *Fluorine in Medicinal Chemistry and Chemical Biology;* Ojima, Iwao, Ed.; John Wiley: Chichester, UK, 2009; pp 213–256.

18. Smits, R.; Cadicamo, C. D.; Burger, K.; Koksch, B. Synthetic Strategies to α-trifluoromethyl and α-difluoromethyl Substituted α-amino Acids. *Chem. Soc. Rev.* **2008,** *37,* 1727–1739.
19. Kukhar, V. P. *Preparation of Fluorine-containing Amino Acids by Methods of Organofluorine Chemistry. Fluorine-Containing Amino Acids;* 1995. 71–111.
20. Platen, T.; Schueler, T.; Tremel, W.; Hoffmann-Roeder, A. Synthesis and Antibody Binding of Highly Fluorinated Amphiphilic MUC1 Glycopeptide Antigens. *Eur. J. Org. Chem.* **2011,** *2011,* 3878–3887. S3878/3871-S3878/3824.
21. Fustero, S.; Sinchez-Rosello, M.; Rodrigo, V.; Sanz-Cervera, J. F.; Piera, J.; Simon-Fuentes, A.; del Pozo, C. Solution-, Solid-phase, and Fluorous Synthesis of β,β-difluorinated Cyclic Quaternary α-amino Acid Derivatives: A Comparative Study. *Chem. Eur. J.* **2008,** *14,* 7019–7029.
22. Cellitti, S. E.; Jones, D. H.; Ryu, Y.; Schultz, P. G.; Geierstanger, B. H. *Site-selective Incorporation of Fluorinated Amino Acids into Proteins Using Translational Systems Including Orthogonal Aminoacyl-tRNA Synthetases and Orthogonal tRNAs* WO2009049223A2, IRM LLC. The Scripps Research Institute: Bermuda, 2009.
23. Mendel, D.; Cornish, V. W.; Schultz, P. G. Site-directed Mutagenesis with an Expanded Genetic Code. *Annu. Rev. Biophys. Biomol. Struct.* **1995,** *24,* 435–462.
24. Furter, R. Expansion of the Genetic Code: Site-directed P-fluoro-phenylalanine Incorporation in *Escherichia coli*. *Protein Sci.* **1998,** *7,* 419–426.
25. Duewel, H. S.; Daub, E.; Robinson, V.; Honek, J. F. Elucidation of Solvent Exposure, Side-chain Reactivity, and Steric Demands of the Trifluoromethionine Residue in a Recombinant Protein. *Biochemistry* **2001,** *40,* 13167–13176.
26. Montanari, V.; Kumar, K. A Fluorous Capping Strategy for Fmoc-based Automated and Manual Solid-phase Peptide Synthesis. *Eur. J. Org. Chem.* **2006,** 874–877.
27. Montanari, V.; Kumar, K. Enabling Routine Fluorous Capping in Solid Phase Peptide Synthesis. *J. Fluorine. Chem.* **2006,** *127,* 565–570.
28. Montanari, V.; Kumar, K. Just Add Water: A New Fluorous Capping Reagent for Facile Purification of Peptides Synthesized on the Solid Phase. *J. Am. Chem. Soc.* **2004,** *126,* 9528–9529.
29. Chiu, H.-P.; Cheng, R. P. Chemoenzymatic Synthesis of (S)-hexafluoroleucine and (S)-tetrafluoroleucine. *Org. Lett.* **2007,** *9,* 5517–5520.
30. Marsh, E. N. G. Fluorinated Proteins: From Design and Synthesis to Structure and Stability. *Acc. Chem. Res.* **2014,** *47,* 2878–2886.
31. Perfetti, R.; Merkel, P. Glucagon-like Peptide-1: A Major Regulator of Pancreatic B-cell Function. *Eur. J. Endocrinol.* **2000,** *143,* 717–725.
32. Nauck, M. A. Glucagon-like Peptide 1 (GLP-1): A Promising Approach and a Novel Treatment for Patients with Type 2 Diabetes. *Int. J. Clin. Pract., Suppl.* **2003,** *138,* 45–52.
33. Gumprecht, J.; Gumprecht, K.; Grzeszczak, W. GLP-1-based Treatment for Type 2 Diabetes. *Diabetol. Dosw. Klin.* **2007,** *7,* 215–219.
34. Meng, H.; Kumar, K. Antimicrobial Activity and Protease Stability of Peptides Containing Fluorinated Amino Acids. *J. Am. Chem. Soc.* **2007,** *129,* 15615–15622.
35. Gottler, L. M.; Ramamoorthy, A. Structure, Membrane Orientation, Mechanism, and Function of Pexiganan – a Highly Potent Antimicrobial Peptide Designed from Magainin. *Biochim. Biophys. Acta, Biomembr.* **2009,** *1788,* 1680–1686.
36. Gottler, L. M.; Lee, H.-Y.; Shelburne, C. E.; Ramamoorthy, A.; Marsh, E. N. G. Using Fluorous Amino Acids to Modulate the Biological Activity of an Antimicrobial Peptide. *ChemBioChem* **2008,** *9,* 370–373.
37. Meng, H.; Clark, G. A.; Kumar, K. *Fluorinated Amino Acids and Biomolecules in Protein Design and Chemical Biology;* 2009. 411–446.

38. Niemz, A.; Tirrell, D. A. Self-association and Membrane-binding Behavior of Melittins Containing Trifluoroleucine. *J. Am. Chem. Soc.* **2001,** *123,* 7407–7413.
39. Bilgicer, B.; Fichera, A.; Kumar, K. A Coiled Coil with a Fluorous Core. *J. Am. Chem. Soc.* **2001,** *123,* 4393–4399.
40. Anderson, J. T.; Toogood, P. L.; Marsh, E. N. G.A. Short and Efficient Synthesis of L-5,5,5,5′,5′,5′-hexafluoroleucine from N-Cbz-L-Serine. *Org. Lett.* **2002,** *4,* 4281–4283.
41. Holmgren, S. K.; Bretscher, L. E.; Taylor, K. M.; Raines, R. T. A Hyperstable Collagen Mimic. *Chem. Biol.* **1999,** *6,* 63–70.
42. Shoulders, M. D.; Kamer, K. J.; Raines, R. T. Origin of the Stability Conferred upon Collagen by Fluorination. *Bioorg. Med. Chem. Lett.* **2009,** *19,* 3859–3862.
43. Shoulders, M. D.; Hodges, J. A.; Raines, R. T. Reciprocity of Steric and Stereoelectronic Effects in the Collagen Triple Helix. *J. Am. Chem. Soc.* **2006,** *128,* 8112–8113.
44. Chattopadhyay, S.; Murphy, C. J.; McAnulty, J. F.; Raines, R. T. Peptides that Anneal to Natural Collagen In Vitro and Ex Vivo. *Org. Biomol. Chem.* **2012,** *10,* 5892–5897.
45. Hodges, J. A.; Raines, R. T. Stereoelectronic and Steric Effects in the Collagen Triple Helix: Toward a Code for Strand Association. *J. Am. Chem. Soc.* **2005,** *127,* 15923–15932.
46. Bejot, R.; Fowler, T.; Carroll, L.; Boldon, S.; Moore, J. E.; Declerck, J.; Gouverneur, V. Fluorous Synthesis of 18F Radiotracers with the [^{18}F]fluoride Ion: Nucleophilic Fluorination as the Detagging Process. *Angew. Chem. Int. Ed.* **2009,** *48,* 586–589.
47. Houston, M. E., Jr.; Honek, J. F. Facile Synthesis of Fluorinated Methionines. *J. Chem. Soc. Chem. Commun.* **1989,** 761–762.
48. Janzen, A. F.; Wang, P. M. C.; Lemire, A. E. Fluorination of Methionine and Methionylglycine Derivatives with Xenon Difluoride. *J. Fluorine. Chem.* **1983,** *22,* 557–559.
49. Honek, J. F. *Fluorinated Methionines as Probes in Biological Chemistry*, Vol. 949; 2007; pp 393–408.
50. Budisa, N.; Pipitone, O.; Siwanowicz, I.; Rubini, M.; Pal, P. P.; Holak, T. A.; Gelmi, M. L. Efforts towards the Design of "Teflon" Proteins: In Vivo Translation with Trifluorinated Leucine and Methionine Analogues. *Chem. Biodivers.* **2004,** *1,* 1465–1475.
51. Moya, I. A.; Westrop, G. D.; Coombs, G. H.; Honek, J. F. Mechanistic Studies on the Enzymatic Processing of Fluorinated Methionine Analogues by *Trichomonas vaginalis* Methionine γ-lyase. *Biochem. J.* **2011,** *438,* 513–521.
52. Tsushima, T.; Ishihara, S.; Fujita, Y. Fluorine-containing Amino Acids and Their Derivatives. 9.1 Synthesis and Biological Activities of Difluoromethylhomocysteine. *Tetrahedron Lett.* **1990,** *31,* 3017–3018.
53. Dannley, R. L.; Taborsky, R. G. Synthesis of DL-S-trifluoromethylhomocysteine (Trifluoromethylmethionine). *J. Org. Chem.* **1957,** *22,* 1275–1276.
54. Wang, D.; Cole, P. A. Protein Tyrosine Kinase Csk-catalyzed Phosphorylation of Src Containing Unnatural Tyrosine Analogues. *J. Am. Chem. Soc.* **2001,** *123,* 8883–8886.
55. Seyedsayamdost, M. R.; Yee, C. S.; Reece, S. Y.; Nocera, D. G.; Stubbe, J. pH Rate Profiles of FnY356-r2s (N = 2, 3, 4) in *Escherichia coli* Ribonucleotide Reductase: Evidence that Y356 Is a Redox-active Amino Acid along the Radical Propagation Pathway. *J. Am. Chem. Soc.* **2006,** *128,* 1562–1568.
56. Wilkins, B. J.; Marionni, S.; Young, D. D.; Liu, J.; Wang, Y.; Di Salvo, M. L.; Deiters, A.; Cropp, T. A. Site-specific Incorporation of Fluorotyrosines into Proteins in *Escherichia coli* by Photochemical Disguise. *Biochemistry* **2010,** *49,* 1557–1559.
57. Gopishetty, B.; Ren, L.; Waller, T. M.; Wavreille, A.-S.; Lopez, M.; Thakkar, A.; Zhu, J.; Pei, D. Synthesis of 3,5-difluorotyrosine-containing Peptides: Application in Substrate Profiling of Protein Tyrosine Phosphatases. *Org. Lett.* **2008,** *10,* 4605–4608.

58. Zheng, H.; Gao, J. Highly Specific Heterodimerization Mediated by Quadrupole Interactions. *Angew. Chem. Int. Ed.* **2010,** *49,* 8635–8639.
59. Robson Marsden, H.; Fraaije, J. G. E.M.; Kros, A. Introducing Quadrupole Interactions into the Peptide Design Toolkit. *Angew. Chem. Int. Ed.* **2010,** *49,* 8570–8572.
60. Cejas, M. A.; Kinney, W. A.; Chen, C.; Leo, G. C.; Tounge, B. A.; Vinter, J. G.; Joshi, P. P.; Maryanoff, B. E. Collagen-related Peptides: Self-assembly of Short, Single Strands into a Functional Biomaterial of Micrometer Scale. *J. Am. Chem. Soc.* **2007,** *129,* 2202–2203.
61. Ryan, D. M.; Doran, T. M.; Nilsson, B. L. Complementary π-π Interactions Induce Multicomponent Coassembly into Functional Fibrils. *Langmuir* **2011,** *27,* 11145–11156.
62. Ueki, H.; Ellis, T. K.; Martin, C. H.; Boettiger, T. U.; Bolene, S. B.; Soloshonok, V. A. Improved Synthesis of Proline-derived Ni(II) Complexes of Glycine: Versatile Chiral Equivalents of Nucleophilic Glycine for General Asymmetric Synthesis of α-Amino Acids. *J. Org. Chem.* **2003,** *68,* 7104–7107.
63. Belokon, Y. N.; Bespalova, N. B.; Churkina, T. D.; Cisarova, I.; Ezernitskaya, M. G.; Harutyunyan, S. R.; Hrdina, R.; Kagan, H. B.; Kocovsky, P.; Kochetkov, K. A.; Larionov, O. V.; Lyssenko, K. A.; North, M.; Polasek, M.; Peregudov, A. S.; Prisyazhnyuk, V. V.; Vyskocil, S. Synthesis of α-Amino Acids via Asymmetric Phase Transfer-Catalyzed Alkylation of Achiral Nickel(II) Complexes of Glycine-Derived Schiff Bases. *J. Am. Chem. Soc.* **2003,** *125,* 12860–12871.
64. Noisier, A. F. M.; Harris, C. S.; Brimble, M. A. Novel Preparation of Chiral α-amino Acids Using the Mitsunobu-Tsunoda Reaction. *Chem. Commun.* **2013,** *49,* 7744–7746.
65. Kukhar, V. P.; Belokon, Y. N.; Soloshonok, V. A.; Svistunova, N. Y.; Rozhenko, A. B.; Kuz'mina, N. A. Asymmetric Synthesis of Organoelement Analogs of Natural Products. Part 12. General Method for the Asymmetric Synthesis of Fluorine-containing Phenylalanines and α-methyl(phenyl)alanines via Alkylation of the Chiral Nickel(II) Schiff's Base Complexes of glycine and Alanine. *Synthesis* **1993,** 117–120.
66. Ojima, I., Ed. *Fluorine in Medicinal Chemistry and Chemical Biology;* Wiley: New York, 2009.
67. Yamaki, Y.; Shigenaga, A.; Tomita, K.; Narumi, T.; Fujii, N.; Otaka, A. Synthesis of Fluoroalkene Dipeptide Isosteres by an Intramolecular Redox Reaction Utilizing N-heterocyclic Carbenes (NHCs). *J. Org. Chem.* **2009,** *74,* 3272–3277.
68. Xiao, J.; Weisblum, B.; Wipf, P. Electrostatic versus Steric Effects in Peptidomimicry: Synthesis and Secondary Structure Analysis of Gramicidin S Analogues with (E)-alkene Peptide Isosteres. *J. Am. Chem. Soc.* **2005,** *127,* 5742–5743.
69. Wipf, P.; Henninger, T. C.; Geib, S. J. Methyl- and (Trifluoromethyl)alkene Peptide Isosteres: Synthesis and Evaluation of Their Potential as β-Turn Promoters and Peptide Mimetics. *J. Org. Chem.* **1998,** *63,* 6088–6089.
70. Meanwell, N. A. Synopsis of Some Recent Tactical Application of Bioisosteres in Drug Design. *J. Med. Chem.* **2011,** *54,* 2529–2591.
71. Black, W. C.; Bayly, C. I.; Davis, D. E.; Desmarais, S.; Falgueyret, J.-P.; Leger, S.; Li, C. S.; Masse, F.; McKay, D. J.; Palmer, J. T.; Percival, M. D.; Robichaud, J.; Tsou, N.; Zamboni, R. Trifluoroethylamines as Amide Isosteres in Inhibitors of Cathepsin K. *Bioorg. Med. Chem. Lett.* **2005,** *15,* 4741–4744.
72. Sani, M.; Volonterio, A.; Zanda, M. The Trifluoroethylamine Function as Peptide Bond Replacement. *ChemMedChem* **2007,** *2,* 1693–1700.
73. Reginster, J. Y.; Neuprez, A.; Beaudart, C.; Lecart, M. P.; Sarlet, N.; Bernard, D.; Disteche, S.; Bruyere, O. Antiresorptive Drugs beyond Bisphosphonates and Selective Oestrogen Receptor Modulators for the Management of Postmenopausal Osteoporosis. *Drugs Aging* **2014,** *31,* 413–424.
74. Piras, M.; Fleming, I. N.; Harrison, W. T. A.; Zanda, M. Linear Trifluoroethylamine RGD Peptidomimetics: Stereoselective Synthesis and Integrin αvβ3 affinity. *Synlett* **2012,** *23,* 2899–2902.

Chapter 5

Organofluorine Pharmaceuticals

Chapter Outline

1. INTRODUCTION

Nearly 20% of the currently marketed pharmaceuticals are organofluorine compounds and, considering the vast number of fluorinated drug candidates that are being tested in clinical trials each year, the number of fluorinated drugs is expected to grow at a rapid pace. Some of the fluorinated drugs such as fluoxetine (Prozac; antidepressant), fluticasone propionate (antiasthma drug), and atorvastatin (cholesterol-lowering drug) are the top-selling, blockbuster drugs with multibillion dollars annual sales.[1] In fact, almost every class of drugs consists of at least one fluorinated blockbuster drug. Due to the highest electronegativity

Organofluorine Compounds in Biology and Medicine. http://dx.doi.org/10.1016/B978-0-444-53748-5.00005-8

of fluorine, combined with its similar van der Waals radius as that of hydrogen, organofluorine compounds show unique biochemical and physiological activity. For example, fluorinated drug candidates usually form relatively more stabilized enzyme–substrate complexes and are relatively more potent than their nonfluorinated analogs. Fluorinated drugs are also relatively more bioavailable and metabolically more stable than their nonfluorinated analogs.

Synthesis and medicinal applications of the fluorinated drugs are reviewed extensively in the literature.[2–5] Fustero, Soloshonok, Liu and coworkers' extensive review describes the detailed synthetic methods, as well as the medicinal uses, for the various classes of fluorine-containing drugs that are approved by US Food and Drug Administration (FDA) during the period 2001–2011.[4] In this chapter, we will focus on the mechanisms of drug action, and outline the synthetic methods for the commonly used FDA-approved drugs.

1.1 Blockbuster Drugs

Several fluorinated pharmaceuticals are routinely being prescribed to treat a variety of diseases. Among the most commonly prescribed fluorinated drugs are fluoroquinolone antibiotics (e.g., Ciprobay), cholesterol-lowering drugs (e.g., Lipitor, ezetimibe, Crestor, and cerivastatin), nonsteroidal antiinflammatory drugs (NSAID; e.g., celecoxib and roflumilast), antimalarials (e.g., Ciprobay), anticancer drugs (e.g., Faslodex and gemcitabine), antidepressants (e.g., fluoxetine), antiviral (e.g., efavirenz), and asthma drugs (e.g., the steroidal drug fluticasone propionate) (Figure 1). Lipitor, also known as atorvastatin, is a member of the statin family of drugs that lower blood cholesterol levels by inhibiting 3-hydroxy-3-methylglutaryl CoA (HMG-CoA) reductase; by inhibiting this enzyme, the initial steps involved in the biosynthesis of cholesterol and other terpenoids are blocked. Lipitor was the number one top-selling drug in 2011 (Pfizer Pharmaceuticals), with annual sales exceeding 7 billion dollars; due to the subsequent expiry of the patent, there has been a massive drop in the revenues for this drug, while the sales of its generic versions have drastically increased.[6,7] GlaxoSmithKline's Advair Diskus (combination drug, consisting of fluticasone propionate and salmeterol), used for treating asthma and chronic obstructive pulmonary disease (COPD), was the fifth top-selling drug with annual sales of over 4.4 billion dollars in 2011, and sixth best-selling drug in 2013. AstraZeneca's antihypercholesterolemia drug Crestor (rosuvastatin), a *p*-fluoroaryl-substituted compound, was listed as the eighth top-selling drug in 2011 and the fifth in 2013. Sitagliptin (Januvia; Merck), an antihyperglycemic drug, used for the treatment of type 2 diabetes mellitus, was the 15th top-selling drug in 2013. Celebrex (celecoxib, Pfizer), a NSAID and the 21st best-selling drug in 2013, is used for the treatment of osteoarthritis and rheumatoid arthritis. Ezetimibe (Zetia; Merck), the 29th best-selling drug in 2013, is used, together with statins, for lowering blood

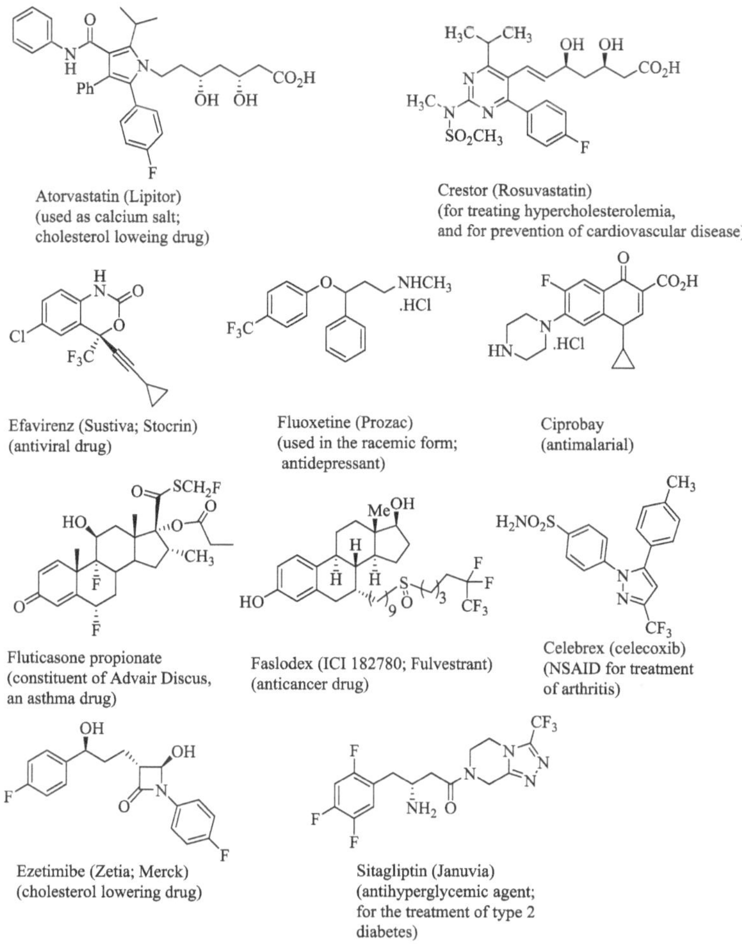

FIGURE 1 Fluorine-containing blockbuster pharmaceuticals; nearly every class of pharmaceuticals has at least one top-selling drug that is an organofluorine compound.

cholesterol levels. Although best-selling rank of the drugs varies year to year, and certain drugs are removed from market periodically due to their adverse side effects, it is remarkable that nearly one-third of the top 100 top-selling drugs are organofluorine compounds (Figure 1).[7]

Historically, 5-fluorouracil, synthesized in 1957, is the first organofluorine pharmaceutical to be used as antimetabolite, and subsequently it became an important anticancer drug after the discovery that it inhibits thymidylate synthase, an enzyme responsible for the intracellular synthesis of thymidine (Chapter 2 provides a detailed discussion of the mechanism of the thymidylate synthase inhibition).[8–10] Due to its high toxicity, analogs of this compound that act as prodrugs and those that release the active drug through intracellular enzymatic catalysis are being developed (vide infra). Representative fluorine-containing drugs and drug candidates, and an overview of their drug action are discussed below. Organofluorine anticancer drugs and drug candidates for neurological disorders are discussed in (chapters 9 and 8, respectively). The pathogenesis of neurological disorders is relatively less thoroughly characterized and has multiple causative factors, complicating the efforts in the drug design.

2. PHARMACEUTICALS WITH *gem*-DIFLUOROMETHYLENE AND TRIFLUOROMETHYL MOIETIES

Many pharmaceuticals having a *gem*-difluoromethylene group are FDA-approved drugs for the treatment of various diseases. These compounds include Faslodex (an anticancer drug), gemcitabine (2′-deoxy-2′,2′-difluorocytidine; an anticancer drug), Tafluprost (attenuates the progression of glaucoma), Maraviroc (an antiretroviral drug used in human immunodeficiency virus (HIV) treatment), pantoprazole (a proton pump inhibitor for treating ulceration of esophagus), roflumilast (for treating COPD), and lubiprostone (for treating irritable bowel syndrome) (Figure 2).

Trifluoromethyl- and *gem*-difluoromethylene groups, in many cases, impart improved lipophilicity and bioavailability to the drug candidates. Due to the increased lipophilicity these drugs molecules exhibit relatively enhanced cell permeability. New synthetic methods for the incorporation of the CF_3 groups

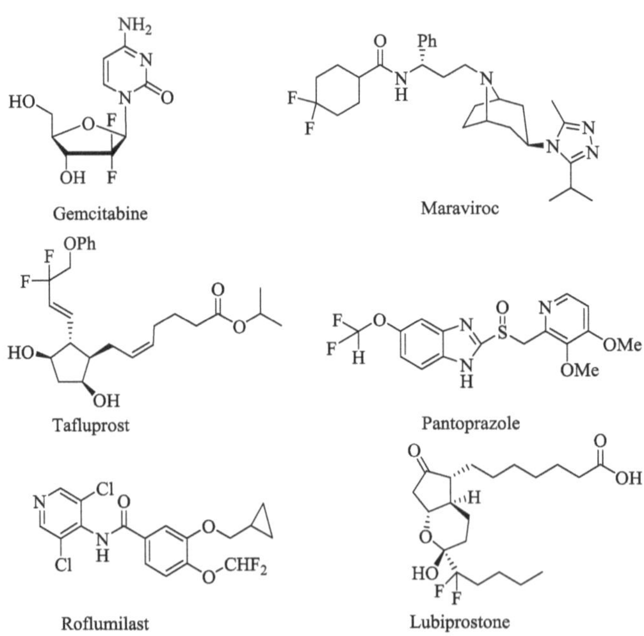

FIGURE 2 Selected pharmaceuticals with a *gem*-difluoromethylene (CF_2) group.

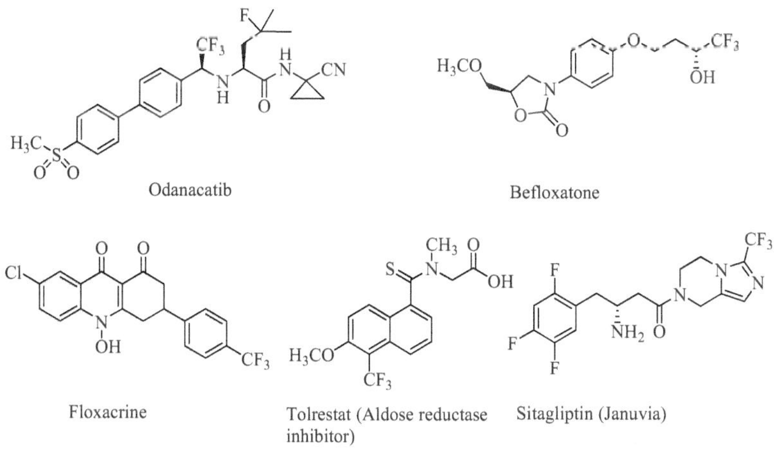

FIGURE 3 Selected pharmaceuticals with a CF_3 moiety.

into organic compounds are being rapidly developed. Examples of the trifluoromethylated top-selling drugs and/or those that have completed clinical trials include efavirenz (Sustiva or Stocrin; antiretroviral therapy for HIV), befloxatone (selective monoamine oxidase-A inhibitor), odanacatib (MK-0822; cathepsin K inhibitor), sitagliptin (an antidiabetic drug), floxacrine (an antimalarial drug), celecoxib (Celebrex; selective COX-2 inhibitor; an NSIAD used to treat osteoarthritis and rheumatoid arthritis), and tolrestat (aldose reductase inhibitor and a potential antidiabetic drug) (Figure 3). Tolrestat was not approved by the FDA to treat diabetes because of its severe liver toxicity. The phase III clinical trials for odanacatib[11] were successfully completed by Merck for treating osteoporosis.

3. ANTIBACTERIALS

3.1 Flurithromycin

Erythromycin is a broad-spectrum macrolide antibiotic, and is effective against multiple strains of bacteria. It is often used for treating bronchitis, Legionnaires' disease, and penicillin-related allergies.[2] Its low stability in the acidic medium makes it ineffective for the treatment of gastritis caused by *Helicobacter pylori*. On the other hand, flurithromycin, a monofluorinated derivative of erythromycin, has higher metabolic stability and relatively more effectiveness, as compared to erythromycin, and therefore is the preferred drug for the treatment of gastritis and other bacterial infections.[12] The increased acid stability of flurithromycin may be rationalized as due to the inductive destabilization of the carbocation intermediate. Further, the fluorine substitution, in case of flurithromycin, also results in increased lipophilicity so that it is relatively more cell permeable and has relatively more bioavailability as compared to that of erythromycin (Figure 4).

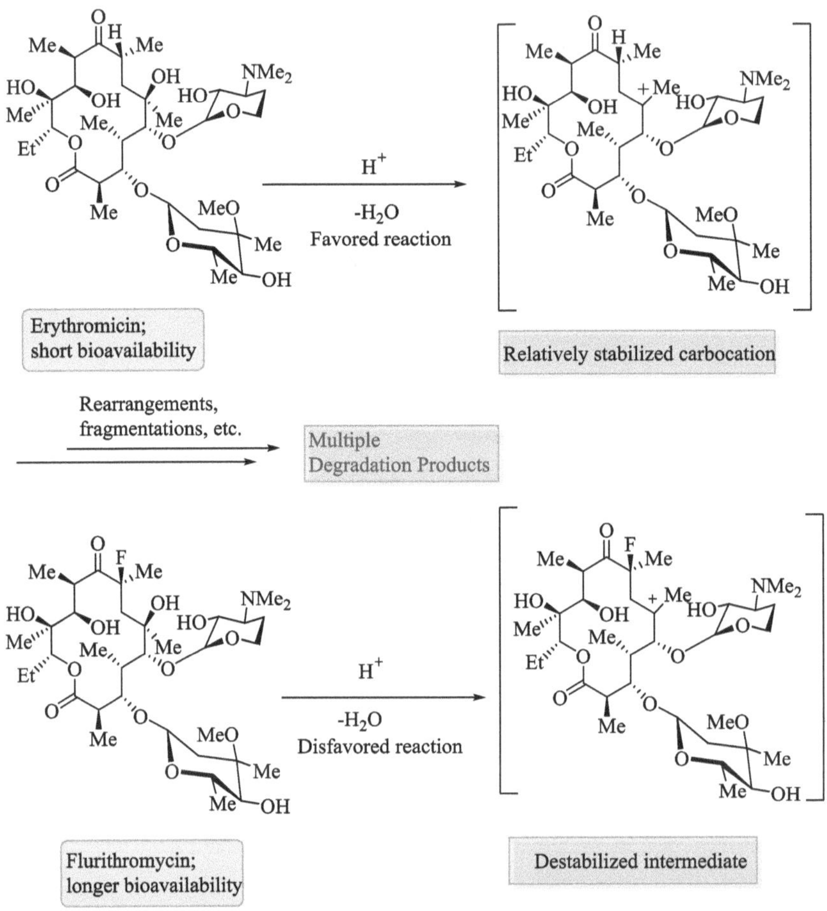

FIGURE 4 Acid-catalyzed degradation of erythromycin and flurithromycin; flurithromycin is relatively less prone to acid-catalyzed degradation because the resulting carbocationic intermediate is relatively destabilized.

3.2 Quinolone Antibiotics

Quinolone antibiotics, the relatively newly developed antibiotics, including the organofluorine compounds such as ofloxacin (Floxin), norfloxacin (Noroxin), ciprofloxacin (Cipro), and moxifloxacin (Avelox) are widely used worldwide. Ofloxacin is a racemic drug, whereas its more active component, the *S*-enantiomer, is marketed as levofloxacin (Levaquin) (Figure 5). Although these fluoroquinolones are broad-spectrum antibiotics and are being used in a broad range of bacterial infections—such as pneumonia and tuberculosis (TB)—in rare cases they may cause adverse effects, which include transient or permanent peripheral neuropathy. The new FDA warnings accordingly advise switching the drugs to the nonfluorinated quinolone antibiotics if these adverse symptoms

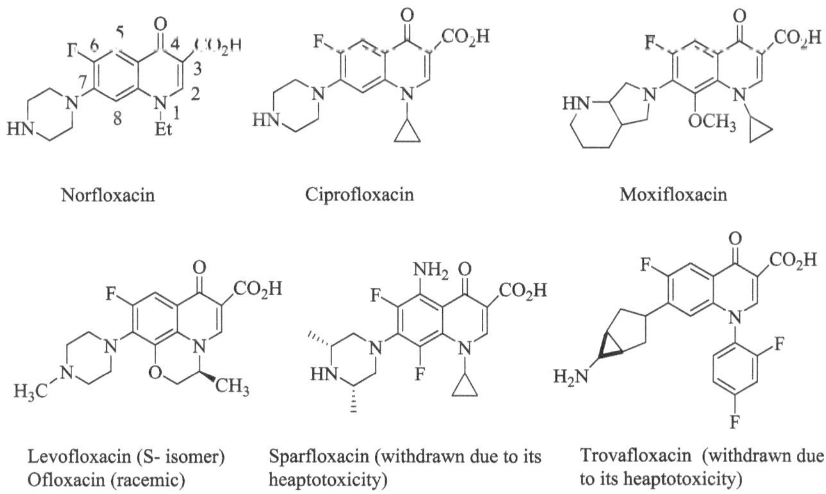

FIGURE 5 Structures of the widely used fluorinated quinolone antibiotics; in rare cases these drugs cause severe adverse effects, which include peripheral neuropathy.

appear upon their usage, unless their benefits of continued treatment outweigh the risks.[13]

The quinolone antibiotics have γ-pyridone ring fused to the benzene ring as a common feature. Further, a fluorine atom at C_6, and a piperidine moiety at C_7 are usually present in these molecules (Figure 5). The quinolone antibiotics, especially ciprofloxacin, ofloxacin (or its chirally pure *S*-isomer, the levofloxacin), and sparfloxacin are currently most commonly used to treat TB cases. The in vitro minimum inhibitory concentrations of these compounds against *Mycobacterium tuberculosis* (TB) are 0.1–0.4 μg/mL.[14] Because these compounds are structurally distinct from the β-lactams or tetracycline antibacterials, they have entirely different bacterial resistance mechanisms and, therefore, are useful as complementary antibacterials for treating β-lactam- or tetracycline-resistant bacterial infections. However, even with these quinolone antibiotics, bacterial resistance is rapidly developing; for example, *Escherichia coli* isolated from turkeys in Great Britain showed resistance to ciprofloxacin in almost one-third of the population tested.[15]

Structure–activity studies of thousands of quinolone compounds reveal that the fluorine at the C_6 improves oral absorption and activity against anaerobes, and the pyrrolidine or piperazine ring at C_7 increases the serum half-life and potency versus the Gram-positive bacteria. The remaining substituents are useful mainly to improve the overall potency.[16] The quinolone antibiotics show adverse side effects including gastrointestinal irritation, and also show drug–drug interactions with NSAIDs, which is mainly attributed to the pyrrolidine or piperazine substituents. Increasing steric bulk of these substituents through *N*-alkylation somewhat ameliorates the latter side effects. As with other classes of pharmaceuticals, some of the FDA-approved quinolone antibiotics show unexpected

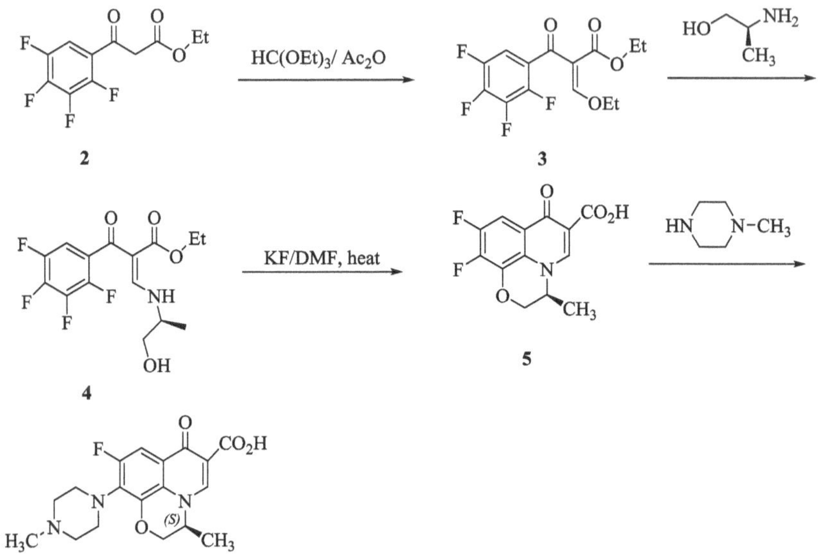

Levofloxacin

FIGURE 6 Synthesis of levofloxacin (*S*-ofloxacin) (Ac_2O = acetic anhydride; DMF = dimethylformamide).

toxicities; trovafloxacin, for example, exerts hepatotoxicity, based on which it has been currently withdrawn from the market.[17] Sparfloxacin has similar toxicity and, therefore, is also withdrawn from the US markets (Figure 5).[18]

3.2.1 Synthesis of Levofloxacin

Ethyl 3-oxo-3-arylpropanoate **2** (obtained through malonic ester synthesis from the 2,3,4,5-tetrafluorobenzoyl chloride (**1**)), upon reaction with triethyl orthoformate gives the vinyl ether **3**, the latter being a masked α-formyl group. Reaction of **3** with (*S*)-2-aminopropanol gives **4**, which upon reaction with a mild base, such as KF, undergoes two tandem cyclization reactions to give **5**. The aromatic nucleophilic substitution of the *para*-F (*para* with respect to the carbonyl group) by *N*-methylpiperazine, followed by hydrolysis of the ester group gives the levofloxacin (Figure 6). The selective nucleophilic displacement of the *ortho*- and *para*-fluorines (with respect to the carbonyl group) in compounds **4** and **5** is facilitated by the electron-withdrawing carbonyl group.[19]

4. ANTIMALARIALS

Malaria is caused by *Plasmodium* parasites—*Plasmodium falciparum. Plasmodium vivax, Plasmodium ovale, Plasmodium malariae*, and *Plasmodium knowles*. Artemisinin and quinine are among the first antimalarial drugs ever used to treat malaria, even in the ancient times. Quinine is an

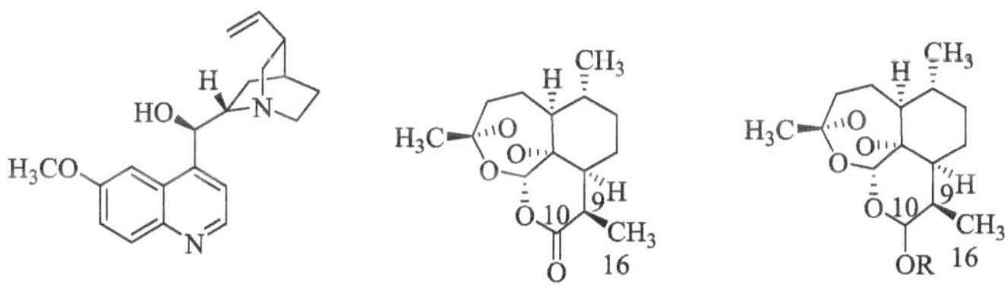

Quinine

Artemisinin

6a; R = CH_3 (artemether);
6b; R = Et (arteether);
6c; R = fluoroalkyl/aryl, e.g.,CF_3, CF_3CH_2, C_6F_5

FIGURE 7 Structures of the widely used first-generation antimalarials (quinine, artemisinin, and its derivatives).

extract of cinchona tree bark, and the bark itself was used to treat malaria as long ago as seventeenth century in Peru, South America. Similarly, artemisinin, an extract of the sweet wormwood (*Artemisia annua*), was in use for treating the malarial fever, hundreds of years even before its structure was elucidated by Chinese scientists in the mid-1970s.[20–22] Artemisinin has poor bioavailability due to its relatively low hydrophilicity and lipophilicity characteristics, and therefore many of its derivatives that retain the 1,2,4-trioxane core have been synthesized and evaluated for their improved efficacy as antimalarial pharmaceuticals. Most of the structural modifications were carried out at the lactone carbonyl carbon (C_{10}), since it is readily functionalizable (Figure 7). Reduction of the lactone carbonyl group gives the hemiacetal, the dihydroartemisinin (DHA), which is readily alkylated to give various acetal derivatives of artemisinin, such as artemether (**6a**; a methyl acetal), Arteether (**6b**; an ethyl acetal), and fluoroalkyl or fluoroaryl acetals (**6c**) (Figure 7).[22] These first-generation artemisinins, developed in the 1980s, had substantial impact on the treatment of malaria. However, these derivatives have relatively short half-life in the body and thereby poor bioavailability, limiting their use as antimalarials. The first-generation artemisinins are, in particular, prone to hydrolysis by proteolytic enzymes at the acetal site to regenerate the hemiacetal form of the DHA, which is rapidly metabolized in the liver by the action of cytochrome P450 enzymes to give the water-soluble DHA-glucuronide derivative (Figure 12; vide infra).[23,24] In order to combat the rapid parasite resistance to antimalarials, World Health Organization's guidelines suggest the use of combination therapy[25] involving the short-acting artemisinin derivatives, which have not yet acquired parasite resistance, and other antimalarials, such as quinine derivatives and various synthetic quinolones that possess longer plasma half-life. Bonnet-Delpon and coworkers have synthesized a variety of fluoroalkyl ethers of DHA (e.g., **6c**), which are relatively poor substrates for cytochrome P450 oxidase so that their bioavailability could be significantly improved.[22–24]

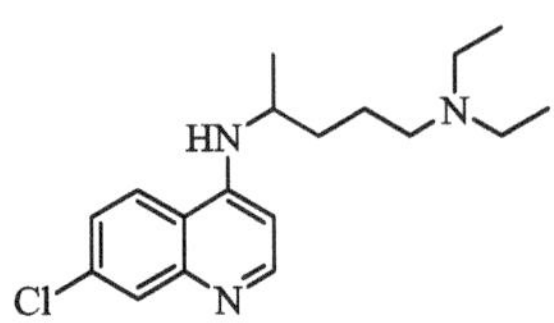

Chloroquine
(a first generation antimalarial drug)

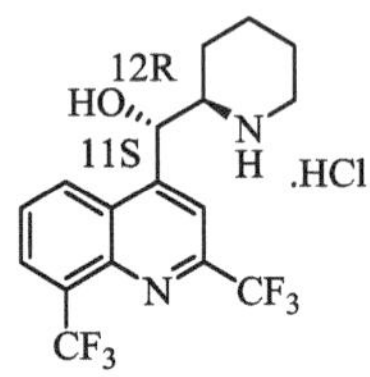

(+)-erythro-Mefloquine HCl
antimalarial

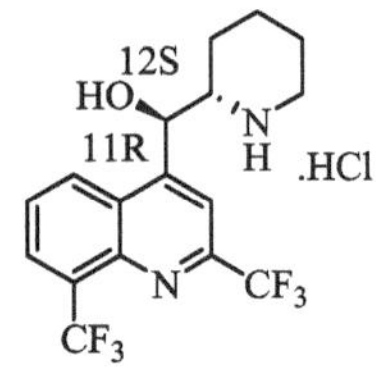

(-)-erythro-Mefloquine HCl
side effect: psychosis

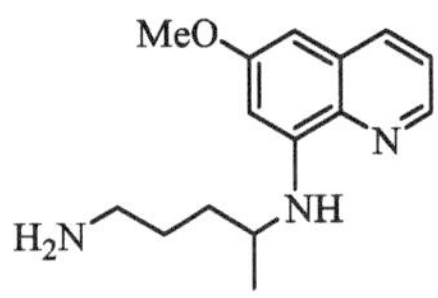

Primaquine

FIGURE 8 Structures of chloroquine, mefloquine (both enantiomers), and primaquine.

Quinine and its chlorinated analogs such as chloroquine and primaquine are typically used not only for the treatment of malaria but also as prophylactics to prevent malarial infections (Figure 8). Due to the evolving resistance of the parasites to these antimalarials, new variants of these drugs are being developed continually. Mefloquine, a bis-trifluoromethylated quinolone derivative, is one of the most successful drugs for treating malaria, and has been approved by the FDA in 1989. It is administered as a racemic mixture, and is marketed as Lariam by Roche Pharmaceuticals. Interestingly, the (+)-(11*S*, 12*R*)-mefloquine HCl is an antimalarial drug, whereas the enantiomeric compound, (−)-(11*R*, 12*S*)-mefloquine HCl, occasionally causes adverse effects, including psychosis and depression in a minor population of the patients, an observation that shows how important it is to use chirally homogeneous drugs. In 2012, the absolute configuration of the active antimalarial diastereomer (+form) has been established as (11*S*, 12*R*) by a combination of nuclear magnetic resonance (NMR), optical rotatory dispersion, and circular dichroism spectroscopy in conjunction with density functional theory calculations.[26] Despite the minor adverse effects, racemic mefloquine is being widely used to treat malaria and is used as a prophylactic agent when traveling to the endemic areas, especially due to its long lifetime of about 2 weeks.

Severe cases of cerebral malaria arise due to the infections of *P. falciparum*. Due to the severity of the symptoms of cerebral malaria, and its resistance to most of the currently available antibacterial drugs, adjunctive therapies, that is, therapies administered in combination with antimalarial drugs, are being explored; activation of γ-peroxisome proliferator-activated receptors (PPARγ) by agonists provides neuronal protection and imparts neuroregenerative properties in cases of central nervous system injury; thus this class of drugs is suitable

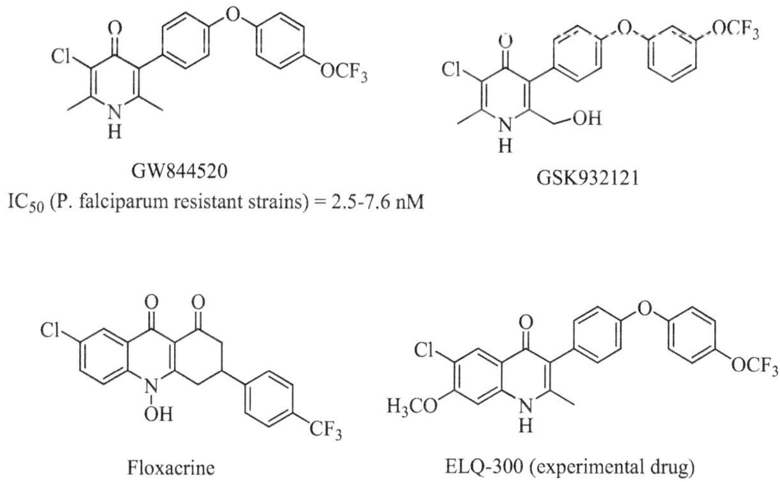

FIGURE 9 Fluorinated antimalarial drugs targeting drug-resistant *P. falciparum*.

as adjunctive therapeutics for treating cerebral malaria. For example, rosiglitazone, a PPARγ agonist and an antidiabetic drug, showed some promise as an adjunctive therapy candidate for malaria and Chagas disease.[27,28]

Several fluorine-substituted pyridine, acridinedione (e.g., floxacrine), and structurally similar acridone derivatives were synthesized and clinically evaluated as antimalarials (Figure 9). Some of these derivatives turned out to be effective drugs against the drug-resistant *P. falciparum* parasite. However, due to the toxicity concerns, their clinical trials have been halted.[29] More and more experimental drugs that evade drug resistance mechanisms of the parasites are currently being developed. One such compound, ELQ-300, evaluated at the Veterans Affairs Medical Center, Portland, Oregon, has the potential to not only cure malaria but also to be used as a prophylactic agent for preventing the incidence of malaria, when exposed to disease-carrying mosquitoes: a 1 mg/kg body weight dose of ELQ-300 cured the infected mice, and a single dose given to noninfected mice protected them from contracting malaria from the plasmodium-infected mosquitoes.[30] ELQ-300 inhibits cytochrome$_{bc1}$ that is responsible for the synthesis of the parasite deoxyribonucleic acid (DNA) building blocks, thereby killing the malarial parasites. This compound has structural similarity with other antimalarials, GW844520 and GSK93212, all having a (trifloromethoxyphenoxyphenyl) moiety appended adjacent to the carbonyl carbon. Some other organofluorine compounds, including GNF156 (an imidazolopiperazine),[31] DSM190 (a triazolopyrimidine),[32] and NITD609 (a spiroindolone) (Figure 10) showed positive results as antimalarials in in vitro studies.[33]

The triazolopyrimidine DSM190 and many other related analogs, including those with an SF_5 moiety on the phenyl ring, were synthesized and tested for

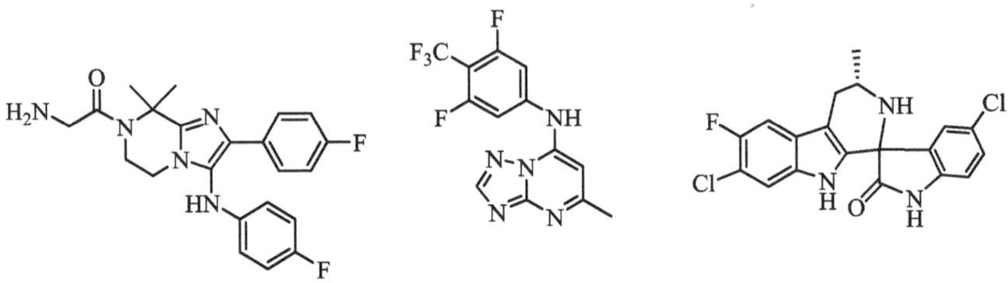

GNF156 DSM190 NITD609

FIGURE 10 Structures of some experimental antimalarial compounds, GNF156, DSM190, and NITD609.

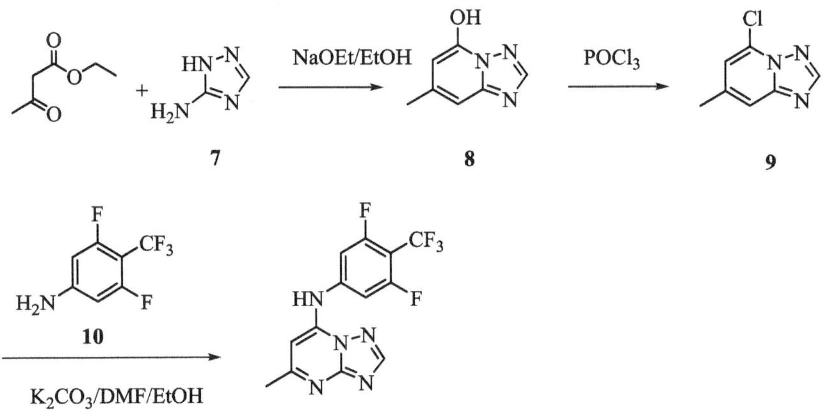

DSM190

FIGURE 11 Synthesis of DSM190, an experimental antimalarial compound.

their antimalarial activity (synthesis of SF_5-mefloquine derivatives is discussed in chapter 3). These compounds show antimalarial activity by inhibiting *P. falciparum* dihydroorotate dehydrogenase (IC_{50} of DSM190 = 0.19 μM), and are currently in the developmental stage.[32]

4.1 Synthesis of DSM190

The triazolopyrimidine and its various analogs were synthesized conveniently in three steps: condensation of 3-amino[1,2,4]triazole **7** with ethyl acetoacetate, followed by dehydroxychlorination of the product **8** using $POCl_3$ gives 7-chloro-[1,2,4]-triazolo[1,5-a]pyrimidine (**9**), which upon aromatic nucleophilic substitution reaction using 3,5-difluoro-4-trifluoromethylaniline (**10**) in the presence of K_2CO_3 gives the DSM190 (Figure 11).[32]

4.2 Fluorinated Artemisinins

The strained endoperoxide group (the 1,2,4-trioxane ring) in artemisinin is primarily responsible for its antimalarial effect. Artemisinin is a fast-acting drug

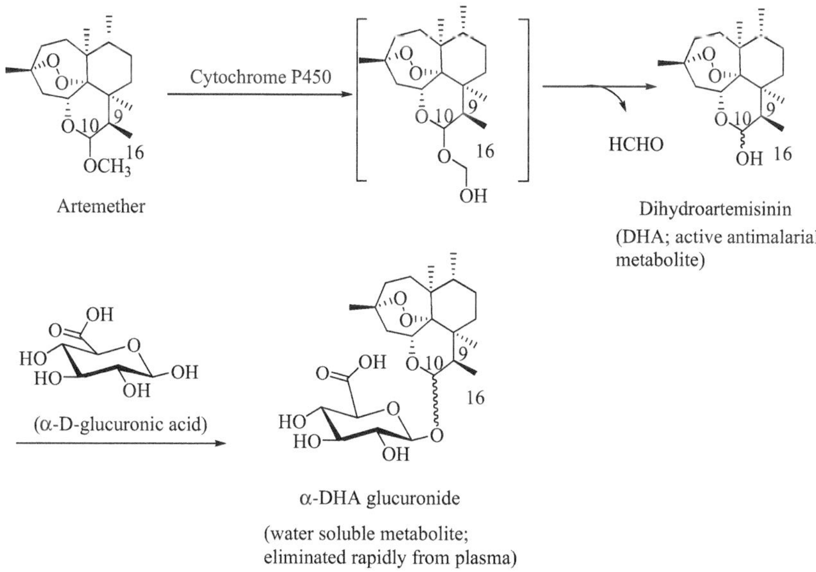

FIGURE 12 Metabolic oxidative degradation of artemether.

in the early stages of parasite development, and is active even in multidrug-resistant *P. falciparum*. However, the short lifetime of this drug under physiological conditions necessitates its use as a combination therapeutic along with other quinine-based drugs, such as mefloquine. Toward the goal of increasing physiological stability of these artemisinin-based drugs, Begue, Bonnet-Delpon, and coworkers have synthesized several fluorinated versions of the artemisinin and its first-generation analogs, and some of these fluorinated versions have relatively increased half-life due to their resistance to oxidation by cytochrome P450 enzymes and also due to their increased hydrolytic stability.[22,34]

The rapid oxidation of the artemether C_{10}-*O*-alkyl group by cytochrome P450 enzymes results in the formation of the hemiacetal, DHA, which undergoes enzyme-catalyzed glucuronation to form the corresponding DHA glucoronide that, being water soluble, is eliminated from the blood plasma at rapid rates (Figure 12). Incorporation of the electron-withdrawing trifluoromethyl groups at C_{10} of dihydroartemisinin or artemether appreciably diminishes their rate of hydrolysis or their CYP-450-catalyzed oxidations, because the strong electron-withdrawing inductive effect of the CF_3 disfavors the formation of the intermediate carbocations, **11** and **12**. This would dramatically increase the plasma stability of the artemisinin analogs, and thereby overcome the physiological instability of the artemisinins (Figure 13).

Begue and coworkers have synthesized the 10-trifluoromethyldihydroartemisinin **13** by reaction of artemisinin with trifluoromethyltrimethylsilane (CF_3SiMe_3) in the presence of tetrabutylammonium fluoride, followed by aqueous workup, in good yield (Figure 14).[35] Reaction of this

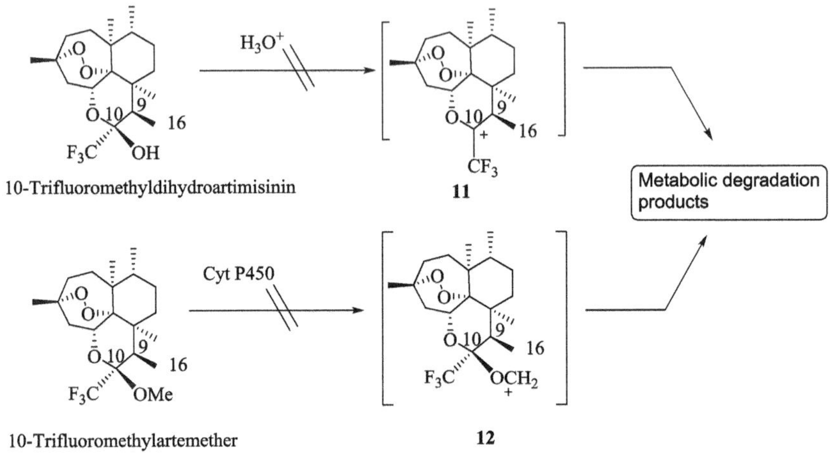

FIGURE 13 Increased hydrolytic and metabolic stability of trifluoromethyl-substituted dihydroartemisinin and artemether; the electron-withdrawing CF_3 group disfavors the formation of the carbocation intermediates.

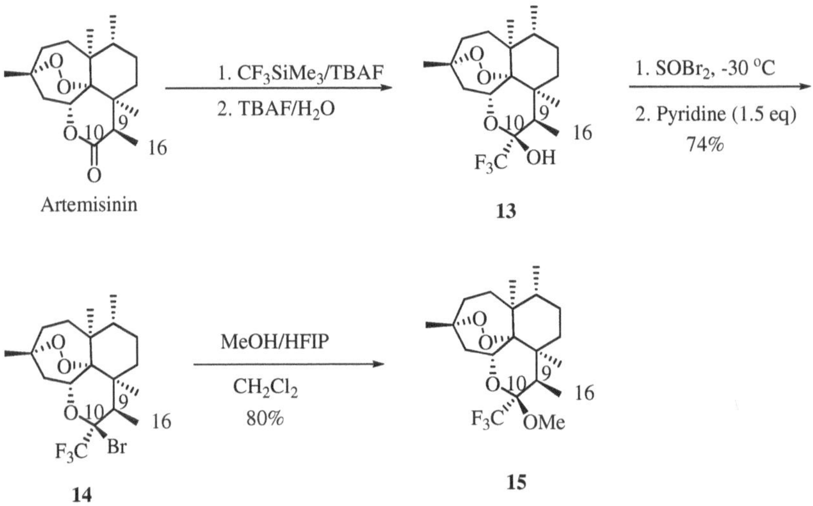

FIGURE 14 Synthesis of 10-trifluoromethylartemether (**15**).

hemiacetal **13** with thionyl bromide, followed by methanolysis gives the acetal, 10-trifluoromethylartemether **15**, in high yield.[36] The hemiacetal **13** is remarkably stable under acidic aqueous conditions ($t_{1/2}$=760h at pH 2 and 37 °C, as compared with 17h for DHA), and has relatively increased half-life of 86min in plasma (cf. 23min for DHA).[23] The trifluoromethyl derivatives **13** and **15** have relatively lower in vivo ED_{50} values in mice infected with *P. falciparum* as compared to their nonfluorinated analogs. The 10-trifluoromethylartemether **15** also shows relatively lower in vitro IC_{50} of 0.8 nM, as compared to an IC_{50}

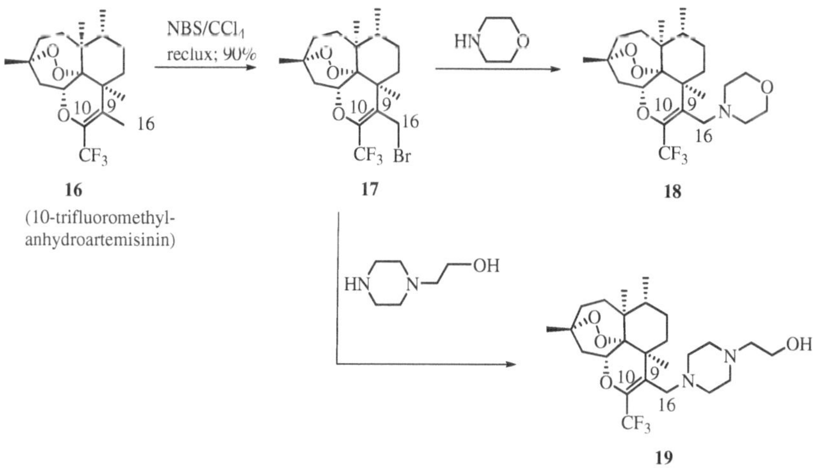

FIGURE 15 Synthesis of water-soluble analogs of 10-trifluoromethylanhydroartemisinin (**16**).

of 3.5 nM for artemether.[23] Thus, these CF_3-substituted analogs of artemisinin have potential applications as orally active antimalarials.

Begue, Bonnet-Delpon, and coworkers have also synthesized various water-soluble CF_3-substituted artemisinin analogs, **18** and **19**, by introducing amino and alcohol functional groups at C_{16} (Figure 15).[24] These compounds could be readily synthesized starting from 10-trifluoromethylanhydroartemisinin (**16**). Allylic bromination of **16**, followed by S_N2 reactions using morpholine and *N*-hydroxyethylpiperazine gives compounds **18** and **19**, respectively. The compounds **18** and **19** have relatively high in vivo antimalarial activities on mice infected with *Plasmodium berghei* as compared to those treated using sodium artesunate (DHA-butanedioic acid monoester, sodium salt). The peperazineethanol derivative, **19**, exhibits relatively high in vivo antimalarial activity: compound **19** at a dose lower than 10 mg/kg cured all the tested *P. berghei*-infected mice, and had 25 times higher bioavailability than the parent artemether drug.[24]

5. ANTIDIABETICS

In spite of the progress in the development of novel and effective drugs to treat diabetes, currently there is no lasting cure for diabetes; antidiabetic drugs are targeted to achieving normal glycemia levels with the long-term goals of preventing the development or progression of the complications of the disease. Most of the drugs are at present focused on type 2 diabetes (noninsulin-dependent diabetes mellitus), with the aim of activating the pancreas to produce more insulin, to decrease the insulin requirements by the body, and to reduce gluconeogenesis by the liver. In severe cases of diabetes, insulin treatment is the only choice available, as is in the case for type 1 diabetes (insulin-dependent diabetes mellitus).

5.1 Sitagliptin

Treatment strategies for the diabetes are increasingly focused on the homeostasis of the glucagon-like peptide 1 (GLP-1), a hormone secreted by α-cells of pancreas and released from the gut after food intake. GLP-1 regulates the levels of insulin based on blood glucose levels, by modulating the concentration of the plasma glucose levels (fluorinated peptide mimetics of GLP-1 are discussed in chapter 4). However, it is metabolically unstable, being rapidly hydrolyzed by the serine protease dipeptidyl peptidase-4 (DPP-4). Therefore, the current trend is directed at finding small-molecule inhibitors of DPP-4 enzyme. Merck's group has identified several β-amino-acid-derived small-molecule inhibitors of DPP-4 enzyme; the most notable one that is currently used as antidiabetic drug is sitagliptin (Januvia) (Figure 16).[37,38]

Sitagliptin ((2*R*)-4-oxo-4-[3-(trifluoromethyl)-5,6-dihydro[1,2,4]triaolo[4,3-α] pyrazin-7(8H)-yl]-1-(2,4,5-trifluorophenyl)butan-2-amine) acts as an antihyperglycemic drug by inhibiting DPP-4 (IC_{50} = 18 nM)[39] and thereby increasing the lifetime of GLP-1. However, the DPP-4 enzymes are also biogenic tumor suppression agents, and inhibition of DPP-4 by sitagliptin or other small molecule inhibitors leads to potential pancreatic cancer risk (pancreatitis) in certain cases. This risk factor, and other adverse effects, may be attenuated by a combination therapy along with the use of other antidiabetic drugs, such as metformin. Such combination therapy of sitagliptin and metformin, generally as their fixed dose combinations, is widely used to treat type 2 diabetes patients. However, the efficacy of this combination therapy, or its precise role in the prevention of the pancreatic cancer, is not clearly established to date, although regulators in Europe concluded that there is little evidence that sitagliptin, with or without metformin, could cause pancreatic inflammation or pancreatic cancer.[40,41] Sitagliptin is generally well tolerated in type 2 diabetes patients.

DPP-4 is also inhibited by (*S*)-valine pyrrolidide. The X-ray crystal structure of this amide bound to DPP-4 was determined at 2.5 Å resolution, which shows that the catalytic site is located in each of the subunits of the homodimeric DPP-4, between the α/β-hydrolase domain and a β-propeller domain.[42] Other small molecule inhibitors, including sitagliptin, are also bound at the same active site of the enzyme, but sitagliptin has opposite orientation of the amide moiety compared

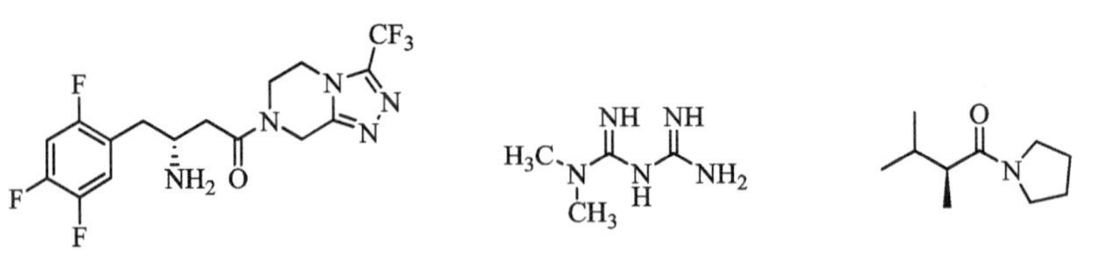

Sitagliptin (Januvia) Metformin (used in combination therapy with sitagliptin) (S)-Valine-pyrrolidide

FIGURE 16 Structures of sitagliptin, metformin, and (*S*)-valine pyrrolidide.

to that of the valine pyrrolidide. The 2,4,5-trifluorophenyl group of sitagliptin is embedded in the S1 hydrophobic pocket (Figures 17 and 18). Sitagliptin analogs with decreased number of the aryl-C-F bonds (e.g., 3,4-difluoro analog) show relatively diminished potency. Interestingly, the trifluoromethyl triazole ring and trifluorophenyl ring of sitagliptin have orthogonal π–π interactions with the Phe357 and Tyr662 aromatic rings, respectively. Thus, the trifluorophenyl and the trifluromethyl triazole rings have decreased electron density at the center of the ring due to the strong electron-withdrawing effect of the three fluorine atoms, and form

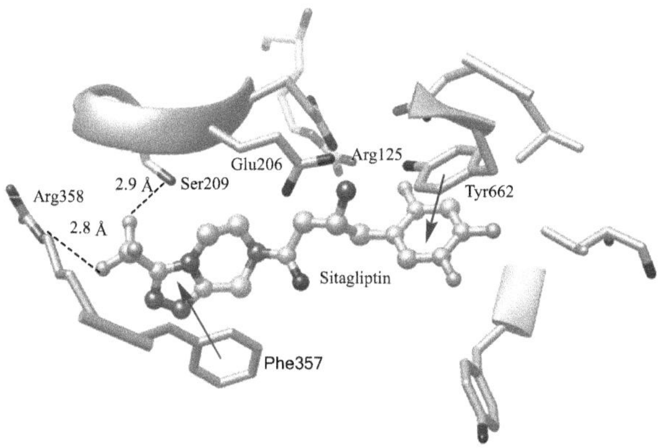

FIGURE 17 Expanded view of the X-ray structure of sitagliptin (ball and stick) bound to DPP-IV (green, F; red, O; blue: N); the fluorine atoms of the trifluoromethyl group have hydrogen bonding interactions with Ser209 and Arg358 side chains. Phe357 and Tyr662 aromatic rings have orthogonal π–π interactions with the electron-deficient triazole and trifluorophenyl rings (shown by arrows), respectively (the structure was created using UCSF Chimera; PDB: 1X70). (For interpretation of the references to color in this figure legend, the reader is referred to the online version of this book.)

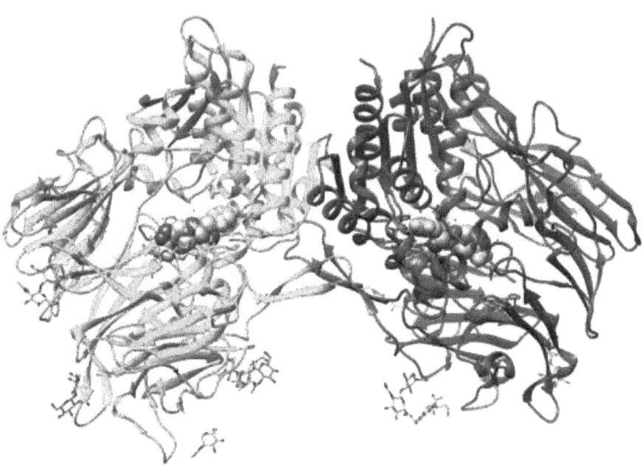

FIGURE 18 X-ray structure of sitagliptin bound to dipeptidyl peptidase-4 (DPP-4). The sitagliptin is shown in CPK model (space filling molecular model) in each of the subunits of the enzyme; the bound *N*-acetylglucosamine molecules are shown as sticks (the structure was created using UCSF Chimera; PDB: 1X70).

π–π donor–acceptor complexes with the neighboring π-electron-rich aromatic rings of Phe357 and Tyr662. The carbonyl oxygen of sitagliptin has long-range hydrogen bonding interaction with Tyr547 through an intervening water molecule, and the 1,2,4-triazole-CF_3 group has hydrogen bonding interactions with the Arg358 and Ser209 side chain residues (Figure 17). The analogs of sitagliptin lacking this CF_3 group show a fourfold decrease in potency. The CF_3 moiety is of the optimal size to fit in this enzyme pocket, and replacement of this group by more crowded substituents results in the loss of the drug potency.[37]

5.2 Synthesis of Sitagliptin

The key intermediate for the synthesis of sitagliptin, compound **23**, was synthesized using Schöllkopf's bis-lactim (**20**) as the chiral auxiliary.[37] The deprotonation of the bis-lactim **20** using *n*-butyllithium gives (2*S*)-2,5-dihydro-3,6-dimethoxy-2-isopropylpyrazin-5-yllithium **21**, which undergoes nucleophilic substitution reaction with 2,4,5-trifluorobenzyl bromide to give compound **22**. Hydrolysis of **22** using aqueous hydrochloric acid, followed by esterification of the carboxy group, *tert*-butyloxycarbonyl (Boc) protection of the amino group, and ester hydrolysis (LiOH) gives (*R*)-*N*-Boc-2,4,5-trifluorophenylalanine **23**. Arndt–Eistert homologation of **23** gives compound **25**, which upon reaction with 3-trifluoromethyl triazolopiperazine (**26**) in the presence of 1-hydroxybenzotriazole (HOBT) and N-(3-dimethylaminopropyl)-N′-ethylcarbodiimide (EDC), followed by workup using methanolic hydrogen chloride and treatment with fumaric acid, gives sitagliptin fumarate, the pharmaceutical ingredient of the drug (Figure 19).[37]

5.3 Aldose Reductase Inhibitors

In the long term, diabetes leads to other complications including diabetic retinopathy, cataracts, diabetic neuropathy, nephropathy, and atherosclerosis. Aldose reductase, a nicotinamide adenine dinucleotide phosphate-dependent oxidoreductase, is involved in glucose metabolism, and it converts glucose into sorbitol, which is abundant in diabetes patients, first identified in the cataractous lenses.[43] Sorbitol is converted to fructose by sorbitol dehydrogenase in a subsequent step. Accumulation of sorbitol in cells leads to diabetic complications.

Aldose reductase pathway is implicated in the onset of the secondary complications of diabetes, and thus most of the drugs to treat these complications are targeted toward aldose reductase inhibitors.[44,45] Pharmaceuticals such as ponalrestat (Prodiax),[46,47] tolrestat (Alredase),[48] Lidorestat,[49] Fidarestat,[50] and Imirestat (AL-1567)[51] are introduced as aldose reductase inhibitors in the treatment of diabetic complications such as diabetic neuropathy and diabetic retinopathy (Figure 20).

Most of the aldose reductases, including tolrestat, ponalrestat, zopolrestat, and zenarestat, however, have severe side effects, and therefore there is a growing demand for the identification of safer and more effective aldose reductase inhibitors for the treatment of diabetic complications.

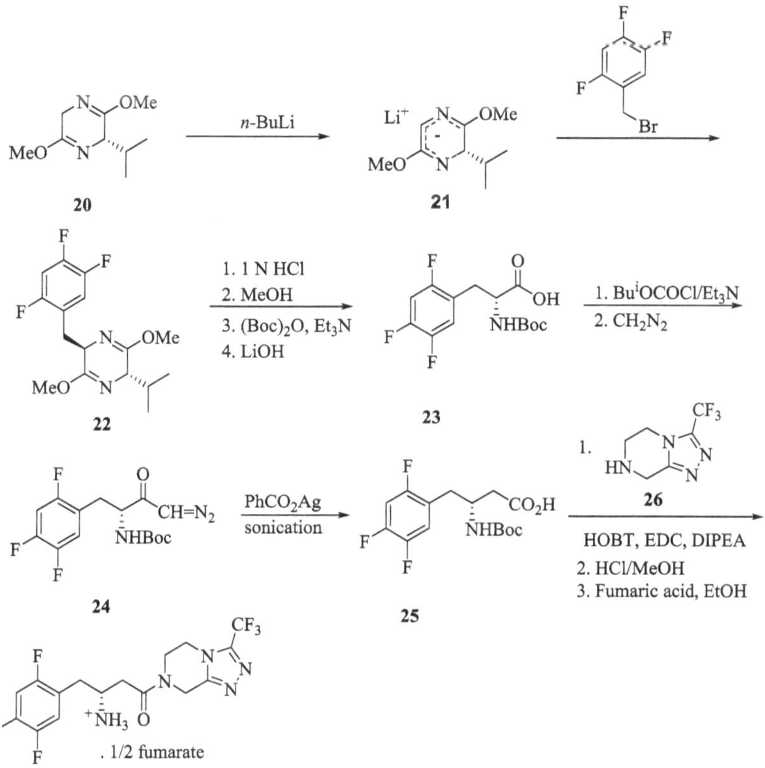

Sitagliptin (Januvia)

FIGURE 19 Synthesis of sitagliptin $(Boc)_2O$ = Bis(tert-butyl) dicarbonate; HOBT = 1-hydroxybenzotriazole; EDC = 1-ethyl-3-(3-dimethylaminopropyl)carbodiimide; DIPEA = diisopropylethylamine).

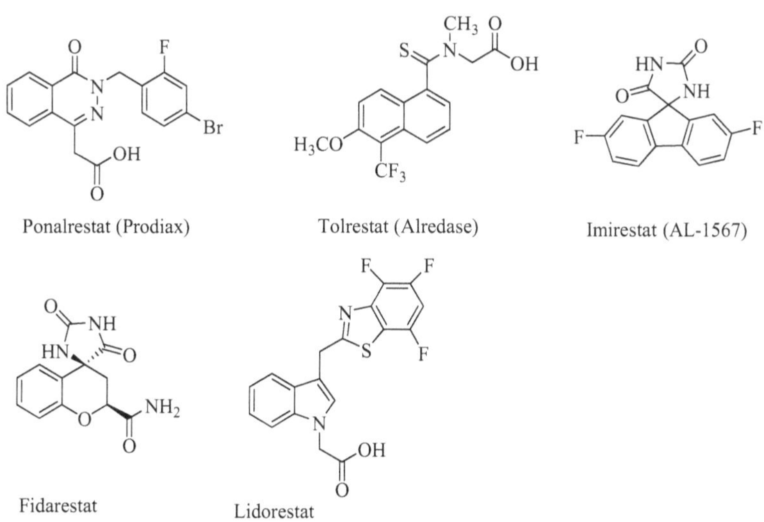

FIGURE 20 Structures of some fluorinated antidiabetic drugs.

6. NONSTEROIDAL ANTIINFLAMMATORY DRUGS

NSAIDs are used as pain killers, analgesics, and antipyretics and some of them such as aspirin (*O*-acetylsalicylic acid) are also finding applications in the prevention of cardiovascular complications. These classes of compounds are inhibitors of cyclooxygenases, COX1 and/or COX2. A majority of NSAIDs are over-the-counter drugs, have a carboxylic acid functional group or other less mildly acidic group, and are nonselective inhibitors of COX1 and COX2. Examples of these drugs are: ibuprofen, naproxen, ketoprofen, aspirin, acetaminophen, fenoprofen, flurbiprofen, oxaprozin, indomethacin, sulindac, tolmetin, etodolac, diclofenac, lumiracoxib, nabumetone, piroxicam, meloxicam, mefenamic acid, and meclofenamic acid (Figure 21). Of these compounds

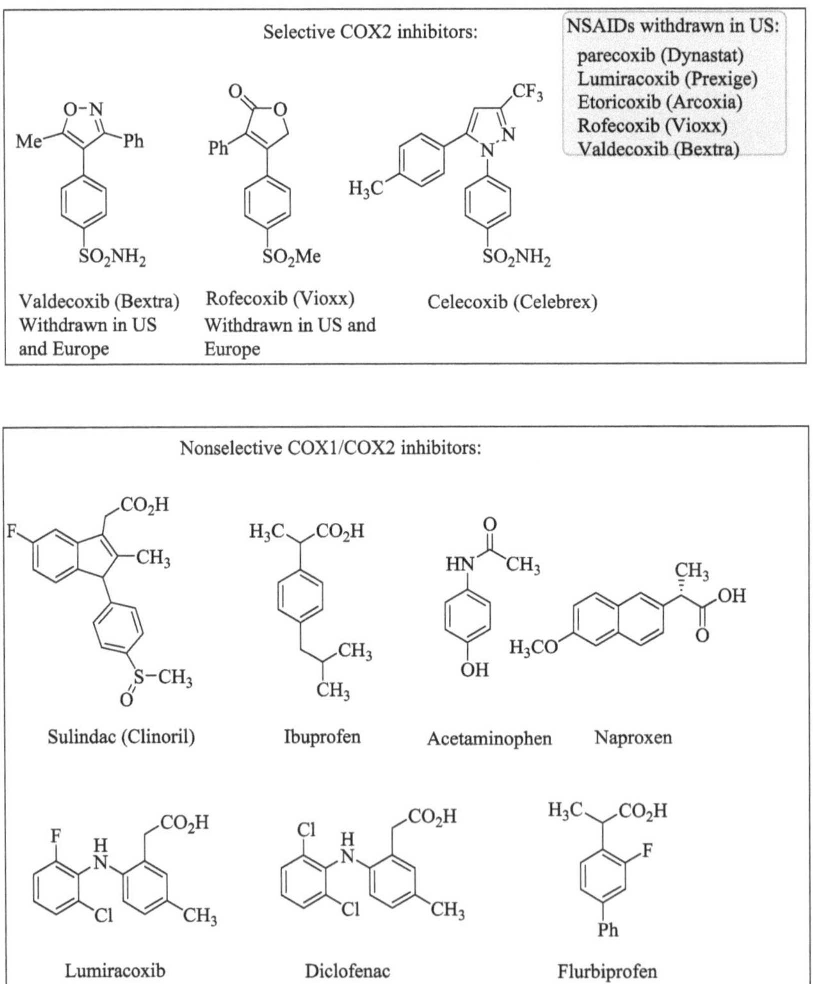

FIGURE 21 Structures of some COX1 and COX2 inhibitors.

flurbiprofen and limracoxib are fluorinated versions. Figure 21 summarizes the structural details of the most widely used NSAIDs.

Cyclooxygenase enzymes (COX1 and COX2) are responsible for the conversion of arachidonic acid into prostaglandins, which are involved in pain, inflammation, and fever. COX2 isoenzyme is predominantly responsible for the pain and inflammation and it is present mainly in the injured tissues.[52] On the other hand, inhibition of COX1 isoenzyme results in adverse effects as this enzyme is responsible for the synthesis of prostaglandins which have important protective role in gastrointestinal and renal functions. Thus nonselective inhibition of COX enzymes by a majority of NSAIDs including ibuprofen, aspirin, acetaminophen and naproxen has adverse effects on gastrointestinal functions resulting in intestinal ulcers and bleeding, especially after prolonged usage. Selective COX2 inhibitors have been developed which include celecoxib, a trifluoromethyl-substituted pyrazole derivative of sulfanilamide.

Inhibition of COX2 isoenzymes also results in adverse cardiovascular and renal complications, although they are relatively less frequent than the gastrointestinal pathologies associated with COX1 inhibition. In order to circumvent the cardiovascular side effects, Aspirin is usually used as a side medication in small doses to prevent thrombosis. Rimon and coworkers, however, showed that celecoxib (Celebra; Celebrex; Pfizer) could reversibly and competitively bind to a subunit of COX1and concluded that the effect of low-dose Aspirin may be blunted when these drugs are used in combination.[53] Further, even with selective COX2 inhibition, gastropathy is still a minor concern. An X-ray structure of celecoxib bound to COX2 shows stabilizing dipolar interactions (3.1 to 3.6 Å) of the CF_3 group with methyl groups of Leu345, Leu517, and guanidinium moiety of Arg106. It is apparent that the highly electronegative fluorine can polarize the C–H bonds of the close-lying side chain amino acids to form hydrogen bond interactions (Figure 22).

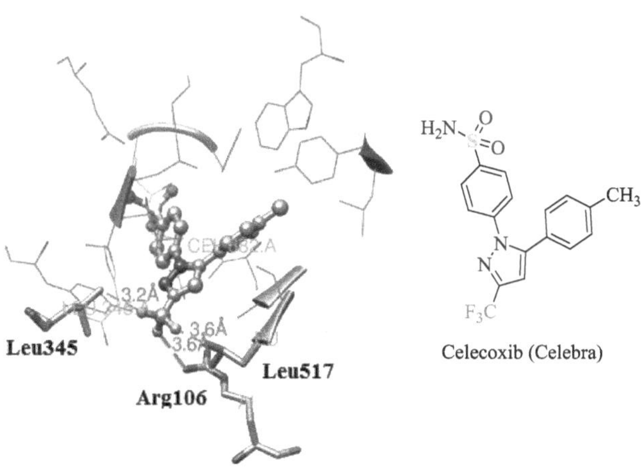

FIGURE 22 X-ray structure of celecoxib (ball and stick) bound to COX2 showing van der Waals interactions of the CF_3 group with terminal methyl groups of Leu 517 and Leu 345, and with guanidinium moiety of Arg106 (the structure was created using UCSF Chimera software; pdb file code: 3LN1).

Some COX2 inhibitors, even when used in the recommended doses for long time, cause severe cardiovascular and cardiorenal effects, and occasionally exhibit serious cutaneous adverse reactions. In view of these observations COX2 inhibitors rofecoxib and valdecoxib have been withdrawn in the United States and Europe in 2004.[54] Thus, currently a blockbuster drug, the trifluoromethyl-group-containing celecoxib (Celebrex) is the only selective COX2 inhibitor that is widely used as NSAID. All these NSAID drugs, including Celebrex, at prolonged or high dosages, have potential cardiovascular, renal, and gastrointestinal complications, and, moreover, their toxicity effect on the elderly population is less clear-cut.[54] Thus, even more selective COX inhibitors are required to minimize their adverse effects, and many challenges remain in the development of pharmaceuticals, especially in the unraveling of the mechanistic details of drug action associated with organofluorine pharmaceuticals.

Diclofenac and lumiracoxib are structurally closely related, selective COX-2 inhibitors that are used for the treatment of inflammatory effects of osteoarthritis and rheumatoid arthritis.[55] The amino group in these compounds activates the *para*-position for electrophilic hydroxylation by cytochrome P450 enzymes (CYP) to give **27**. Further CYP-catalyzed oxidation of this compound results in the formation of *p*-benzoquinone imine **29**, a toxic metabolite (Figure 23).[55] The Michael additions of the amino, hydroxyl, and thiol side chain substituents on the liver proteins to compound **29**, followed by fluoride ion elimination, results in the formation of protein adducts **31**. These protein adducts eventually result in liver necrosis, when formed in large concentrations. Similar reactions of *p*-benzoquinone-imines derived from acetaminophen that result in acute hepatotoxicity are well established.[56] However, when formed in reasonably small concentrations, the antioxidant glutathione undergoes Michael additions with these reactive metabolites and attenuates their toxicity effects. Human whole blood assay shows that lumiracoxib is 515 times more selective to COX2 than COX1. Lumiracoxib also reduces the incidence of gastroduodenal erosions, as compared to naproxen.[57]

(*S*)-Flurbiprofen, a fluorine-containing NSAID, is a relatively less widely used pharmaceutical. Although (*S*)-flurbiprofen is the major inhibitor of the cyclooxygenases, currently this drug is administered as a racemic mixture. Unfortunately it causes gastrointestinal toxicity with continued use as is the common case with a majority of the nonsteroidal antiinflammatory agents.

7. CORTICOSTEROIDS

Corticosteroids, sometimes called steroid drugs, are widely used as topical antiinflammatory agents. These drugs are also often abused by athletes as they can perform while injured. However, the use of corticosteroids is associated with excitatory effects, and may lead to addiction.

Fluorinated corticosteroids, triamcinolone and dexamethasone, are used as antiinflammatory agents, and in long-term care for rheumatoid arthritis. Fluprednisolone is used as topical antiinflammatory agent, and fluticasone

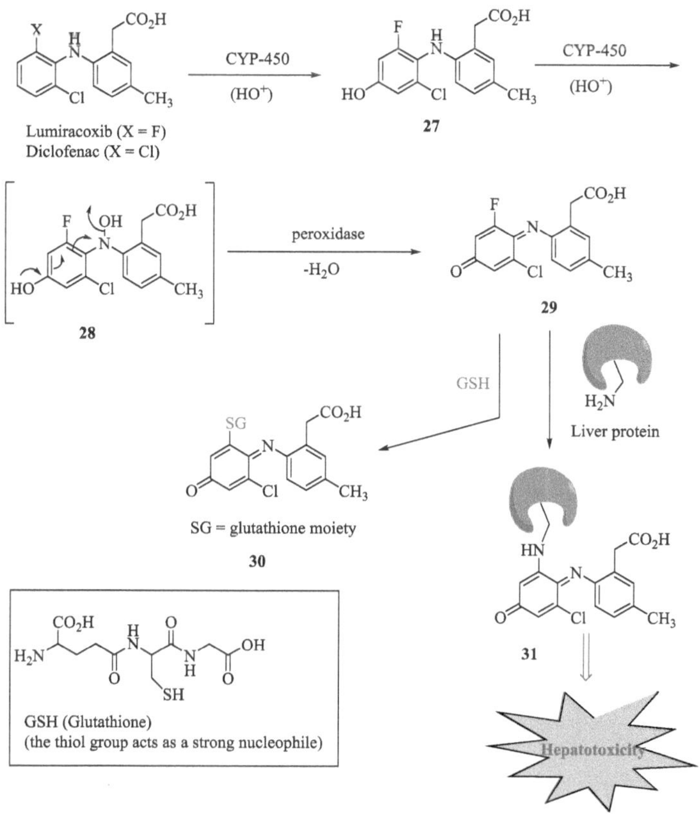

FIGURE 23 Schematic representation of toxic metabolite formation from the NSAIDs diclofenac and lumiracoxib, and their competitive trapping by glutathione (GSH).

propionate (Flonase), a glucocorticosteroid, is a widely prescribed antiinflammatory agent for the treatment of allergic rhinitis and asthma (Figure 24).

A relatively more metabolically stable analog of fluticasone propionate, fluticasone furoate has been introduced as a novel drug which is more effective than dexamethasone or fluticasone propionate in the intranasal treatment of seasonal and perennial allergic rhinitis.[58,59] All these corticosteroids are lipophilic and exert their antiinflammatory effect by binding to the endogeneous ligand binding sites of the glucocorticoid receptors as agonists. The relative binding affinities for the active sites of glucocorticoid receptor is estimated as, 1:18:29, for dexamethasone, fluticasone propionate, and fluticasone furoate, respectively. Further, the protein–ligand interactions are essentially the same for all these three compounds as shown by their X-ray crystal structures of the ligands bound to the receptors.[59,60] From the relative binding constants, it is evident that the hydrogen bonding interactions due to the terminal hydroxyl group of the 17-hydroxyacetyl group of dexamethasone are more than compensated in case of the fluticasone furoate by the strong electrostatic interactions of the fluorine

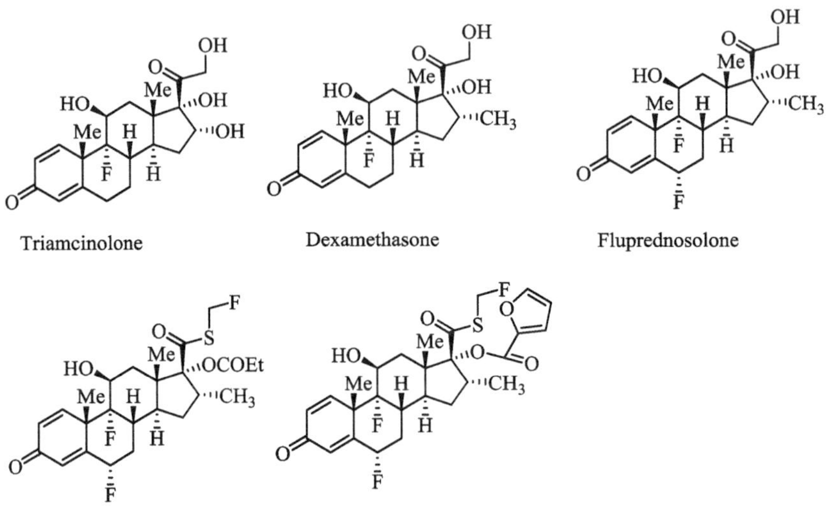

Triamcinolone Dexamethasone Flupredosolone

Fluticasone propionate Fluticasone furoate

FIGURE 24 Structures of some corticosteroid drugs.

with the neighboring side chain C–H bonds (Phe749 and Thr739); by the polar interactions with the Phe625 backbone amide moiety; or by the relatively weak hydrogen bonding interactions with the Asn564 side chain amide N–H bond. Further, there are favorable hydrophobic interactions of the 17α-substituent (i.e., furanoate) in case of the fluticasone furanoate (Figure 25). Similar interactions are also present for fluticasone propionate at the enzyme active site.

The 17α-pocket accommodates the propionate and furanosyl moieties, and the hydrophobic stabilization achieved is relatively higher for the furanosyl moiety than for the propionate group as revealed by the relatively higher binding constant for fluticasone furanoate. The C6 and C9 fluorines in these compounds presumably give improved metabolic stability or optimal lipophilicity for the drug bioavailability.

7.1 Synthesis of the Fluorinated Corticosteroids

Synthesis of the fluorinated corticosteroids can be conveniently achieved using electrophilic fluorination of the steroidal enol acetates, followed by S_N2 ring opening of the epoxide using HF. For example, reaction of the steroidal dienol acetate **32** with Selectfluor, an electrophilic fluorinating reagent, results in the stereoselective formation of 6-fluoroderivative, **33**. A subsequent trans-diaxial ring opening of the steroidal 9,11-epoxide in **33** by aqueous HF reagent gives the corresponding 9-fluoro derivative, the fluprednisolone. The aqueous HF used in this reaction also affects the hydrolysis of the 17α-acetoxy group[61] (Figure 26). The 17α-OH group in fluprednisolone can be further transformed into various ester derivatives (e.g., furanoate or propanoate derivatives).

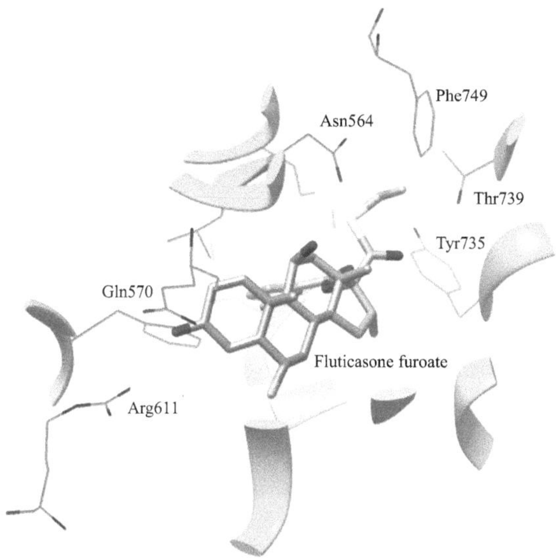

FIGURE 25 Expanded view of the X-ray crystal structure of fluticasone furanoate bound to glucocorticoid receptor ligand binding domain, showing the active site binding interactions; the fluorine atoms have either polar interactions or weak C–H hydrogen bonding interactions with the neighboring amino acid residues. Arg611, Gln570, and Asn564 are within the hydrogen bonding distance to the carbonyl oxygens of the ligand (the structure was created using UCSF Chimera; PDB3CLD).

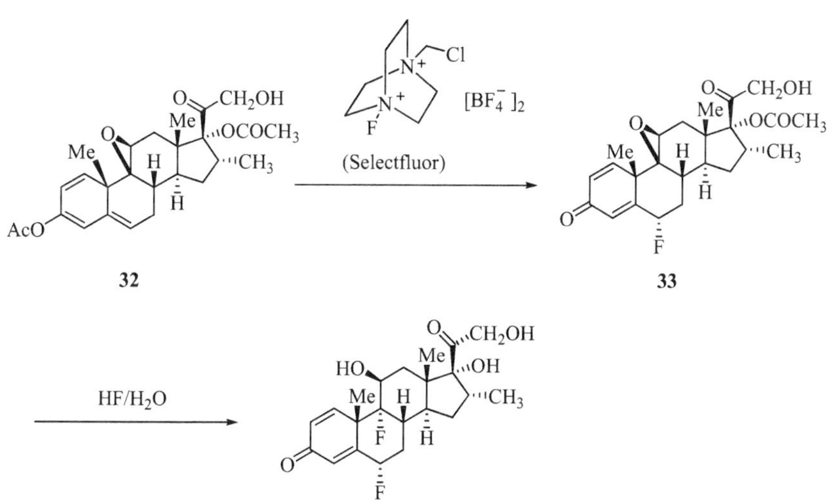

FIGURE 26 General synthetic route for fluorinated corticosteroids.

8. ANTIVIRAL DRUGS

Epidemics of influenza virus, and its ever-persistent growing resistance to pharmaceuticals, prompted the innovative development of new versions of antiviral drugs. Unlike most antibacterial drugs, antiviral drugs do not destroy the viral cells but inhibit their growth and proliferation.

8.1 Efavirenz

Drugs targeting HIV infections are mostly aimed at reverse transcriptase inhibitors which will stop or postpone indefinitely the replication of the virus. Bristol-Myers Squibb's anti HIV drug, (*S*)-6-chloro-4-(2-cyclopropylethynyl)-1,4-dihydro-4-(trifluoromethyl)-2H-3,1-benzoxazin-2-one, marketed as efavirenz, (also called as Sustiva or Stocrin) is a reverse transcriptase inhibitor. Efavirenz binds to the enzyme away from the active site, altering the active site conformations and thereby resulting in enzyme inhibition. Combination therapy including protease inhibitors and other nucleoside and nonnucleoside reverse transcriptase inhibitors are found to be effective strategies to control HIV infections with great success.[62] In Efavirenz, the trifluoromethyl group helps to increase the acidity of the cyclic carbamate amide group and thereby its hydrogen bonding interactions are appropriately modulated for efficient binding.[63] The three fluorine atoms of the CF_3 group are within about 3.4 Å away from the neighboring Val79, Val106, and Tyr188 side chains, and thus have weak van der Waals attractions with the C–H bonds of their side chains (Figure 27). In the most common mutant form of the enzyme, the Lys103 is replaced by asparagine

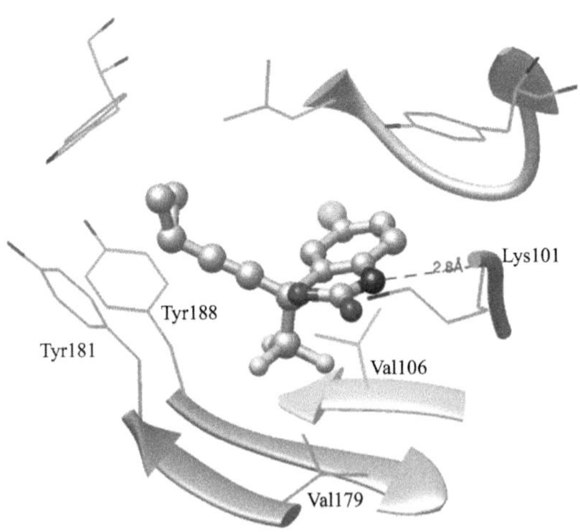

FIGURE 27 X-ray structure of efavirenz (ball and stick) bound to the wild-type HIV-1 reverse transcriptase; the CF_3 group increases the acidity of the neighboring NH moiety so that it has strong hydrogen bonding interactions with the lysine backbone carbonyl oxygen. The three fluorines are within ~3.4 Å of the V106, V179, and Tyr 188 side chain C–H bonds. Efavirenz is also active against the mutated form of the enzyme, in which the Lys 103 is replaced by asparagine (K103N); (PDB: 1IKW; the structure was created using UCSF Chimera software).

(K103N). Efavirenz is also active against the mutated form, and has essentially similar active site interactions. Of the two possible enantiomers, only the (*S*)-isomer is active antiviral agent.

Several synthetic methods for the enantioselective synthesis of the *S*-efavirenz have been developed.[64–70] Most of these synthetic methods are based on the asymmetric propargylation of aryl trifluoromethyl ketones; a catalytic version of these reactions has also been reported.[64] The key step of one of the improved synthesis of this drug involves the enantioselective trifluoromethylation (using Ruppert–Prakash reagent CF_3SiMe_3) in the presence of a catalytic amount of *N*-(3,5-bis-trifluoromethylphenyl)-cinchonidine as the chiral auxiliary[71]; the enantioselectivities for these reactions were in the range of 50%, and a single recrystallization of the product in the next step increased the enantiomeric purity to 94% (at a cost of overall yield). The synthetic scheme for this method is outlined in Figure 28.

Trifluoromethylation of the propargyl ketone **34** in the presence of the chiral cinchonidine-derived catalyst **35** gave **36** in high yield, with 50% enantioselectivity. However, metal-catalyzed reduction of the nitro group in **36**,

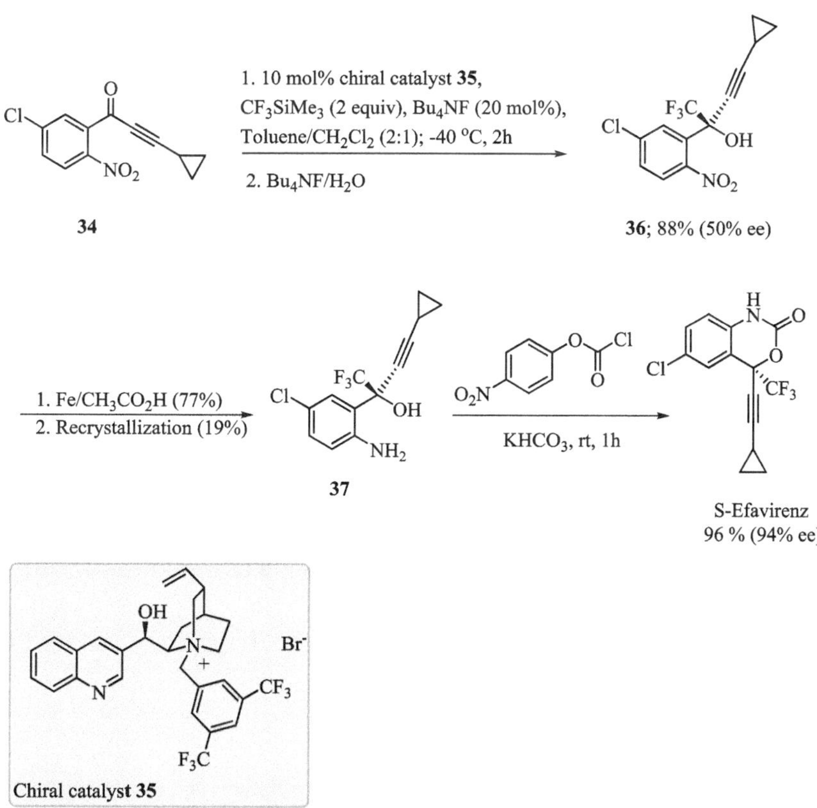

FIGURE 28 Synthesis of (*S*)-efavirenz.

followed by recrystallization of the resulting amine **37** improved the enantioselectivity to 94%. Reaction of **37** with *p*-nitrophenyl chloroformate afforded the cyclic carbamate, *S*-efavirenz, in high yield (Figure 28). Enantioselective trifluoromethylation using a dimeric cinchonine derivative as the chiral catalyst, on the other hand, afforded the *R*-isomer, with ~96% enantioselectivity.

8.2 Tipranavir

Tipranavir (Boehringer Ingelheim Pharmaceuticals) is used as an antiviral drug to treat acquired immunodeficiency syndrome (AIDS) cases. It is an orally bioavailable nonpeptidic class of drug for the inhibition of the aspartic HIV protease, with an inhibition constant (K_i) of 8 pM.[72] It is administered along with ritonavir, a peptide-based drug, in order to improve its pharmacokinetics. There are, however, reports of adverse side effects of tipranavir, such as intracranial hemorrhage.[73] Ritonavir in combination with darunavir, another nonpeptidic drug, is also being used to treat AIDS (Figure 29).[74]

8.3 Trifluridine

Trifluridine (5-trifluoromethyluridine; trifluorothymidine) is used for the treatment of herpes infection cornea (herpetic keratitis, superficial or stromal).[75] Trifluridine is an irreversible thymidylate synthase inhibitor, and thereby suppresses DNA synthesis. It is also likely that the strong electron-withdrawing trifluoromethyl group dramatically reduces the basicity of the adjacent carbonyl group so that its base pairing capacity, and thereby viral replication, is halted. Similar decrease of

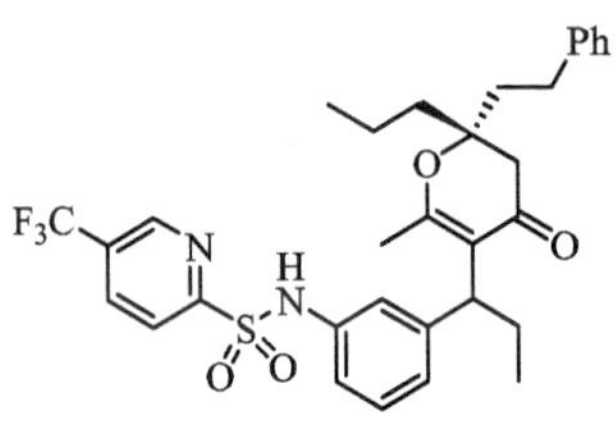

Tipranavir; protease inhibitor
(used to treat HIV)

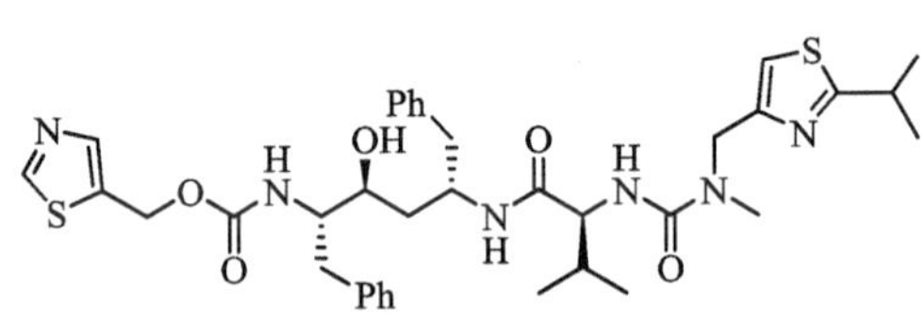

Ritonavir

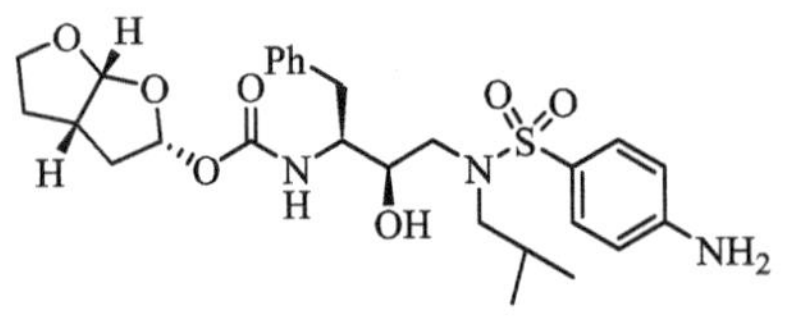

(-)Darunavir

FIGURE 29 Structures of tipranavir, ritonavir, and (−)darunavir, antiviral drugs to treat AIDS.

nucleophilicity of the neighboring hydroxy group by the adjacent CF_2 group in case of gemcitabine results in the suppression of cancer cell DNA synthesis (nucleoside-based cancer therapeutics are discussed in chapter 9).[76]

Synthesis of trifluridine can be effected under free radical condition, by the reaction of uridine (**38**) with Togni's reagent (**39**) in the presence of catalytic amounts of Cu(I) salts (Figure 30).[77] In-situ-generated CF_3Cu (using CF_3I and copper powder), upon reaction with 5-iodouridine (**40**), also gives the trifluridine in moderate yields.[78]

The reaction of 3′,5′-di-*O*-acetyl-5-iodo-2′-deoxyuridine (**41**) with methyl (fluorosulfonyl)difluoroacetate in the presence of CuI, or copper metal, gives 2′-deoxy-3′,5′-di-*O*-acetyl-5-trifluoromethyluridine (**42**) in moderate yields.[79] The latter compound **42** may be hydrolyzed under mild reaction conditions to give the trifluridine (Figure 31).

8.4 Fluorinated Nucleosides as Antibacterials and Antiviral Compounds

Nucleocidin (4′α-fluoro-5′-*O*-sulfamoyladenosine), the only known naturally occurring carbohydrate or nucleoside, is an antitrypanosomal antibiotic. It was first isolated from a soil microbe, *Streptomyces calvus* in 1957.[76] The structural

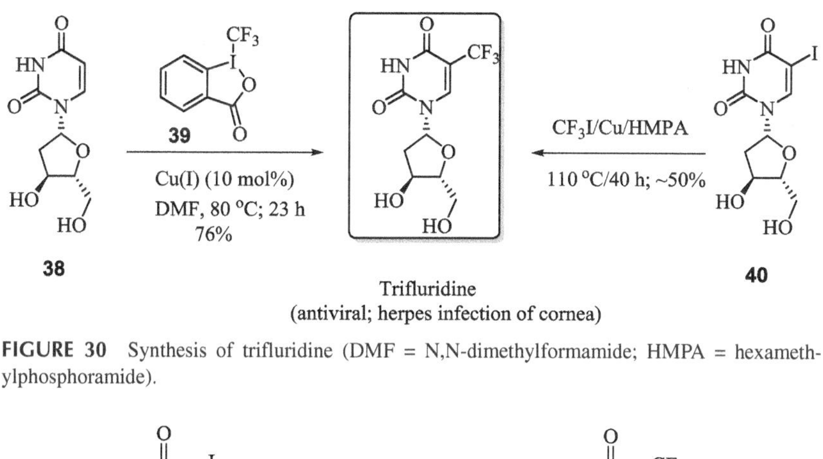

38

40

Trifluridine
(antiviral; herpes infection of cornea)

FIGURE 30 Synthesis of trifluridine (DMF = N,N-dimethylformamide; HMPA = hexamethylphosphoramide).

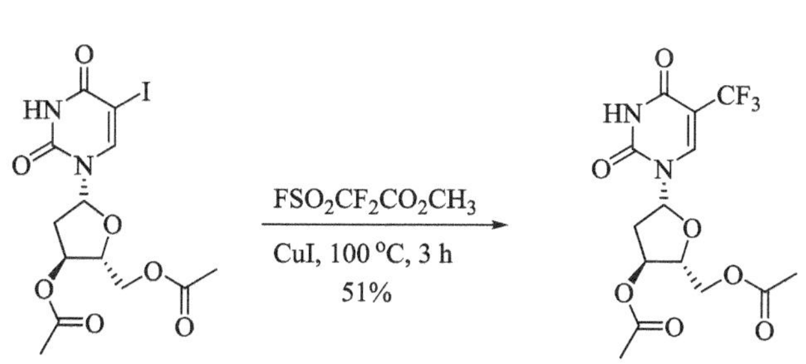

41 **42**

FIGURE 31 Synthesis of 2′-deoxy-3′,5′-di-*O*-acetyl-5-trifluoromethyluridine, **42** (DMF = N,N-dimethylformamide; HMPA = hexamethylphosphoramide).

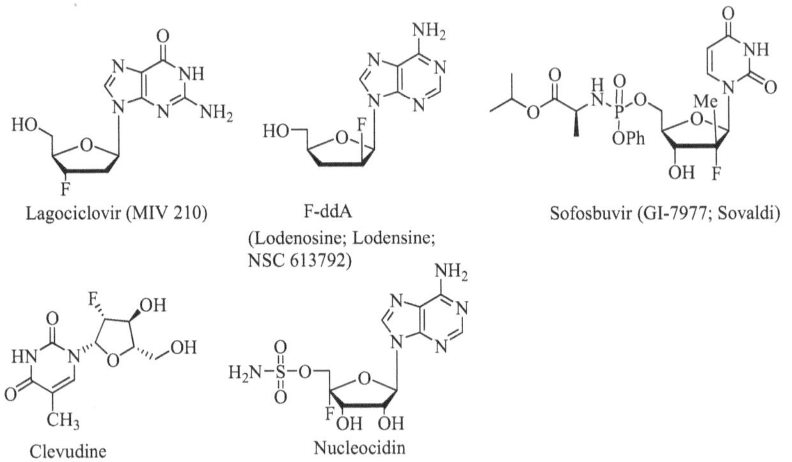

FIGURE 32 Structures of fluorinated nucleosides Lagociclovir, F-ddA, sofosbuvir, clevudine, and nucleocidin.

details of nucleocidin have been established using high-resolution mass spectrometry and ^{1}H, ^{13}C, and ^{19}F NMR spectroscopy.[80] The significant structural feature that distinguishes it from other nucleoside drugs is in the presence of a C4′-α-fluorine atom on ribose moiety.

Since the discovery of nucleocidin as an antibiotic, numerous fluorinated nucleosides have been synthesized and explored as pharmaceutically interesting candidates.[81] The notable examples include the widely used pharmaceuticals clevudine, an antiviral drug for the treatment of hepatitis B (anti-HBV); lagociclovir (3′-α-fluoro-2′,3′-dideoxyguanosine), an antiviral drug for the treatment of HBV[82]; and 2′-β-fluoro-2′,3′-dideoxyadenosine (F-ddA) (Figure 32). F-ddA exhibits synergistic antiviral effects when used in combination with other major classes of anti-HIV drugs, including nucleoside reverse transcriptase inhibitors, nonnucleoside reverse transcriptase inhibitors, and protease inhibitors. It is especially useful in combination therapy involving nonnucleoside drug (e.g., nevirapine and costatolide), nucleoside analogs such as AZT, and protease inhibitors (e.g., ritonavir and nelfinavir).[83] However, its clinical trials were abandoned due to safety concerns.

Sofosbuvir (GI-7977; Sovaldi; Gilead Sciences) is another example of a nucleoside analog that was approved by the FDA in 2013 for treating hepatitis C infections. This compound serves as a prodrug that is transformed into the active drug, the 2′-fluoro-2′-methyluridine triphosphate, through a series of intracellular enzyme-catalyzed reactions (Figure 33). It inhibits NS5B ribonucleic acid (RNA)-dependent RNA polymerase, and thereby prevents RNA synthesis in the viral cells.[84] The sofosbuvir-derived 2′-fluoro-2′-methyluridine triphosphate mimics the natural substrate uridine triphosphate, so that it is competitively incorporated into the growing hepatitis C virus RNA polymerase, and

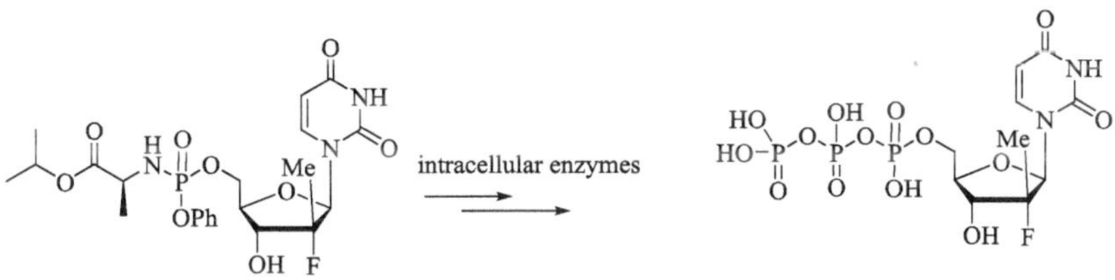

FIGURE 33 The enzymatic conversion of the prodrug sofosbuvir to the active drug, 2′-fluoro-2′-methyluridine triphosphate.

once it is incorporated into the RNA polymerase, the RNA polymerase reaction is terminated (chain termination).

9. ANTIDEPRESSANT DRUGS

The Na^+- and Cl^--dependent serotonin (5-hydroxytryptamine, 5-HT) transporter is expressed on human neuronal, platelet, placental, and pulmonary membranes. The serotonin receptor in the brain seems to be the principal site of action of the antidepressant drugs.[85] Several halogenated drugs, such as Prozac (*N*-methyl-3-[4-(trifluoromethyl)phenoxy]-3-phenylpropanamine, fluoxetine) are agonists for these receptors.

Prozac (fluoxetine) is a widely used antidepressant drug.[2] It was approved by the FDA in 1987 for treating depression and in 1994 for treating obsessive-compulsive disorder and bulimia. Its annual sales exceed 1 billion dollars, and due to its widespread use it is also of environmental concern.[86] Its mechanism of action involves inhibiting the reuptake of serotonin (5-HT) by nerve terminals, allowing the continued activation of specific receptors by the neurotransmitter (serotonin). The CF_3 group imparts fivefold enhanced potency for this drug as compared to its nonfluorinated analog. Unlike other selective serotonin reuptake inhibitors (SSRI), fluoxetine increases the extracellular concentration of not only serotonin but also dopamine and norepinephrine.[87]

Due to the sterically crowded CF_3 group, fluoxetine adopts a conformation favoring its more efficient binding to the serotonin transporter (SERT) proteins, the transmembrane G-protein-coupled receptors.[88] Using a bacterial homolog of SERT, leucine transporter, it has been demonstrated through X-ray crystal structure determination that all the serotonin reuptake inhibitors, including fluoxetine, bind at the same region of the protein (halogen binding pocket). The CF_3 group of fluoxetine has dipolar van der Waals F···HC interactions with Leu29, Leu25, and Tyr108, and also has similar nonbonding interactions with Gly26 and Arg30 (Figure 34). Both the *R*- and *S*-isomers of this compound have the same type of interactions with these amino acids.[89] Based on this, it was concluded that the specificity of the SERT for SSRI depends mostly on the dipolar interactions of the halogens with the protein's halogen binding pocket.[89]

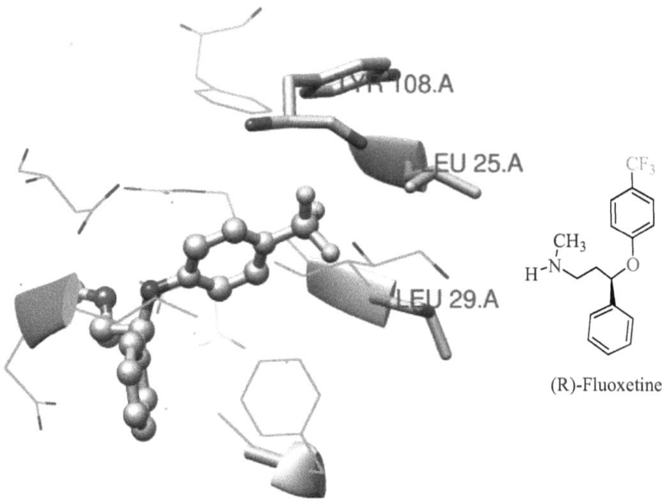

FIGURE 34 (*R*)-fluoxetine (ball and stick) bound to the leucine transporter, a *SERT homolog; the CF_3* fluorine atoms (green) have van der Waals F···H–C dipolar interactions with Leu25, Leu29, and Tyr108 side chains (PDB: 3GWV; the structure was created using UCSF Chimera software). (For interpretation of the references to color in this figure legend, the reader is referred to the online version of this book.)

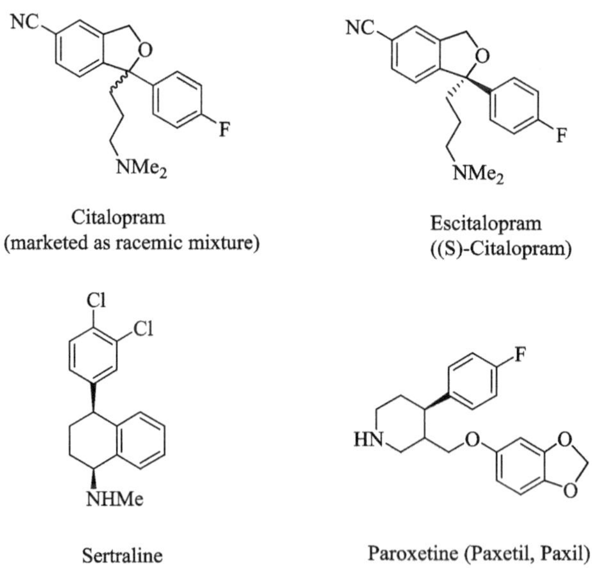

FIGURE 35 Structures of some serotonin reuptake inhibitors.

Other serotonin reuptake inhibitors include citalopram, sertraline, and paroxetine (Figure 35). The (*S*)-isomer of citalopram (escitalopram) is the pharmacologically active agent, even though the drug is prescribed in the racemic form.[90] Escitalopram is remarkably more effective than the *R*-citalopram in its binding to the serotonin reuptake proteins, but the *R*-isomer may modulate the binding affinity of the *S*-isomer, and therefore the drug is sold as a racemic mixture.[91]

Unlike Prozac, paroxetine is a selective inhibitor for G-protein-coupled receptor kinase 2 (GRK2) in vivo and in living cells. GRK2 is a therapeutic target for the treatment of heart failure, and thus paroxetine and related pharmaceuticals that are potent and specific GRK2 inhibitors may have important applications for cardiovascular diseases.[92]

10. ANTIHYPERCHOLESTEROLEMIA DRUGS

Hypercholesterolemia is the outcome of the disorders associated with lipid metabolism. Patients with hypercholesterolemia have elevated levels of cholesterol and low-density lipoproteins (LDLs) in plasma. The relatively high cholesterol levels in blood plasma results in atherosclerosis, in which case the arterial walls are thickened due to the plaque, consisting of mostly cholesterol, fats, and calcium. This cholesterol-induced atherosclerosis is a major cause of stroke and ischemia. The cholesterol and LDL levels could, in principle, be controlled by modifying the dietary food intake. As an alternative, there are many drugs that are marketed to lower the cholesterol and LDL levels in plasma, which interfere with lipid metabolism in the liver. Statin drugs are shown to selectively inhibit the HMG-CoA enzyme, which is involved in the early stages of lipid metabolism. All statins have the characteristic β,δ-dihydroxycarboxylic acid (or lactone) unit. Currently, several statin drugs are available for this treatment, including atorvastatin, fluvastatin (Lescol), lovastatin (Mevacor), pitavastatin (Livalo), pravastatin (Pravachol), simvastatin (Zocor), and rosuvastatin (Crestor). Atorvastatin (Lipitor) and ezetimibe (Figure 36) are among the best known cholesterol-lowering drugs currently in the market.

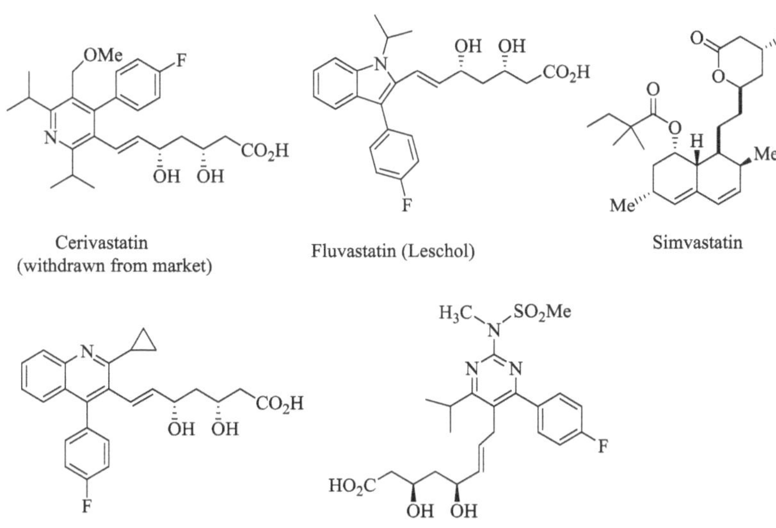

FIGURE 36 Structures of some of the widely used fluorinated statins and Simvastatin.

10.1 Ezetimibe

One of the often popularized, leading cholesterol-lowering drugs in the market is ezetimibe (1-(4-fluorophenyl)-(3*R*)-[3-(4-fluorophenyl)-(3*S*)-hydroxypropyl]-(4*S*)-(4-hydroxyphenyl)-2-azetidinone), an azetidinone derivative, developed by Schering-Plough and approved by the FDA in 2002.[93,94] This drug blocks biliary excretion of cholesterol into the intestine by selectively binding to the hepatic Niemann-Pick C1 like-1 (NPC1L1) transmembrane protein, which is highly expressed in the small intestine and in the liver, and which is involved in the cellular trafficking of cholesterol and other lipids.[95] The NPC1L1-knockout mice show substantial decrease of cholesterol absorption into the small intestine, an effect comparable with that of ezetimibe-treated mice.[95] Thus, through selective inhibition of the hepatic NPC1L1, ezetimibe, as in the case of the other statins, helps in lowering the overall cholesterol and other lipid content in plasma, although its mechanism for lowering plasma cholesterol is distinct from that of the statins: whereas statins lower cholesterol levels by intervening in the early stages of cholesterol biosynthesis, ezetimibe lowers the cholesterol levels by inhibiting intestinal absorption of cholesterol.

The discovery of ezetimibe demonstrates the principles of rational drug design and structure–activity correlations. Thus, Schering-Plough's (now Merck) SCH 48461, which was developed in the early1990s, is also a potent cholesterol absorption inhibitor, but is metabolically unstable as it has four metabolically degradable sites: two of the *p*-methoxy groups in SCH 48461 are prone to demethylation during metabolism to give the corresponding phenols, and the secondary benzyl group and phenyl-*para*-carbon are susceptible to enzymatic hydroxylation (Figure 37).

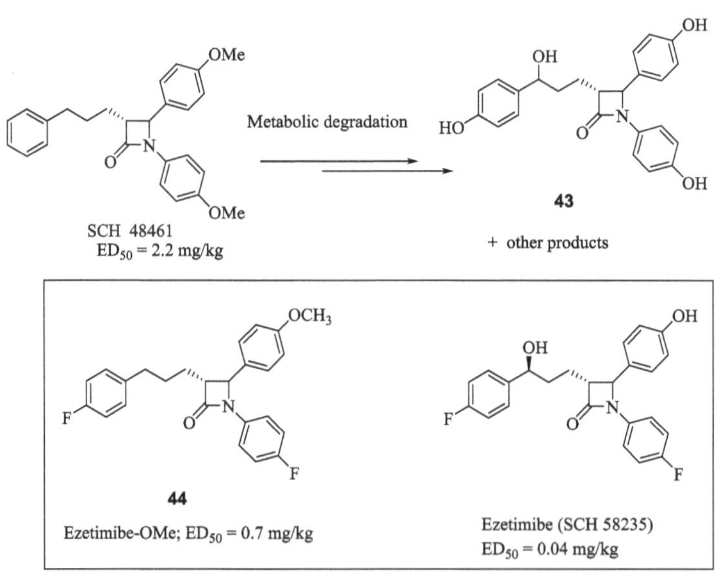

FIGURE 37 Structures of ezetimibe and its analogs SCH 48461 and ezetimibe-OMe; the metabolic oxidation of the aryl ring in SCH 48461 is attenuated by fluorine substitution in ezetimibe and its ether analog ezetemibe-OMe.

Because of the relative ease of the formation of these secondary metabolites the potency of the drug is dramatically reduced (ED_{50} = 2.2 mg/kg). Blocking of these sites by incorporating a hydroxy moiety and fluorine, and modifying the reactivity of the two other aromatic rings gave metabolically stable ezetimibe, which is 50-fold more potent (ED_{50} = 0.04 mg/kg). On the other hand, retaining the benzylic carbon and one of the methoxy groups gives an ezetimibe analog (ezetimibe-OMe; **44**), with relatively lower potency than ezetimibe (ED_{50} = 0.7 mg/kg), although it exhibits relatively improved potency as compared to SCH 48461 (ED_{50} = 2.2 mg/mg).[96] Several similar derivatives of ezetimibe were synthesized and evaluated for their potency, safety, and effectiveness, out of which ezetimibe was chosen as the final optimal drug candidate, and was approved by the FDA, in 2002, as a cholesterol-lowering drug that inhibits intestinal absorption of cholesterol.

The synthetic strategy for ezetimibe and its various derivatives is shown in Figure 38.[96] The 2+2 cycloaddition of the ketene, generated in situ from 2-methoxycarbonylpropanoyl chloride **45**, and diarylimine **46** gave racemic mixture of the azetidinones **47**. Chromatographic resolution using Chiralcel-OD (cellulose tris(3,5-dimethylphenylcarbamate), coated on silica gel) column, followed by ester hydrolysis, conversion to acid chloride, and Pd(0)-mediated Negishi coupling with *p*-fluorophenylzinc bromide, afforded the azetidinone

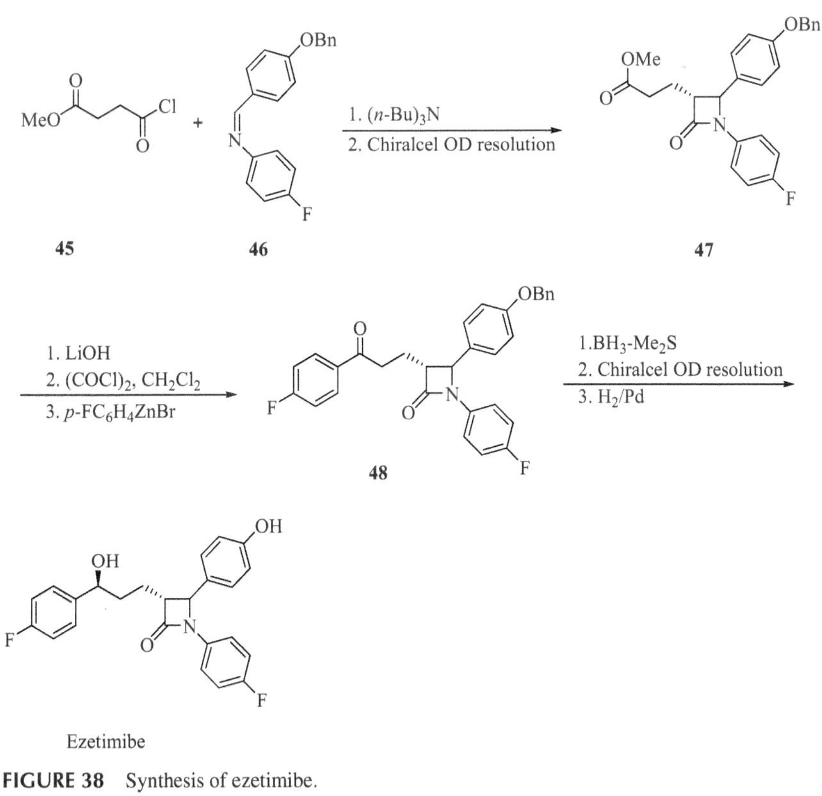

FIGURE 38 Synthesis of ezetimibe.

48. Reduction of the benzylic ketone moiety of the latter compound by borane dimethyl sulfide, followed by Chiralcel-OD resolution, and deprotection of the *O*-benzyl group by catalytic hydrogenation, provided the ezetimibe in high yields and with high optical purity.

10.2 Lomitapide

Lomitapide (Juxtapid; Lojuxta; Aegerion Pharmaceuticals) is being used for treating patients with homozygous familial hypercholesterolemia, and is approved by the FDA (in 2012) and European Medicines Agency (in 2013) (Figure 39). This compound inhibits microsomal triglyceride transfer protein, thereby preventing the biosynthesis of the LDL cholesterol. It is used in conjunction with statins for patients with high-risk hypercholesterolemia or for individuals who are intolerant to the first-line therapy with statins.[97]

11. ANTICOAGULATING (BLOOD-THINNING) AGENTS

11.1 Ticagrelor

Ticagrelor (Brilinta; AstraZeneca) acts as an antagonist, binding reversibly, to the $P2Y_{12}$ receptor, a main constituent of the blood platelets (Figure 40). $P2Y_{12}$

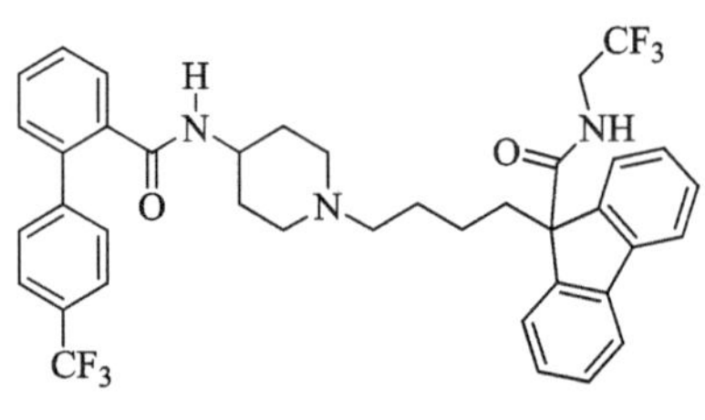

Lomitapide (Juxtapid)

FIGURE 39 Structure of lomitapide.

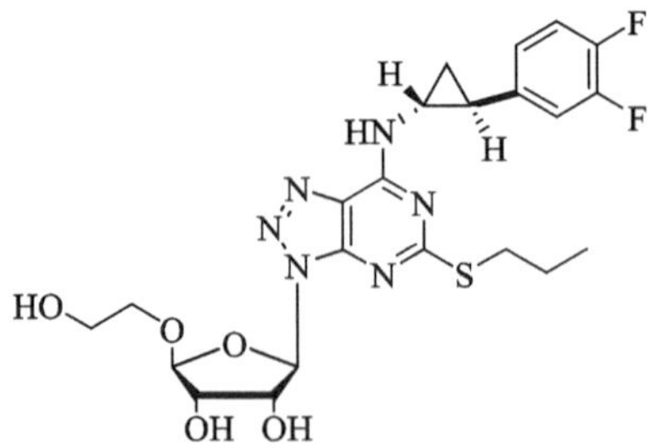

Ticagrelor

FIGURE 40 Structure of ticagrelor.

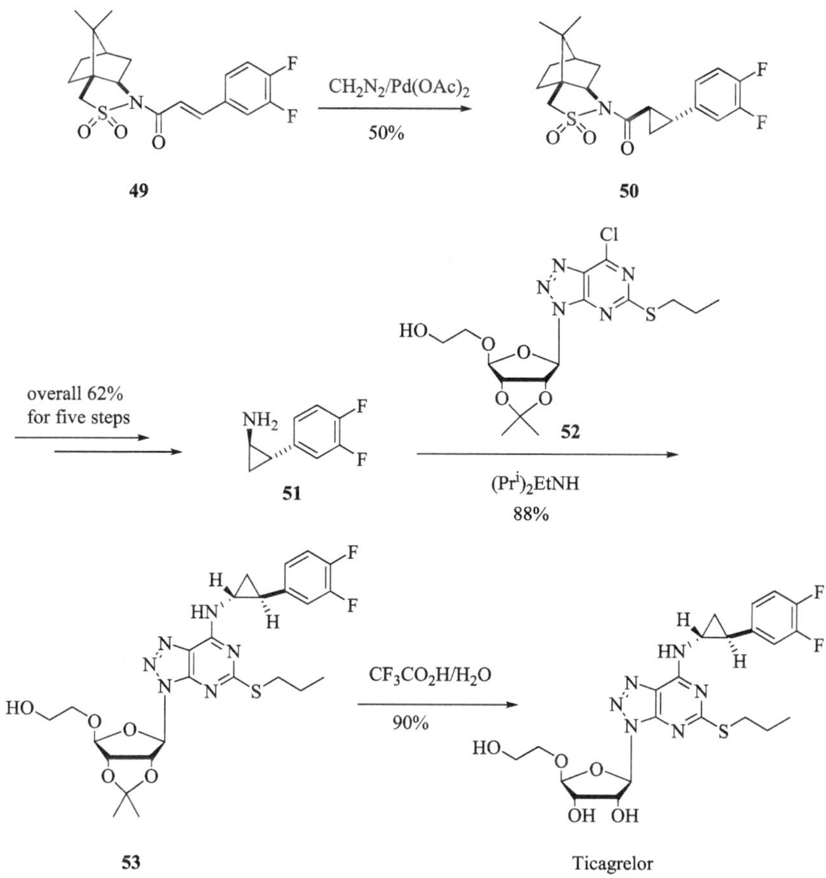

FIGURE 41 AstraZenica's synthesis of ticagrelor.

protein belongs to the family of the G-protein-coupled purinergic receptors that bind to adenosine diphosphate (ADP) to initiate cell signaling, thereby controlling blood clotting. Ticagrelor was approved by the FDA in 2011 as a blood-thinning agent for reducing the rate of thrombotic cardiovascular events.

Ticagrelor inhibits platelet activation and aggregation by reversibly binding to the $P2Y_{12}$ receptor at a site distinct from the ADP binding site (antagonist), so that the conformational changes of the receptor associated with the normal binding of ADP are inactivated. Thus, the inhibition of the ADP-induced activation of this G-protein, and thereby signaling, depends on the effective concentration of the ticagrelor at the binding site.[98]

AstraZeneca's synthetic strategy for the ticagrelor (Figure 41) involves the diastereoselective cyclopropanation of the Oppolzer's sultam derivative **49**, followed by the amide hydrolysis and Curtius degradation (62% overall yield) to the chirally pure (*E*)-2-(3,4-difluorophenyl)cyclopropylamine

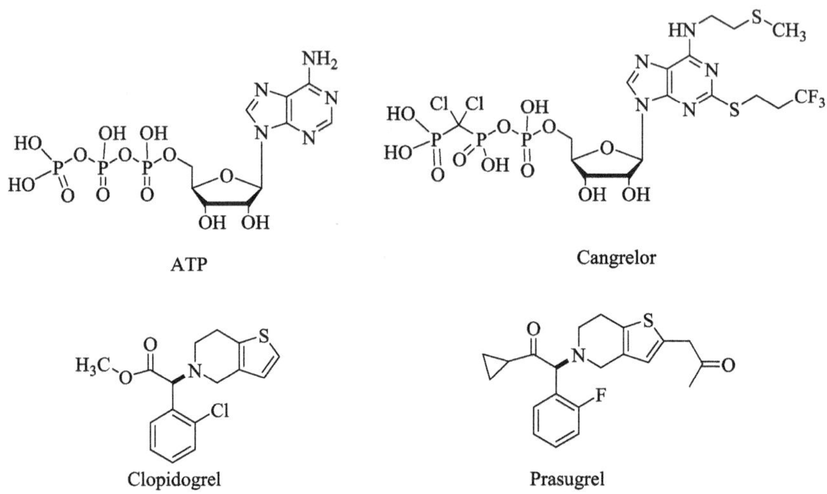

FIGURE 42 Structures of adenosine triphosphate (ATP), cangrelor, clopidogrel, and prasugrel.

51. Compound **51**, upon reaction with **52**, affords **53** (through the aromatic nucleophilic substitution of the aryl chlorine), which after removal of the acetal-protecting group by mild acid-catalyzed hydrolysis (aqueous trifluoroacetic acid) gives ticagrelor.[99]

11.2 Other Inhibitors of $P2Y_{12}$ Receptors

Slight structural changes can result in dramatic alterations in the agonist or antagonist properties of small molecules. ADP, for example is an agonist of the $P2Y_{12}$ receptor, whereas adenosine triphosphate (ATP) is an antagonist for the same receptor ($pIC_{50}=3.5$).[99] Thus, the terminal phosphate group of ATP plays an important role in its antagonist properties. Cangrelor was designed as a hydrolytically stable ATP mimic to act as antagonist of the $P2Y_{12}$ receptor (Figure 42). The dichloromethyl phosphonate moiety in cangrelor makes it relatively hydrolytically stable. Cangrelor is a relatively more efficient and selective $P2Y_{12}$ receptor antagonist, and is suitable for intravenous injection.

Ticagrelor (*vide supra*) is an orally active reversible $P2Y_{12}$ receptor antagonist, and is an improved version of Cangrelor in its antagonistic properties. Ticagrelor is relatively more hydrolytically stable than Cangrelor as the "triphosphate" moiety of Cangrelor is replaced by hydroxyethyl mimic in ticagrelor, and has over 100-fold greater binding affinity for the $P2Y_{12}$ receptor than Cangrelor. Unlike Cangrelor or ticagrelor, the commonly used antithrombotic agents, clopidogrel and prasugrel (Figure 42) are metabolized to active drugs that bind irreversibly at the ADP-binding site of the $P2Y_{12}$ receptor, and thereby irreversibly inactivate the protein for signaling (Figure 43). This irreversible

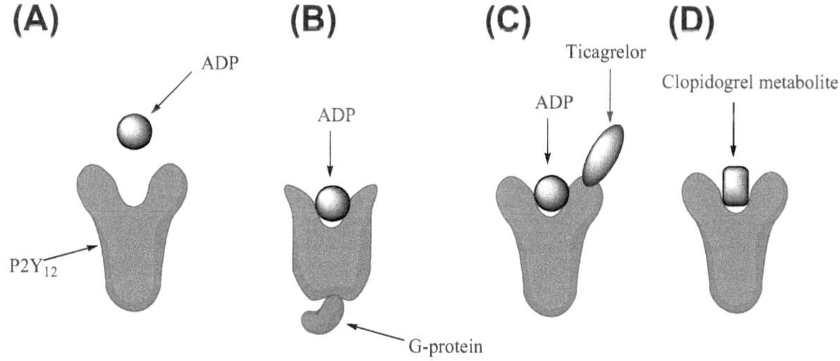

FIGURE 43 A schematic view of $P2Y_{12}$ receptor: before binding to the ligand adenosine diphosphate (ADP) (A); after binding to ADP, resulting in conformational change of the receptor and G-protein activation (B); after reversibly binding to ticagrelor and ADP (C); and after binding of the metabolite of clopidogrel (D). Reversible binding of ticagrelor locks the conformation of the receptor so that simultaneous binding of ADP to the receptor has no net effect on conformational change and thereby no signaling; on the other hand, clopidogrel metabolite irreversibly binds at the ADP-binding region, rendering the receptor inactive throughout the lifetime of the platelet.[98]

inactivation of the receptor persists until the affected platelets are replaced by new ones, and may be associated with the increased risk of bleeding in certain patients.[98]

12. SUMMARY AND OUTLOOK

About one-fifth of the currently marketed pharmaceuticals are organofluorine compounds. Some of the top-selling drugs such as atorvastatin, Crestor, efavirenz, fluoxetine, Ciprobay, fluticasone propionate, Faslodex, ezetimibe, Celebrex, and sitagliptin consist of fluoroaryl moiety, *gem*-difluoromethylene moiety, or a trifluoromethyl moiety. Fluoxetine, a serotonin reuptake inhibitor is a widely used antidepressant drug. In this drug molecule, the trifluoromethyl moiety forces the molecule to adopt a favorable conformation to the SERT proteins. Fluorinated drugs such as atorvastatin and ezetimibe are among the top-selling antihypercholesterolemia drugs.

Sitagliptin, one of the widely used antidiabetic drugs, is a DPP-4 inhibitor. At the enzyme active site, the trifluoromethyl-substituted triazole and trifluorophenyl ring of sitagliptin, due to their decreased electron density, are involved in orthogonal π–π interactions with the neighboring phenylalanine and tyrosine residues. Several fluorinated drugs (e.g., aldose reductase inhibitors) are being developed for treating secondary complications of diabetes, such as diabetic neuropathy. Fluorinated nonsteroidal anti inflammatory drugs such as valdecoxib and celecoxib (Celebra) are selective COX2 inhibitors. Corticosteroids such as fluticasone furanoate and dexamethasone are widely used antiinflammatory agents. The fluorinated corticosteroids are relatively more lipophilic than are their nonfluorinated versions, and exert their antiinflammatory effect by binding as agonists to the gluococorticoid receptors.

Antibacterials, such as flurithromycin have enhanced bioavailability and metabolic stability due to the presence of a fluorine substituent at the metabolically vulnerable site. Mefloquine, a trifluoromethyl-substituted quinine analog, is one of the most successful drugs for treating malaria. Antimalarials such as floxacrine, consisting of a trifluoromethylphenyl moiety, are used to treat drug-resistant malarial infections. Fluorinated analogs of artemisinins are explored as alternative antimalarials for drug-resistant infections. The bioavailability as well as the chemical and metabolic stabilities of the trifluoromethyl-substituted analogs of the first-generation artemisinins is significantly increased. Many fluorinated antiviral drugs are being developed to combat viral infections, including HIV. The antiviral drug *S*-efavirenz is a reverse transcriptase inhibitor, and is administered as a chiral drug to treat HIV infections. Antiviral drug trifluridine (trifluorothymidine) is an irreversible thymidylate synthase inhibitor, and thereby suppresses viral DNA synthesis. Fluorinated nucleoside analogs such as F-ddA and sofosbuvir are some of the other commonly used antiviral drugs. Novel therapeutics are being developed at a rapid pace than ever before, largely due to the emerging advances in synthetic organofluorine chemistry.

REFERENCES

1. O'Hagan, D. Fluorine in Health Care: Organofluorine Containing Blockbuster Drugs. *J. Fluorine Chem.* **2010,** *131,* 1071–1081.
2. Purser, S.; Moore, P. R.; Swallow, S.; Gouverneur, V. Fluorine in Medicinal Chemistry. *Chem. Soc. Rev.* **2008,** *37,* 320–330.
3. Filler, R.; Saha, R. Fluorine in Medicinal Chemistry: A Century of Progress and a 60-Year Retrospective of Selected Highlights. *Future Med. Chem.* **2009,** *1,* 1370.
4. Wang, J.; Sanchez-Rosello, M.; Acena, J. L.; del Pozo, C.; Sorochinsky, A. E.; Fustero, S.; Soloshonok, V. A.; Liu, H. Fluorine in Pharmaceutical Industry: Fluorine-Containing Drugs Introduced to the Market in the Last Decade (2001–2011). *Chem. Rev.* **2014,** *114,* 2432–2506.
5. Kirk, K. L. Fluorination in Medicinal Chemistry: Methods, Strategies, and Recent Developments. *Org. Process Res. Dev.* **2008,** *12,* 305–321.
6. Anonymous. Patent Watch. *Nat. Rev. Drug Discov.* **2013,** *12,* 14–15.
7. http://www.Drugs.com/stats/top100.
8. Heidelberger, C.; Chaudhuri, N. K.; Danneberg, P.; Mooren, D.; Griesbach, L.; Duschinsky, R.; Schnitzer, R. J.; Pleven, E.; Scheiner, J. Fluorinated Pyrimidines, a New Class of Tumour-Inhibitory Compounds. *Nature* **1957,** *179,* 663–666.
9. Ghoshal, K.; Jacob, S. T. An Alternative Molecular Mechanism of Action of 5-Fluorouracil, a Potent Anticancer Drug. *Biochem. Pharmacol.* **1997,** *53,* 1569–1575.
10. Graham, J. S.; Cassidy, J. Adjuvant Therapy in Colon Cancer. *Expert Rev. Anticancer Ther.* **2012,** *12,* 99–109.
11. Novinec, M.; Lenarcic, B. Cathepsin K: A Unique Collagenolytic Cysteine Peptidase. *Biol. Chem.* **2013,** *394,* 1163–1179.

12. Mabe, S.; Eller, J.; Champney, W. S. Structure-Activity Relationships for Three Macrolide Antibiotics in *Haemophilus influenzae*. *Curr. Microbiol.* **2004,** *49,* 248–254.
13. http://www.fda.gov/Drugs/DrugSafety/ucm365050.htm.
14. Berning, S. E. The Role of Fluoroquinolones in Tuberculosis Today. *Drugs* **2001,** *61,* 9–18.
15. Gosling, R. J.; Clouting, C. S.; Randall, L. P.; Horton, R. A.; Davies, R. H. Ciprofloxacin Resistance in *E. coli* Isolated from Turkeys in Great Britain. *Avian Pathol.* **2012,** *41,* 83–89.
16. Domagala, J. M. Structure-Activity and Structure-Side-Effect Relationships for the Quinolone Antibacterials. *J. Antimicrob. Chemother.* **1994,** *33,* 685–706.
17. Stepan, A. F.; Walker, D. P.; Bauman, J.; Price, D. A.; Baillie, T. A.; Kalgutkar, A. S.; Aleo, M. D. Structural Alert/Reactive Metabolite Concept as Applied in Medicinal Chemistry to Mitigate the Risk of Idiosyncratic Drug Toxicity: A Perspective Based on the Critical Examination of Trends in the Top 200 Drugs Marketed in the United States. *Chem. Res. Toxicol.* **2011,** *24,* 1345–1410.
18. Psaty Bruce, M. Clinical Trial Design and Selected Drug Safety Issues for Antibiotics Used to Treat Community-Acquired Pneumonia. *Clin. Infect. Dis.* **2008,** *47* (Suppl. 3), S176–S179.
19. Mitscher, L. A.; Chu, D. T. *Process for Preparation of Racemic and Optically Active Ofloxacin and Related Derivatives;* Abbott Laboratories: USA, 1988. US4777253A.
20. Cui, L.-W.; Su, X.-Z. Discovery, Mechanisms of Action and Combination Therapy of Artemisinin. *Expert Rev. Anti Infect. Ther.* **2009,** *7,* 999–1013.
21. Butler, A. R.; Khan, S.; Ferguson, E. A Brief History of Malaria Chemotherapy. *J R Coll Physicians Edinb* **2010,** *40,* 172–177.
22. Begue, J.-P.; Bonnet-Delpon, D. *Antimalarial fluoroartemisinins: increased metabolic and chemical stability;* In *Fluorine in Medicinal Chemistry and Chemical Biology;* Ojima, I., Ed.; John Wiley & Sons Ltd.: Chichester, UK, 2009; p 141–163.
23. Magueur, G.; Crousse, B.; Charneau, S.; Grellier, P.; Begue, J.-P.; Bonnet-Delpon, D. Fluoroartemisinin: Trifluoromethyl Analogues of Artemether and Artesunate. *J. Med. Chem.* **2004,** *47,* 2694–2699.
24. Grellepois, F.; Chorki, F.; Ourevitch, M.; Charneau, S.; Grellier, P.; McIntosh, K. A.; Charman, W. N.; Pradines, B.; Crousse, B.; Bonnet-Delpon, D.; Begue, J.-P. Orally Active Antimalarials: Hydrolytically Stable Derivatives of 10-Trifluoromethyl Anhydrodihydroartemisinin. *J. Med. Chem.* **2004,** *47,* 1423–1433.
25. Wiseman, V.; Kim, M.; Mutabingwa, T. K.; Whitty, C. J. M. Cost-Effectiveness Study of Three Antimalarial Drug Combinations in Tanzania. *PLoS Med.* **2006,** *3,* e373.
26. Schmidt, M.; Sun, H.; Rogne, P.; Scriba, G. K. E.; Griesinger, C.; Kuhn, L. T.; Reinscheid, U. M. Determining the Absolute Configuration of (+)-Mefloquine HCl, the Side-Effect-Reducing Enantiomer of the Antimalaria Drug Lariam. *J. Am. Chem. Soc.* **2012,** *134,* 3080–3083.
27. Hovsepian, E.; Penas, F.; Mirkin, G. A.; Goren, N. B. Role of PPARs in *Trypanosoma cruzi* Infection: Implications for Chagas Disease Therapy. *PPAR Res.* **2012**, http://dx.doi.org/10.1155/2012/528435; *Chem. Abstr.* 2012:332513.
28. Serghides, L. The Case for the Use of PPARγ Agonists as an Adjunctive Therapy for Cerebral Malaria. *PPAR Res.* **2012**, http://dx.doi.org/10.1155/2012/513865; *Chem. Abstr.* 512011:793222.
29. Barton, V.; Fisher, N.; Biagini, G. A.; Ward, S. A.; O'Neill, P. M. Inhibiting Plasmodium Cytochrome Bc1: A Complex Issue. *Curr. Opin. Chem. Biol.* **2010,** *14,* 440–446.
30. Nilsen, A.; LaCrue, A. N.; White, K. L.; Forquer, I. P.; Cross, R. M.; Marfurt, J.; Mather, M. W.; Delves, M. J.; Shackleford, D. M.; Saenz, F. E.; Morrisey, J. M.; Steuten, J.; Mutka, T.; Li, Y.; Wirjanata, G.; Ryan, E.; Duffy, S.; Kelly, J. X.; Sebayang, B. F.; Zeeman, A.-M.; Noviyanti, R.; Sinden, R. E.; Kocken, C. H. M.; Price, R. N.; Avery, V. M.; Angulo-Barturen, I.; Jimenez-Diaz,

M. B.; Ferrer, S.; Herreros, E.; Sanz, L. M.; Gamo, F.-J.; Bathurst, I.; Burrows, J. N.; Siegl, P.; Guy, R. K.; Winter, R. W.; Vaidya, A. B.; Charman, S. A.; Kyle, D. E.; Manetsch, R.; Riscoe, M. K. Quinolone-3-Diarylethers: A New Class of Antimalarial Drug. *Sci. Transl. Med.* **2013,** *5,* 177ra137. 114 pp.

31. Chatterjee, A. K. Cell-Based Medicinal Chemistry Optimization of High-Throughput Screening (HTS) Hits for Orally Active Antimalarials. Part 1: Challenges in Potency and Absorption, Distribution, Metabolism, Excretion/Pharmacokinetics (ADME/PK). *J. Med. Chem.* **2013,** *56,* 7741–7749.
32. Gujjar, R.; El Mazouni, F.; White, K. L.; White, J.; Creason, S.; Shackleford, D. M.; Deng, X.; Charman, W. N.; Bathurst, I.; Burrows, J.; Floyd, D. M.; Matthews, D.; Buckner, F. S.; Charman, S. A.; Phillips, M. A.; Rathod, P. K. Lead Optimization of Aryl and Aralkyl Amine-Based Triazolopyrimidine Inhibitors of *Plasmodium falciparum* Dihydroorotate Dehydrogenase with Antimalarial Activity in Mice. *J. Med. Chem.* **2011,** *54,* 3935–3949.
33. Rottmann, M.; McNamara, C.; Yeung, B. K. S.; Lee, M. C. S.; Zou, B.; Russell, B.; Seitz, P.; Plouffe, D. M.; Dharia, N. V.; Tan, J.; Cohen, S. B.; Spencer, K. R.; Gonzalez-Paez, G. E.; Lakshminarayana, S. B.; Goh, A.; Suwanarusk, R.; Jegla, T.; Schmitt, E. K.; Beck, H.-P.; Brun, R.; Nosten, F.; Renia, L.; Dartois, V.; Keller, T. H.; Fidock, D. A.; Winzeler, E. A.; Diagana, T. T. Spiroindolones, a Potent Compound Class for the Treatment of Malaria. *Science* **2010,** *329,* 1175–1180.
34. Begue, J.-P.; Bonnet-Delpon, D. Fluoroartemisinins: Metabolically More Stable Antimalarial Artemisinin Derivatives. *ChemMedChem* **2007,** *2,* 608–624.
35. Abouabdellah, A.; Begue, J.-P.; Bonnet-Delpon, D.; Gantier, J.-C.; Truong, T. T. N.; Truong, D. T. Synthesis and In vivo Antimalarial Activity of 12α-Trifluoromethyl-Hydroartemisinin. *Bioorg. Med. Chem. Lett.* **1996,** *6,* 2717–2720.
36. Magueur, G.; Crousse, B.; Ourevitch, M.; Begue, J.-P.; Bonnet-Delpon, D. Preparation of 10-Trifluoromethyl Artemether and Artesunate. Influence of Hexafluoropropan-2-ol on Substitution Reaction. *J. Org. Chem.* **2003,** *68,* 9763–9766.
37. Kim, D.; Wang, L.; Beconi, M.; Eiermann, G. J.; Fisher, M. H.; He, H.; Hickey, G. J.; Kowalchick, J. E.; Leiting, B.; Lyons, K.; Marsilio, F.; McCann, M. E.; Patel, R. A.; Petrov, A.; Scapin, G.; Patel, S. B.; Roy, R. S.; Wu, J. K.; Wyvratt, M. J.; Zhang, B. B.; Zhu, L.; Thornberry, N. A.; Weber, A. E. (2R)-4-Oxo-4-[3-(Trifluoromethyl)-5,6-Dihydro[1,2,4]Triazolo[4,3-a]Pyrazin-7(8H)-yl]-1-(2,4,5-Trifluorophenyl)Butan-2-Amine: A Potent, Orally Active Dipeptidyl Peptidase IV Inhibitor for the Treatment of Type 2 Diabetes. *J. Med. Chem.* **2005,** *48,* 141–151.
38. Johnson, K. M. S.; Schurr, K. Sitagliptin: A DPP-4 Inhibitor for the Treatment of Type 2 Diabetes Mellitus. *Clin. Med. Insights Ther.* **2011,** *3,* 81–92.
39. Farnham, W. B. Silicon Mediated Reactions in Organofluorine Chemistry. *Synth. Fluorine Chem.* **1992,** 247–257.
40. Plosker, G. L. Sitagliptin: A Review of Its Use in Patients with Type 2 Diabetes Mellitus. *Drugs* **2014,** *74,* 223–242.
41. Matveyenko, A. V.; Dry, S.; Cox, H. I.; Moshtaghian, A.; Gurlo, T.; Galasso, R.; Butler, A. E.; Butler, P. C. Beneficial Endocrine but Adverse Exocrine Effects of Sitagliptin in the Human Islet Amyloid Polypeptide Transgenic Rat Model of Type 2 Diabetes: Interactions with Metformin. *Diabetes* **2009,** *58,* 1604–1615.
42. Burton, D. J. Organometallics in Synthetic Organofluorine Chemistry. *Synth. Fluorine Chem.* **1992,** 205–226.
43. Carey, F. Diabetic Complications: Aldose Reductase Inhibitors. *Emerging Drugs* **1996,** *1,* 83–101.
44. Kador, P. F. The Role of Aldose Reductase in the Development of Diabetic Complications. *Med. Res. Rev.* **1988,** *8,* 325–352.

45. Oates, P. J.; Mylari, B. L. Aldose Reductase Inhibitors: Therapeutic Implications for Diabetic Complications. *Expert Opin. Invest. Drugs* **1999,** *8,* 2095–2119.
46. Faes, T. J.; Yff, G. A.; DeWeerdt, O.; Lanting, P.; Heimans, J. J.; Bertelsmann, F. W. Treatment of Diabetic Autonomic Neuropathy with an Aldose Reductase Inhibitor. *J. Neurol.* **1993,** *240,* 156–160.
47. Soulis-Liparota, T.; Cooper, M. E.; Dunlop, M.; Jerums, G. The Relative Roles of Advanced Glycation, Oxidation and Aldose Reductase Inhibition in the Development of Experimental Diabetic Nephropathy in the Sprague-Dawley Rat. *Diabetologia* **1995,** *38,* 387–394.
48. Bravi, M. C.; Pietrangeli, P.; Laurenti, O.; Basili, S.; Cassone-Faldetta, M.; Ferri, C.; De Mattia, G. Polyol Pathway Activation and Glutathione Redox Status in Non-Insulin-Dependent Diabetic Patients. *Metab. Clin. Exp.* **1997,** *46,* 1194–1198.
49. Van Zandt, M. C.; Jones, M. L.; Gunn, D. E.; Geraci, L. S.; Jones, J. H.; Sawicki, D. R.; Sredy, J.; Jacot, J. L.; DiCioccio, A. T.; Petrova, T.; Mitschler, A.; Podjarny, A. D. Discovery of 3-[(4,5,7-Trifluorobenzothiazol-2-yl)Methyl]Indole-*N*-Acetic Acid (Lidorestat) and Congeners as Highly Potent and Selective Inhibitors of Aldose Reductase for Treatment of Chronic Diabetic Complications. *J. Med. Chem.* **2005,** *48,* 3141–3152.
50. Fournier, B.; Bendeif, E.-E.; Guillot, B.; Podjarny, A.; Lecomte, C.; Jelsch, C. Charge Density and Electrostatic Interactions of Fidarestat, an Inhibitor of Human Aldose Reductase. *J. Am. Chem. Soc.* **2009,** *131,* 10929–10941.
51. Griffin, B. W.; McNatt, L. G.; Chandler, M. L.; York, B. M. Effects of Two New Aldose Reductase Inhibitors, Al-1567 and Al-1576, in Diabetic Rats. *Metab. Clin. Exp.* **1987,** *36,* 486–490.
52. Roth, S. H. Nonsteroidal Anti-inflammatory Drug Gastropathy: New Avenues for Safety. *Clin. Interv. Aging* **2011,** *6,* 125–131.
53. Rimon, G.; Sidhu Ranjinder, S.; Lauver, D. A.; Lee Jullia, Y.; Sharma Narayan, P.; Yuan, C.; Frieler Ryan, A.; Trievel Raymond, C.; Lucchesi Benedict, R.; Smith William, L. Coxibs Interfere with the Action of Aspirin by Binding Tightly to One Monomer of Cyclooxygenase-1. *Proc. Natl. Acad. Sci. U.S.A.* **2010,** *107,* 28–33.
54. Richy, F.; Rabenda, V.; Mawet, A.; Reginster, J. Y. Flurbiprofen in the Symptomatic Management of Rheumatoid Arthritis: A Valuable Alternative. *Int. J. Clin. Pract.* **2007,** *61,* 1396–1406.
55. Huq, F. Molecular Modelling Analysis of the Metabolism of Lumiracoxib. *Int. J. Pure Appl. Chem.* **2007,** *2,* 377–385.
56. Moyer, A. M.; Fridley, B. L.; Jenkins, G. D.; Batzler, A. J.; Pelleymounter, L. L.; Kalari, K. R.; Ji, Y.; Chai, Y.; Nordgren, K. K. S.; Weinshilboum, R. M. Acetaminophen-NAPQI Hepatotoxicity: A Cell Line Model System Genome-Wide Association Study. *Toxicol. Sci.* **2011,** *120,* 33–41.
57. Rordorf, C. M.; Choi, L.; Marshall, P.; Mangold, J. B. Clinical Pharmacology of Lumiracoxib: A Selective Cyclo-Oxygenase-2 Inhibitor. *Clin. Pharmacokinet.* **2005,** *44,* 1247–1266.
58. Sorbera, L. A.; Serradell, N.; Bolós, J. Fluticasone Furoate: Glucocorticoid Receptor Agonist Intranasal Corticosteroid. *Drugs Future* **2007,** *32,* 12–16.
59. Biggadike, K.; Bledsoe, R. K.; Hassell, A. M.; Kirk, B. E.; McLay, I. M.; Shewchuk, L. M.; Stewart, E. L. X-ray Crystal Structure of the Novel Enhanced-Affinity Glucocorticoid Agonist Fluticasone Furoate in the Glucocorticoid Receptor-Ligand Binding Domain. *J. Med. Chem.* **2008,** *51,* 3349–3352.
60. Valotis, A.; Högger, P. Human Receptor Kinetics and Lung Tissue Retention of the Enhanced-Affinity Glucocorticoid Fluticasone Furoate. *Respir. Res.* **2007,** *8.*
61. Qiu, X.-L.; Yue, X.; Qing, F.-L. Fluorine-Containing Chiral Drugs. *Chiral Drugs* **2011,** 195–251.
62. Adkins, J. C.; Noble, S. Efavirenz. *Drugs* **1998,** *56,* 1055–1064.
63. Rabel, S. R.; Patel, M.; Sun, S.; Maurin, M. B. Electronic and Resonance Effects on the Ionization of Structural Analogues of Efavirenz. *AAPS PharmSci* **2001,** *3,* 1.

64. Chinkov, N.; Warm, A.; Carreira, E. M. Asymmetric Autocatalysis Enables an Improved Synthesis of Efavirenz. *Angew. Chem.* **2011,** *123,* 3014–3018.
65. Choudhury, A.; Moore, J. R.; Pierce, M. E.; Fortunak, J. M.; Valvis, I.; Confalone, P. N. In situ Recycling of Chiral Ligand and Surplus Nucleophile for a Noncatalytic Reaction: Amplification of Process Throughput in the Asymmetric Addition Step of Efavirenz (DMP 266). *Org. Process Res. Dev.* **2003,** *7,* 324–328.
66. Grabowski, E. J. J. Enantiopure Drug Synthesis: From Methyldopa to Imipenem to Efavirenz. *Chirality* **2005,** *17,* S249–S259.
67. Pierce, M. E.; Parsons, R. L., Jr.; Radesca, L. A.; Lo, Y. S.; Silverman, S.; Moore, J. R.; Islam, Q.; Choudhury, A.; Fortunak, J. M. D.; Nguyen, D.; Luo, C.; Morgan, S. J.; Davis, W. P.; Confalone, P. N.; Chen, C. Y.; Tillyer, R. D.; Frey, L.; Tan, L.; Xu, F.; Zhao, D.; Thompson, A. S.; Corley, E. G.; Grabowski, E. J. J.; Reamer, R.; Reider, P. J. Practical Asymmetric Synthesis of Efavirenz (DMP 266), an HIV-1 Reverse Transcriptase Inhibitor. *J. Org. Chem.* **1998,** *63,* 8536–8543.
68. Tan, L.; Chen, C.; Tillyer, R. D.; Grabowski, E. J. J.; Reider, P. J. A novel, highly enantioselective ketone alkynylation reaction mediated by chiral zinc aminoalkoxides. *Angew. Chem. Int. Ed.* **1999,** *38,* 711–723.
69. Thompson, A.; Corley, E. G.; Huntington, M. F.; Grabowski, E. J. J.; Remenar, J. F.; Collum, D. B. Lithium Ephedrate-Mediated Addition of a Lithium Acetylide to a Ketone: Solution Structures and Relative Reactivities of Mixed Aggregates Underlying the High Enantioselectivities. *J. Am. Chem. Soc.* **1998,** *120,* 2028–2038.
70. Thompson, A. S.; Corley, E. G.; Huntington, M. F.; Grabowski, E. J. J. Use of an Ephedrine Alkoxide to Mediate Enantioselective Addition of an Acetylide to a Prochiral Ketone: Asymmetric Synthesis of the Reverse Transcriptase Inhibitor L-743,726. *Tetrahedron Lett.* **1995,** *36,* 8937–8940.
71. Kawai, H.; Kitayama, T.; Tokunaga, E.; Shibata, N. A New Synthetic Approach to Efavirenz through Enantioselective Trifluoromethylation by Using the Ruppert-Prakash Reagent. *Eur. J. Org. Chem.* **2011,** *2011,* 5959–5961.
72. Turner, S. R.; Strohbach, J. W.; Tommasi, R. A.; Aristoff, P. A.; Johnson, P. D.; Skulnick, H. I.; Dolak, L. A.; Seest, E. P.; Tomich, P. K.; Bohanon, M. J.; Horng, M.-M.; Lynn, J. C.; Chong, K.-T.; Hinshaw, R. R.; Watenpaugh, K. D.; Janakiraman, M. N.; Thaisrivongs, S. Tipranavir (PNU-140690): A Potent, Orally Bioavailable Nonpeptidic HIV Protease Inhibitor of the 5,6-Dihydro-4-Hydroxy-2-Pyrone Sulfonamide Class. *J. Med. Chem.* **1998,** *41,* 3467–3476.
73. Chrysos, G.; Gerakari, S.; Stasini, F.; Kokkoris, S.; Kourousis, D.; Velegraki, A. *Intracranial Haemorrhage Possibly Related to Tipranavir in an HIV-1 Patient with Cryptococcal Meningitis;* Infectious Diseases Unit, Tzaneion General Hospital of Piraeus: Athens, Greece. *J. Infect.* **2008,** *57(1),* 85–87.
74. Berhan, A.; Berhan, Y. Virologic Response to Tipranavir-Ritonavir or Darunavir-Ritonavir Based Regimens in Antiretroviral Therapy Experienced HIV-1 Patients: A Meta-Analysis and Meta-Regression of Randomized Controlled Clinical Trials. *PLoS ONE* **2013,** *8,* e60814.
75. Schlosser, M. Parametrization of Substituents: Effects of Fluorine and Other Heteroatoms on OH, NH, and CH Acidities. *Angew. Chem. Int. Ed.* **1998,** *37,* 1496–1513.
76. Maruyama, T.; Ikejiri, M.; Izawa, K.; Onishi, T. Synthesis and Biological Activity of Fluorinated Nucleosides. In *Fluorine in Medicinal Chemistry and Chemical Biology;* Ojima, I., Ed.; Wiley, 2009; pp 165–198.
77. Kukhar, V. P. Preparation of Fluorine-Containing Amino Acids by Methods of Organofluorine Chemistry; In *Fluorine-Containing Amino Acids*; Kukhar, V. P., Soloshonok, V. A., Eds.; Wiley: Chichester, UK, 1995; pp 71–111.

78. Kobayashi, Y.; Kumadaki, I.; Yamamoto, K. Simple Synthesis of Trifluoromethylated Pyrimidine Nucleosides. *J. Chem. Soc. Chem. Commun.* **1977,** 536–537.
79. Shimizu, T.; Asai, T.; Morisawa, Y. *Method for Preparation of 5-(Trifluoromethyl)Uracil Derivatives by Trifluoromethylation of 5-Halogenated Uracils with Difluoro(Fluorosulfonyl)Acetic Acid Esters;* Asahi Glass Co., Ltd.: Japan, 1999. JP11246590A.
80. Morton, G. O.; Lancaster, J. E.; Van Lear, G. E.; Fulmor, W.; Meyer, W. E. Structure of Nucleocidin. III. Revised Structure. *J. Am. Chem. Soc.* **1969,** *91,* 1535–1537.
81. Qiu, X.-L.; Xu, X.-H.; Qing, F.-L. Recent Advances in the Synthesis of Fluorinated Nucleosides. *Tetrahedron* **2010,** *66,* 789–843.
82. Brodszki, M.; Backstrom, B.; Horvath, K.; Larsson, T.; Malmgren, H.; Pelcman, M.; Wahling, H.; Wallberg, H.; Wennerberg, J. Synthesis of the Hepatitis B Nucleoside Analogue Lagociclovir Valactate. *Org. Process Res. Dev.* **2011,** *15,* 1027–1032.
83. Buckheit, R. W., Jr.; Russell, J. D.; Pallansch, L. A.; Driscoll, J. S. Anti-Human Immunodeficiency Virus Type 1 (HIV-1) Activity of 2′-Fluoro-2′,3′-Dideoxyarabinosyladenine (F-ddA) Used in Combination with Other Mechanistically Diverse Inhibitors of HIV-1 Replication. *Antiviral Chem. Chemother.* **1999,** *10,* 115–119.
84. Belema, M.; Lopez, O. D.; Bender, J. A.; Romine, J. L.; St. Laurent, D. R.; Langley, D. R.; Lemm, J. A.; O'Boyle, D. R., II; Sun, J.-H.; Wang, C.; Fridell, R. A.; Meanwell, N. A. Discovery and Development of Hepatitis C Virus NS5A Replication Complex Inhibitors. *J. Med. Chem.* **2014,** *57,* 1643–1672.
85. Ramamoorthy, S.; Bauman, A. L.; Moore, K. R.; Han, H.; Yang-Feng, T.; Chang, A. S.; Ganapathy, V.; Blakely, R. D. Antidepressant- and Cocaine-Sensitive Human Serotonin Transporter: Molecular Cloning, Expression, and Chromosomal Localization. *Proc. Natl. Acad. Sci. U.S.A.* **1993,** *90,* 2542–2546.
86. Mennigen, J. A.; Stroud, P.; Zamora, J. M.; Moon, T. W.; Trudeau, V. L. Pharmaceuticals as Neuroendocrine Disruptors: Lessons Learned from Fish on Prozac. *J. Toxicol. Environ. Health Part B* **2011,** *14,* 387–412.
87. Bymaster, F. P.; Zhang, W.; Carter, P. A.; Shaw, J.; Chernet, E.; Phebus, L.; Wong, D. T.; Perry, K. W. Fluoxetine, but Not other Selective Serotonin Uptake Inhibitors, Increases Norepinephrine and Dopamine Extracellular Levels in Prefrontal Cortex. *Psychopharmacology (Berl.)* **2002,** *160,* 353–361.
88. Roman, D. L.; Walline, C. C.; Rodriguez, G. J.; Barker, E. L. Interactions of Antidepressants with the Serotonin Transporter: A Contemporary Molecular Analysis. *Eur. J. Pharmacol.* **2003,** *479,* 53–63.
89. Zhou, Z.; Zhen, J.; Karpowich, N. K.; Law, C. J.; Reith, M. E. A.; Wang, D. N. Antidepressant Specificity of Serotonin Transporter Suggested by Three LeuT-SSRI Structures. *Nat. Struct. Mol. Biol.* **2009,** *16,* 652–657.
90. Hyttel, J.; Boegesoe, K. P.; Perregaard, J.; Sanchez, C. The Pharmacological Effect of Citalopram Resides in the (S)-(+)-Enantiomer. *J. Neural Transm. Gen. Sect.* **1992,** *88,* 157–160.
91. Combs, S.; Kaufmann, K.; Field, J. R.; Blakely, R. D.; Meiler, J. Y95 and E444 Interaction Required for High-Affinity S-Citalopram Binding in the Human Serotonin Transporter. *ACS Chem. Neurosci.* **2011,** *2,* 75–81.
92. Thal, D. M.; Homan, K. T.; Chen, J.; Wu, E. K.; Hinkle, P. M.; Huang, Z. M.; Chuprun, J. K.; Song, J.; Gao, E.; Cheung, J. Y.; Sklar, L. A.; Koch, W. J.; Tesmer, J. J. G. Paroxetine Is a Direct Inhibitor of G Protein-Coupled Receptor Kinase 2 and Increases Myocardial Contractility. *ACS Chem. Biol.* **2012,** *7,* 1830–1839.
93. Clader, J. W. The Discovery of Ezetimibe: A View from Outside the Receptor. *J. Med. Chem.* **2004,** *47,* 1–9.

94. Shah, P.; Westwell, A. D. The Role of Fluorine in Medicinal Chemistry. *J. Enzyme Inhib. Med. Chem.* **2007,** *22,* 527–540.
95. Tang, W.; Jia, L.; Ma, Y.; Xie, P.; Haywood, J.; Dawson, P. A.; Li, J.; Yu, L. Ezetimibe Restores Biliary Cholesterol Excretion in Mice Expressing Niemann-Pick C1-Like 1 Only in Liver. *Biochim. Biophys. Acta Mol. Cell Biol. Lipids* **2011,** *1811,* 549–555.
96. Rosenblum, S. B.; Huynh, T.; Afonso, A.; Davis, H. R., Jr.; Yumibe, N.; Clader, J. W.; Burnett, D. A. Discovery of 1-(4-Fluorophenyl)-(3R)-[3-(4-Fluorophenyl)-(3S)-Hydroxypropyl]-(4S)-(4-Hydroxyphenyl)-2-Azetidinone (SCH 58235): A Designed, Potent, Orally Active Inhibitor of Cholesterol Absorption. *J. Med. Chem.* **1998,** *41,* 973–980.
97. Noto, D.; Cefalu, A. B.; Averna, M. R. Beyond Statins: New Lipid Lowering Strategies to Reduce Cardiovascular Risk. *Curr. Atheroscler. Rep.* **2014,** *16,* 1–10.
98. Husted, S.; van Giezen, J. J. J. Ticagrelor: The First Reversibly Binding Oral P2Y12 Receptor Antagonist. *Cardiovasc. Ther.* **2009,** *27,* 259–274.
99. Springthorpe, B.; Bailey, A.; Barton, P.; Birkinshaw, T. N.; Bonnert, R. V.; Brown, R. C.; Chapman, D.; Dixon, J.; Guile, S. D.; Humphries, R. G.; Hunt, S. F.; Ince, F.; Ingall, A. H.; Kirk, I. P.; Leeson, P. D.; Leff, P.; Lewis, R. J.; Martin, B. P.; McGinnity, D. F.; Mortimore, M. P.; Paine, S. W.; Pairaudeau, G.; Patel, A.; Rigby, A. J.; Riley, R. J.; Teobald, B. J.; Tomlinson, W.; Webborn, P. J. H.; Willis, P. A. From ATP to AZD6140: The Discovery of an Orally Active Reversible P2Y12 Receptor Antagonist for the Prevention of Thrombosis. *Bioorg. Med. Chem. Lett.* **2007,** *17,* 6013–6018.

Chapter 6

Organofluorine Anesthetics

Chapter Outline

1. INTRODUCTION

One of the most important applications of organofluorine compounds in medicinal chemistry is their use as inhalation anesthetics. Each year over 20 million people, in the United States alone, undergo surgical operations using general anesthetics, and thus there is an increasing demand to discover and identify safer and more effective anesthetics that could be rapidly eliminated from the body after the surgical procedure. In this respect, the organofluorine-based volatile inhalation anesthetics complement the state-of-the-art intravenous anesthetics such as propofol. The fluorinated inhalation anesthetics that are currently used in hospital rooms have rapid onset and offset anesthetic properties. Typically, the organofluorine-based anesthetics are very small molecules, having one to four carbons. The major fluorinated anesthetics currently in market are halothane (Fluothane), isoflurane (Forane), enflurane (Ethrane, also called as compound 374), desflurane (Suprane), and sevoflurane (Ultane, Sevorane, Sojourn) (Figure 1). None of the currently available general anesthetics (both intravenous and inhalation type) are completely free of toxic effects, as each of these compounds exerts

Organofluorine Compounds in Biology and Medicine. http://dx.doi.org/10.1016/B978-0-444-53748-5.00006-X

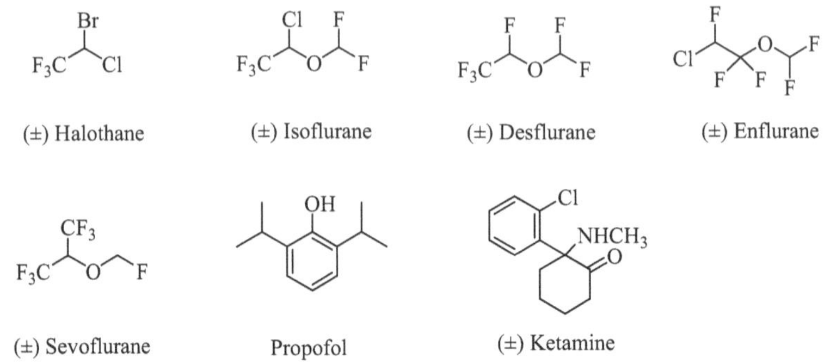

FIGURE 1 Structures of fluorinated inhalation anesthetics, and intravenous anesthetics propofol and ketamine.

its unique adverse effects on the patients. Therefore, selection of a given anesthetic is irrelevant in most surgeries and is based mainly on the expertise of the resident anesthesiologists, although the use of halothane is discouraged in some countries due to its relatively high hepatotoxicity in some susceptible cases.[1,2] Animal models show that the toxicities of the inhalation anesthetics (organofluorine-derived compounds) are dose and exposure time dependent, and some of them may have neurodegenerative effects (e.g., isoflurane).[3] However, their complete toxicity effects on human patients are not clearly elucidated despite their worldwide use.

Organofluorine compounds are ideally suited for use as inhalation anesthetics from the point of view of anesthetic administration. That is, organofluorine compounds, in general, are relatively nonflammable and have relatively higher volatility characteristics, as compared to their nonfluorinated analogs. Compound volatility is an important criterion for the design of inhalation anesthetics. The volatile organofluorine compounds have obvious advantages, because they would be able to pass through lungs in the vapor phase prior to reaching the brain through circulation. Further, due to the high strength of the C–F bond, organofluorine-based anesthetics are also metabolically inert. In the early stages of development, various fluoroalkyl ethers were introduced as viable alternatives to the first-generation anesthetics—which include diethyl ether, chloroethane, cyclopropane, and chloroform—due to their high flammability and adverse side effects. These compounds, especially, chloroform, were used on patients for over 100 years until the mid-1950s, and are no longer used as anesthetics. The new era in inhalation general anesthetics began with the discovery of halothane as a safe and effective inhalation anesthetic in 1955.[4]

Fluorinated inhalation general anesthetics, such as halothane, isoflurane, desflurane, enflurane, and sevoflurane, were developed as relatively safer and more efficient alternatives to other nonfluorinated intravenous anesthetics (e.g., propofol (2,6-diisopropylphenol) and ketamine (2-(2-chlorophenyl)-2-(methylamino)cyclohexanone)), and are currently widely used as general anesthetics in the hospitals worldwide.[5] Among these fluorinated inhalation anesthetics, halothane is less commonly used due to its relatively higher postoperative toxicity effects. Occasionally, combinations of various anesthetics are often used together in order to produce more effective anesthesia that generally involves the combined effects of unconsciousness, amnesia, insensitiveness, and analgesic effects. In principle, fluorinated inhalation anesthetics provide all these effects without the need of additional pharmacological agents or other coanesthetics, although several other drugs are coadministered to bring about effective muscle relaxation and cardiovascular response. There are some reports of carbon monoxide evolution during the administration of the organofluorine anesthetics. Small amount of carbon monoxide is generated from the partial degradation of these anesthetics by strong bases ($Ca(OH)_2$, NaOH, KOH), which are constituents of soda lime canisters that are used in the anesthetic setup; soda lime is used to capture the patients' exhaled carbon dioxide.[6] Alternate CO_2 sequestrating agents devoid of these strong bases, such as molecular sieves, may avoid this complication.[7]

Despite the advances in the anesthesiology in modern medicine, it is still under debate as to whether the inhalation anesthetics are neuroprotective or neurotoxic. There are some studies indicating that they show neuroprotection when administered at relatively shorter times, while other studies show that they are neurotoxic, especially when continuously administered for over 2 h.[8–10] A clinical study on the possible hepatotoxicity of the inhalation anesthetics, isoflurane, desflurane, and sevoflurane, revealed that drug-induced hepatotoxicity occurs in a significant number of patients, that is, nearly a quarter of the 1556 patients examined developed significant liver injury, and about 3% of the patients had abnormal postoperative liver biochemistry, presumably due to the inhalation anesthetics.[11] This exemplifies the necessity of safety precautions that need to be taken during the administration of these anesthetics.

2. MECHANISM OF ACTION OF INHALATION ANESTHETICS

Due to the wide variations in the structures of the commonly used anesthetic compounds, ranging from nonvolatile steroids to volatile fluorinated ethers, there may be divergent mechanisms of action for each class of these compounds. According to an earlier hypothesis for the mechanism of anesthetic action, called as Meyer–Overton rule, the anesthetic efficiency is based on the lipid solubility of anesthetics; that is, the higher the lipid solubility, the greater is the anesthetic effect. According to this hypothesis, it was assumed that the general anesthetics exert their action by lipid bilayer perturbation within the cell membrane, which is then transmitted to a membrane protein.[12] Because the potency of the anesthetics increases in proportion to their solubility in olive oil, it was assumed that the site of anesthetic action is within the cell membrane. It was later recognized that due to the use of low levels of the anesthetics during the anesthesia, the perturbation of the lipid bilayer membranes is minimal and, therefore other sites, especially the protein–lipid interface, and ion channels at the synapse may be more relevant for exerting the anesthetic properties. There is now general agreement that the general anesthetic compounds exert their action by binding to proteins rather than to the lipid bilayer membranes, although the exact mechanism of their action is still in debate.[13] Among other experimental techniques, the ^{19}F nuclear magnetic resonance (NMR) studies, fluorescence spectroscopy, isothermal titration calorimetry studies, and photoaffinity labeling techniques helped to identify the approximate binding sites of anesthetic molecules with the protein targets.[14,15] Based on these mechanistic studies, it is now generally accepted that the favored sites of action of anesthetics on the central nervous system are a variety of plasma membrane proteins, such as Cys-loop superfamily of ligand-gated ion channels (LGICs), nicotinic acetylcholine (nACh) receptors, N-methyl-D-aspartate (NMDA) receptors, and potassium ion channels.[14] The volatile inhalation anesthetics, being nonpolar, are only capable of forming weak van der Waals' interactions with the target proteins, and due to these experimental challenges, direct X-ray structural studies of the protein–ligand interactions are reported for only a few cases.

Anesthetics bind to human serum albumin (HSA) competitively with fatty acids and/or some of the pharmaceuticals, and therefore HSA is the principal transporter of anesthetics to the neuronal cells. The X-ray structures of HSA bound to propofol and halothanes show similarity in their binding sites, and these anesthetics compete with fatty acid binding sites on the protein. Binding of these anesthetics minimally alters the structural or conformational characteristics of the proteins (Figure 2).[16] Halothane molecules are surrounded by different sets of amino acid residues at the two binding regions; in one of these binding regions, the fluorine atoms have van der Waals interactions with the side chain C–H bonds of Phe403, Leu453, and Val433 and polar interaction with the backbone carbonyl carbons of Leu430, whereas at the second site, the fluorine atoms have polar interactions with the backbone carbonyls of Arg209, Asp324, and Glu354 (Figure 2). At clinically relevant concentrations, about 50% of the halothane and 80% of the propofol are bound to the HSA.[16]

2.1 (G-Protein)-Coupled Receptor Binding Sites for Anesthetics

Guanine nucleotide-binding protein (G-protein)-coupled receptors (GPCR), consisting of seven transmembrane α-helices connected by six loops of varying length, are involved in the activation of G-proteins and thereby play a key role in the cellular signal transduction pathways. GPCRs are the primary targets of the inhalation anesthetics, neurotransmitters, hormones, and numerous pharmaceuticals. The prototypical GPCR bovine rhodopsin reversibly binds to halothane through noncovalent interactions, mainly at the ligand (retinal) binding cavity in the transmembrane core of the receptor

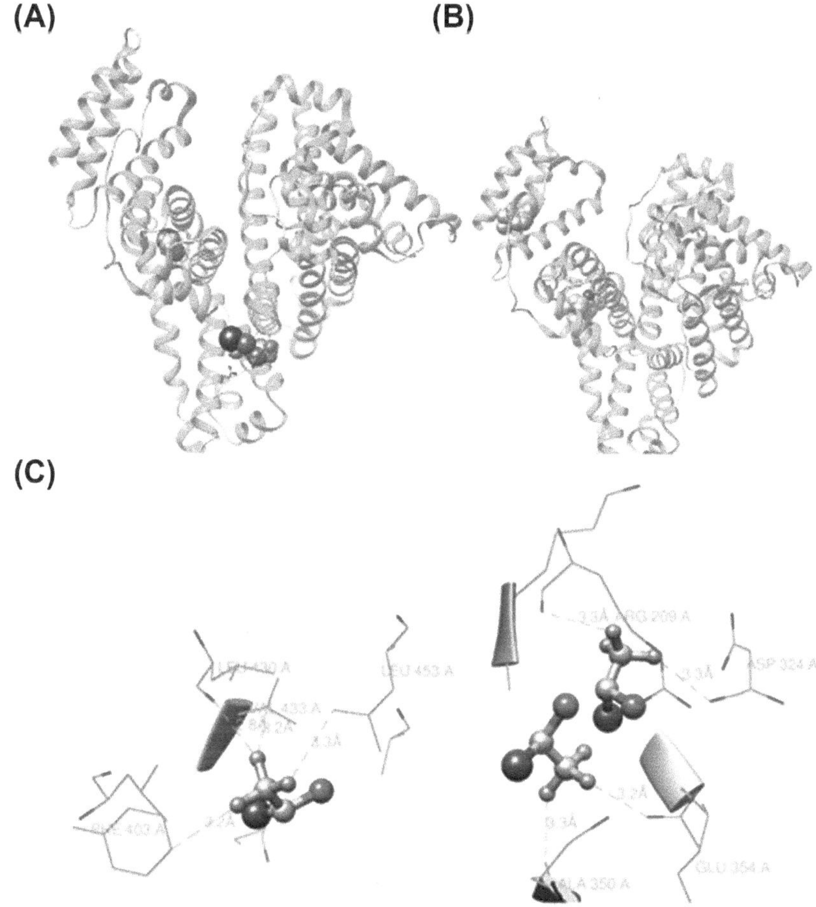

FIGURE 2 X-ray structures for the human serum albumin (HSA) bound to three molecules of halothane (spheres; A) and two molecules of propofol (spheres; B), and expanded view of the halothane–protein interactions at the two binding regions (C); these anesthetic molecules have van der Waals′ interactions with the surrounding nonpolar as well as polar amino acids with minimal change of structural features at the binding regions; the structures were generated using UCSF Chimera software; PDB: 1E7B (halothane) and 1E7A (propofol).

(Figure 3). This hydrophobic ligand binding cavity is closer to Trp265, as shown by tryptophan fluorescence quenching by halothane.[17] It was also observed that halothane competes with retinal in its binding to rhodopsin, and based on this it was suggested that anesthetic action may be due to the disruption of signal transduction through competition with endogenous ligands (such as retinal) in the GPCR core.[17] Due to the similarity of the surface volumes, all the fluorinated inhalation anesthetics that are currently in use may bind at the same transmembrane cavity through noncovalent interactions.

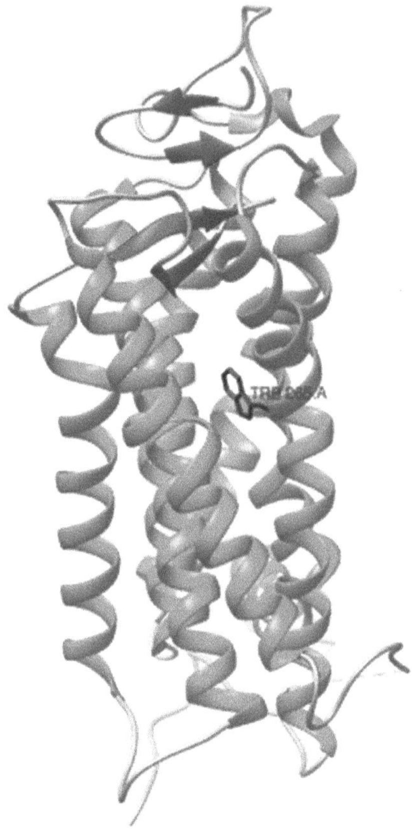

FIGURE 3 X-Ray structure of bovine rhodopsin showing the halothane (ligand) binding cavity in the transmembrane binding core at Trp265; the structure was generated using UCSF Chimera software; PDB ID: 1F88.

Molecular dynamics simulations show that halothane binds to the hydrophobic cavity of the nicotinic receptors close to M2–M3 loop, and this binding modulates the conformational changes of the protein at the vicinity, and these effects are transmitted to the channel gate, thereby altering the protein functions.[18] These molecular dynamics simulations are also in accordance with the NMR (^{19}F and Nuclear Overhauser Effect NMR (NOE)) experiments, which show that the anesthetic compounds are localized mostly at the membrane–water interface and reversibly alter the protein function.[14]

2.2 LGIC Receptors in the Binding of Anesthetics

The LGIC receptors are activated by extracellular ligands that include small molecules such as glycine, serotonin, γ-aminobutyric acid (GABA), and nicotine. They are broadly classified into four superfamilies: (1) the Cys-loop superfamily that binds to cationic ligands such as nicotine and 5-hydroxytryptamine (5-HT), as well as anionic ligands such as GABA; (2) the glutamate receptors that bind to NMDA, kainate

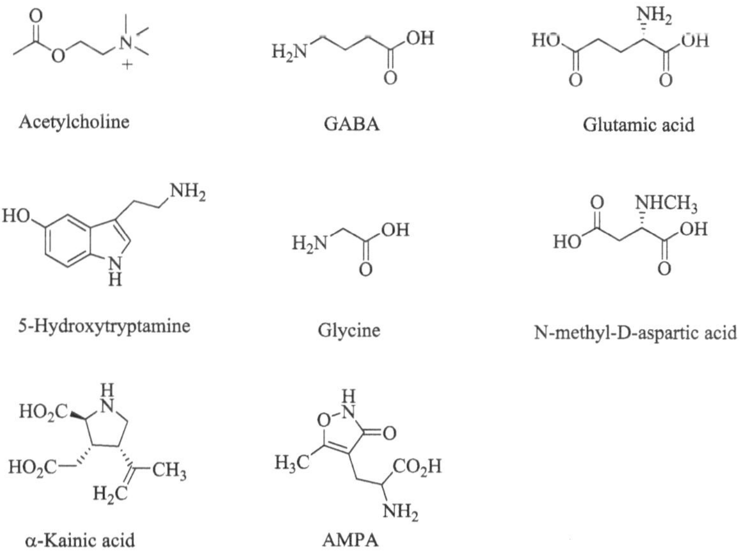

FIGURE 4 Examples of small molecule, endogenous neurotransmitters.

(2-carboxy-4-(1-methylethenyl)-3-pyrrolidineacetic acid), and α-amino-3-hydroxy-5-methyl-4-isoxazolepropionic acid (AMPA); (3) the transient receptor potential channels; and (4) adenosine-triphosphate-gated channels. The Cys-loop superfamily receptors play a key role in the physiological functions and their mutations lead to various diseases, and therefore these receptors provide important targets for therapeutics. nACh receptors, a subclass of the Cys-loop superfamily receptors, are expressed in neuronal and muscle cells and are important for synaptic transmission.[19,20] Small molecule neurotransmitters, acting either as agonists or as antagonists, activate the ion channels by binding at the allosteric sites, which are topologically distinct from the orthosteric binding sites (that is, the active sites where endogenous ligands bind). The endogenous ligands include acetylcholine, GABA, glutamic acid, 5-HT, glycine, NMDA, and kainate (Figure 4).

$GABA_A$ receptors, members of the pentameric ligand-gated ion channels (pLGICs) superfamily, bind to neurotransmitters at the orthosteric binding site to modulate the opening of the ion channel (gate). These receptors are pentameric transmembrane protein aggregates, formed by the assembly of five identical subunits, and are frequently called Cys-loop receptors since their extracellular domains consist of two cysteine residues connected by a disulfide bridge. These receptors contain multiple allosteric binding sites that bind to a variety of pharmaceuticals and general anesthetics. An X-ray crystal structure of bromoform bound to the *Erwinia chrysanthemi* LGIC reveals involvement of three distinct binding sites for anesthetics;

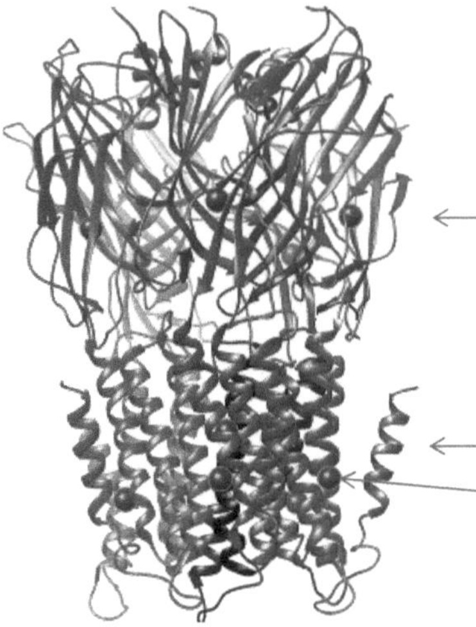

FIGURE 5 pLGIC in complex with bromoform (red spheres); each of the five chains is shown in a different color. Anesthetics bind to pLGIC in the transmembrane and extracellular sites, as well as at the channel pore (UCSF software; PDB: 3ZKR). (For interpretation of the references to color in this figure legend, the reader is referred to the online version of this book.)

one site is at the channel pore, as is common with many other pLGICs. A second binding site is located in the lower part of the transmembrane region and the third is in the extracellular domain (Figure 5).[21] Anesthetics, such as bromoform and chloroform bind in the hydrophobic region of the channel pore.

The anesthetics, through their binding at the allosteric sties on the $GABA_A$ receptors, enhance GABA-mediated chloride ion currents at the receptor. Enflurane requires two specific amino acid residues in $GABA_A$-α2 receptors, namely, Ser270 and Ala291 in the transmembrane domains TM2 and TM3 for the positive modulation of the $GABA_A$ receptors.[22] isoflurane, on the other hand, is insensitive to the presence of these amino acid residues for its positive $GABA_A$ modulating effect; site-specific mutation of these amino acids by other amino acids has little or no effect on the clinically relevant concentrations of isoflurane.[22]

Inhalation anesthetics, in addition to binding to ion channels in the central nervous system, also modulate the immune functions via direct interaction with leukocyte targets, such as integrin lymphocyte function-associated antigen-1 (LFA-1).[23] The X-ray crystal structure of isoflurane with LFA-1 shows hydrophobic interactions of the terminal CF_3 fluorine atoms with side chains of Ile235, Tyr257, Lys287, and Ile259. The CF_2 fluorine atoms have hydrogen bonding type of polar interactions with Tyr307, with a fluorine to aryl oxygen distance of 3.2 Å, and the chlorine atom also has weak hydrogen bonding interactions with Glu301 hydroxy group as can be seen from the X-ray crystal structure (Figure 6). Although racemic isoflurane was used

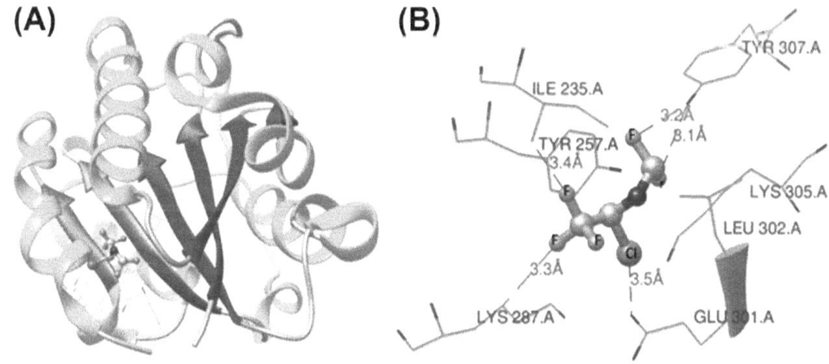

FIGURE 6 X-ray crystal structure of isoflurane bound to integrin LFA-1 (A) and its expanded view at the binding region (B); the structure was generated using UCSF Chimera software; PDB ID: 3F78.

in this crystal structure determination, the S-isomer selectively binds to the protein active site.[23]

3. CHIRAL ANESTHETICS

As described above, much progress has been made towards elucidating the mechanism of anesthetic action of the fluorinated compounds, and it is now generally accepted that the anesthetic effect is due to the reversible binding of the anesthetics to the proteins, such as LGIC and GPCR. Thus, enantiomerically pure compounds are expected to show relatively more selective binding to proteins, and thereby are expected to exert minimal adverse effects. However, currently, fluorinated anesthetics are administered as racemic mixtures due to the technological issues related to the industrial manufacture of the enantiomerically pure compounds. The future trend in this area may be focused on the development of the chiral anesthetics, available through either asymmetric synthesis or chiral resolution of the racemic mixtures.[24,25] The chirally homogeneous anesthetics may prove to be more efficient in some cases. For example, in rodents, S-isoflurane is more potent than the racemic mixture, whereas the enantiomers of desflurane show no difference in their activity, although one of the enantiomers helps in quicker recovery from anesthesia.[26,27]

4. ENVIRONMENTAL IMPACT OF FLUORINATED ANESTHETICS

The fluorinated inhalation anesthetics are greenhouse gases, and there is some concern about the use of these compounds as anesthetic agents. A 1 h use of desflurane, for example, is equivalent to 235–470 miles of driving.[28] However, the use of these anesthetics is justified in view of their medical necessity. Future technologies may be able to recycle the exhaled anesthetic gases and thus reduce the environmental impact.

5. SYNTHESIS AND TOXICITY OF FLUORINATED ANESTHETICS

5.1 Halothane

One of the earliest fluorine containing anesthetics halothane, 1-bromo-1-chloro-2,2,2-trifluoroethane ($CF_3CHBrCl$; trade named as Fluothane), had superior nonflammability characteristics, and fewer side effects as compared with other nonfluorinated anesthetics, and proved immediate success in the hospitals in the mid-1950s.[4]

5.1.1 Synthesis of Halothane

The ICI (Imperial Chemical Industries) production of halothane involves free radical bromination of 1-chloro-2,2,2-trifluoroethane (**2**), which is prepared by the reaction of HF with trichloroethylene (**1**) in the presence of antimony pentachloride or antimony trichloride (Figure 7).[29] The electrophilic addition of HF to trichloroethylene (**1**) in the presence of the Lewis acid catalyst is accompanied by the regioselective chloride to fluoride anion exchange (i.e., Lewis-acid-mediated S_N1 reactions) of the initially formed 1,2,2-trichloro-2-fluoroethane to give 1-chloro-3,3,3-trifluoroethane (**2**).

Ramig and coworkers have synthesized enantiomerically pure (S)-halothane in a multistep reaction sequence starting from 1,1,2-trichloro-3,3,3-trifluoropropene (**3**), which is transformed in three convenient steps into diastereomeric amide mixture **5** that could be separated by high-performance liquid chromatography (HPLC) (Figure 8).[30,31] The purified enantiomer **6** was reacted with conc. H_2SO_4 to give the amide **8**, which, upon hydrolysis in aqueous acid afforded the chirally pure carboxylic acid **10**. The final key step in the synthesis of the (S)-halothane involves the thermal anionic decarboxylation of **10**. The enantioselectivity of this decarboxylation depends on the base used. The strychnine salt of **10** upon thermal decarboxylation gave the (S)-halothane in 97.5% enantiomeric purity, whereas use of triethylamine or other bases under these reaction conditions gave much lower enantioselectivities. Unfortunately, the synthesis of **10** involves HPLC separation of the diastereomeric mixture of **5**, limiting the large-scale synthesis of the chirally pure halothane using this procedure.

5.1.2 Hepatotoxicity of Halothane

Halothane, despite making a breakthrough entry into the safe-acting inhalation anesthetics, is susceptible to induce some adverse effects, including hepatic necrosis. This prompted the development of alternative fluorinated anesthetics, and currently use of halothane is limited to certain countries.

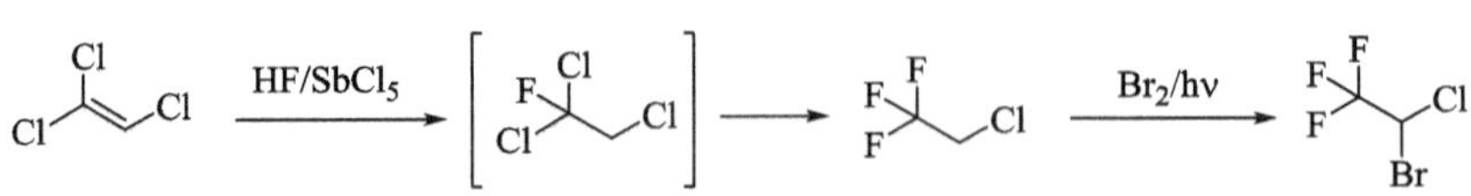

1 **2** Halothane

FIGURE 7 ICI synthesis of halothane.

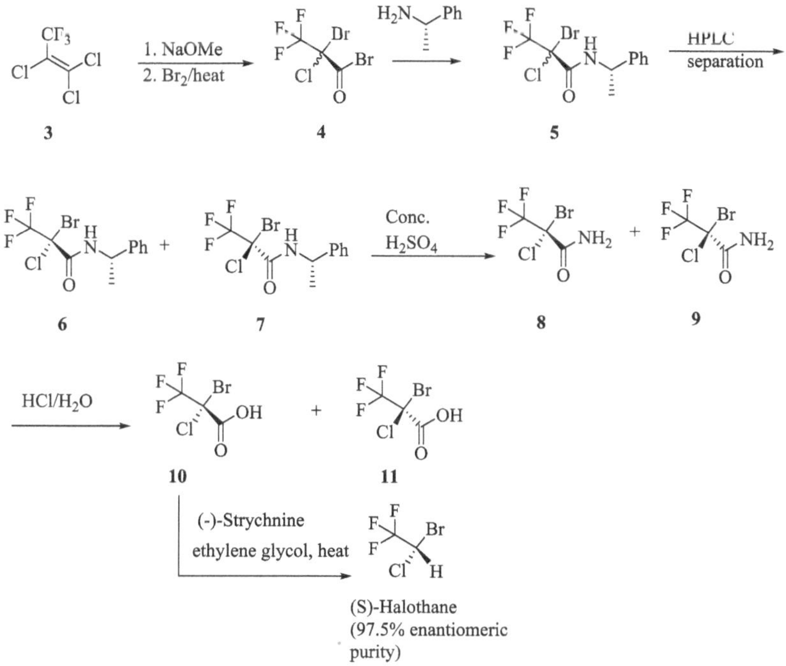

FIGURE 8 Synthesis of chirally pure (S)-halothane.

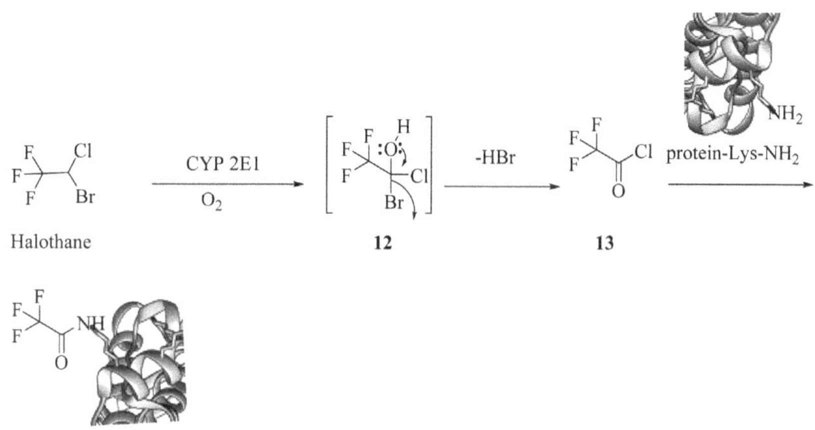

14; N-trifluoroacetyl-Lys-protein

FIGURE 9 Mechanism of oxidative degradation of halothane and protein modification.

The hepatotoxicity of halothane can be rationalized as shown in Figure 9. Halothane undergoes oxidative metabolism by cytochrome P450 (CYP) enzymes[32]; the oxidation of halothane by CYP2E1 and CYP2A6 enzymes gives trifluoroacetyl chloride **13** as the major metabolite, which reacts with N-terminal amino groups of liver proteins, including CYP2E1, to give the N-trifluoroacetylated protein adduct, **14**; the resulting protein modification leads to the formation of neoantigens, which elicit immune response in susceptible patients, eventually causing liver necrosis and fatal consequences in rare cases of patients (Figure 9).[1,33,34] CYP2E1 inhibitor disulfiram diminishes halothane oxidation in vivo, supporting the involvement of CYP2E1 in the oxidation of halothane.[32,35] The protein modifications resulting from the halothane adducts presumably results in hapten formation, leading to the induction of immune response and hepatitis. The clinical features of the halothane-induced hepatitis are consistent with immune-mediated adverse drug reactions, as demonstrated in a murine model.[34] Similar mechanisms are operative for hepatitis associated with enflurane, isoflurane, and desflurane. Enzyme-linked immunosorbent assay (ELISA) also showed that the halothane reactivity to hepatic proteins from rats is significantly greater than that of enflurane and desflurane.[32] The relatively lower toxicity associated with the latter anesthetics may be due to their relatively lower rates of metabolic bioactivation so that the reactive acid chloride **13** is formed at significantly lower levels than in the case of halothane. A vast majority of patients anesthetized with enflurane or desflurane do not sustain liver injury as compared with those anesthetized with halothane.[1]

Under low oxygen concentrations, halothane undergoes reductive metabolism by CYP2A6 and CYP3A4 enzymes in human liver microsomes to an unstable 1-chloro-2,2,2-trifluoroethyl free radical that serves as an initiator for lipid peroxidation, with consequent adverse effects on respiratory and circulatory functions.[36] Further, this free radical is also a source of the volatile metabolites, chlorodifluoroethene and chlorotrifluoroethane that may show other adverse effects.

5.1.3 Chemical Degradation of Halothane

Halothane is somewhat unstable to NaOH in soda lime (used in the sterilization procedures for the drug) and undergoes dehydrofluorination to give 2-bromo-2-chloro-1,1-difluoroethylene (**15**). The latter product, however, reacts rapidly and nonenzymatically with sulfur nucleophiles, such as intracellularly abundant glutathione (GSH) to form the stable adduct **16**, and is depleted from the human body immediately after anesthetic administration to patients (Figure 10).[6] This GSH conjugate **16**, however, is potentially nephrotoxic, as observed in animal models. Similar dehydrohalogenations are also observed for other fluorinated

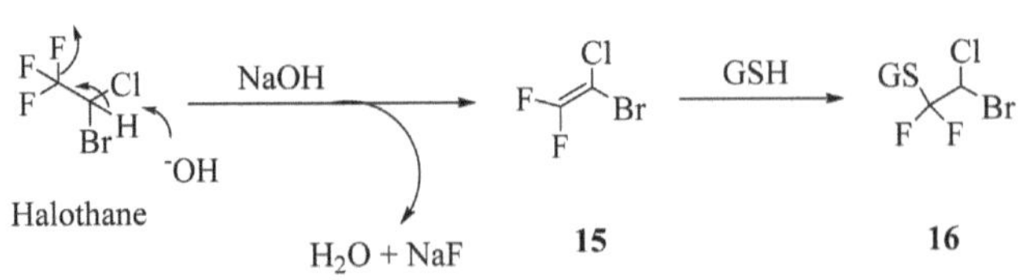

FIGURE 10 Dehydrofluorination of halothane and sequestration of the intermediate by glutathione (GSH = glutathione).

inhalation anesthetics, including sevoflurane, isoflurane, and desflurane. The dehydrohalogenation may be avoided by replacing soda lime by other materials such as molecular sieves to remove the carbon dioxide in the anesthetic procedure.[7]

5.2 Isoflurane

Isoflurane (2-chloro-2-(difluoromethoxy)-1,1,1-trifluoroethane, Forane), a general inhalation anesthetic (marketed by Anaquest since 1981), is a chiral molecule as is the case with other fluorinated inhalation anesthetics. Isoflurane as well as other fluorinated anesthetics are currently administered as racemic mixtures, although it has been demonstrated that there is a significant difference in the anesthetic properties of the enantiomers.[37]

5.2.1 Synthesis of Isoflurane

Isoflurane is a regioisomer of enflurane, an alternative inhalation anesthetic, and is produced on industrial scale in two steps: reaction of trifluoroethanol with in-situ-generated difluorocarbene (F_2C:), formed through the deprotonation of chlorodifluoromethane, followed by spontaneous elimination of the fluoride anion gives 2,2,2-trifluoroethyl difluoromethyl ether **17**, which upon free radical halogenation using aqueous Cl_2 in the presence of UV light, at low temperature, gives isoflurane (Figure 11).[38] The byproduct HCl formed in the last step remains in the aqueous layer and is easy to separate. Isoflurane is currently marketed by three major pharmaceutical companies—Abbott, Eli Lilly, and Pharmacia.[39]

The racemic isoflurane can be resolved by gas chromatographic (GC) separation using a chiral GC column consisting of a cyclodextrin as the stationary phase. In this way (S)-(+)-isoflurane was obtained in over 98% enantiomeric excess.[40,41] Due to the difficulties involved in the scale-up of this procedure, this anesthetic, as in the case of the other fluorinated anesthetics, is currently used in the racemic form.

Both enantiomers of isoflurane were synthesized in chirally pure form starting from the racemic (1-chloro-2,2,2-trifluoroethoxy)difluoroacetic acid (**18**), which is readily synthesized from 2,2,2-trifluoroethanol (Figure 12). The resolution of **18** into the chirally pure compounds **20** and **21** was achieved using (+) dehydroabietylamine (**19**) as the resolving agent. Thermal decarboxylation of **20** and **21** gave the (R)-isoflurane and (S)-isoflurane, respectively, with over

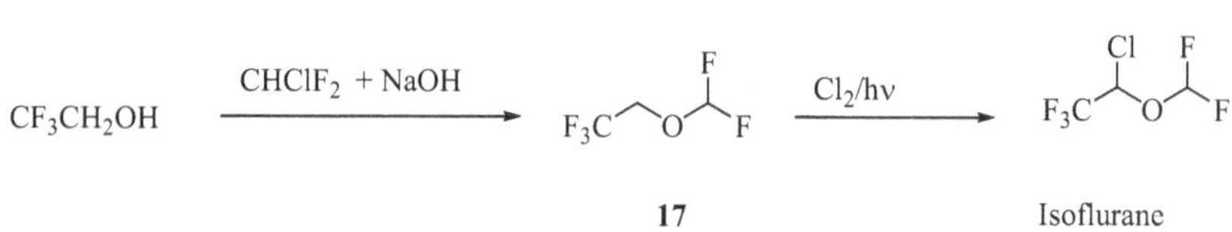

FIGURE 11 Industrial-scale synthesis of isoflurane.

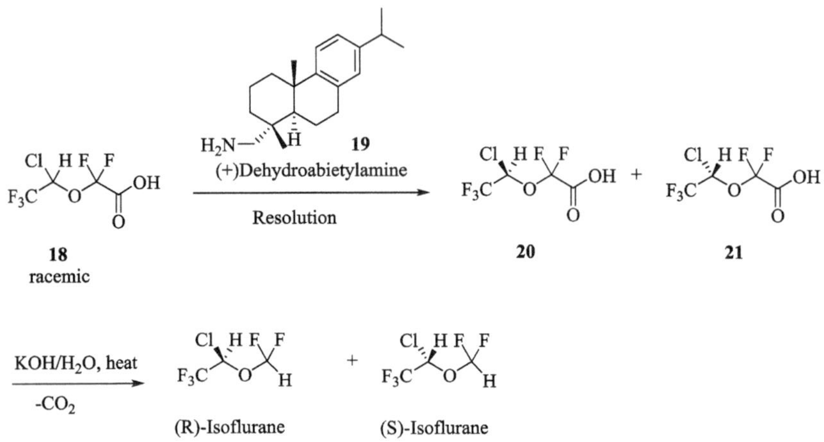

FIGURE 12 Synthesis of chirally pure (R)- and (S)-isoflurane.

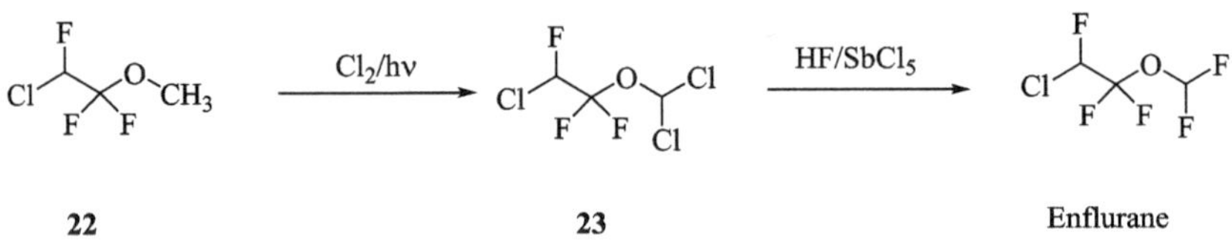

FIGURE 13 Industrial-scale synthesis of enflurane.

99% enantiomeric excess.[42] However, this synthetic route would not be practical for the large-scale synthesis of the chirally pure anesthetic because half of the material is wasted during the resolution step.

5.3 Enflurane

Enflurane (2-chloro-1-(difluoromethoxy)-1,1,2-trifluoroethane, Ethrane), a regioisomer of isoflurane, is synthesized on industrial scale by free radical chlorination of methoxy-1,1,2-trifluoro-2-chloroethane (**22**), followed by halogen (chloride to fluoride) exchange in the presence of $HF/SbCl_5$ (Figure 13).[43]

Enflurane is oxidatively metabolized in the liver microsomes by CYP2E1 enzymes to give the unstable intermediate, **24**, which upon dehydrohalogenation gives the corresponding acid chloride, **25**. This acid chloride reacts rapidly with liver proteins to give the protein adducts (**26**) (Figure 14). Similar mechanisms operate with isoflurane and desflurane.[32] The liver protein adducts **26** generate immune response, which upon subsequent exposure to these adducts leads to toxicity and eventual cellular death at large enough concentrations. ELISA experiments showed that the hepatotoxicity resulting from this acylated protein modification is the greatest for halothane among the fluorinated ether anesthetics.[32]

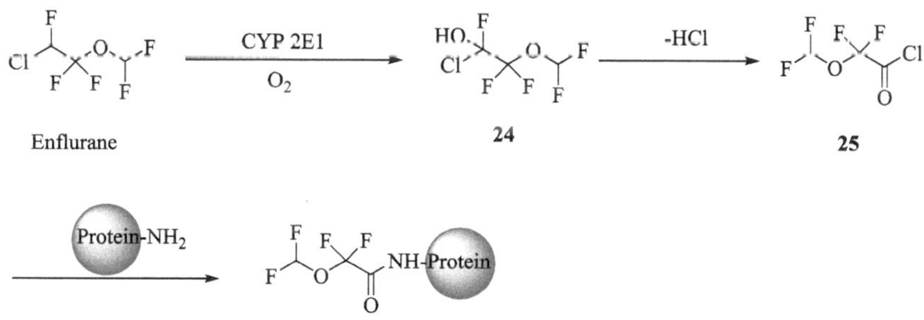

26; Enflurane-protein adduct

FIGURE 14 Oxidative metabolism of enflurane and formation of its protein adduct.

5.4 Sevoflurane

The racemic sevoflurane (1,1,1,3,3,3-hexafluoro-2-(fluoromethoxy)propane) was developed as an inhalation anesthetic by Maruishi in Japan and licensed to Abbott.[25] Currently, two major manufacturers in the United States supply sevoflurane: Abbott markets it under the trade name Ultane, and Baxter Health Care Corporation provides the generic product. The sevoflurane sold by these two pharmaceutical companies is rated as therapeutically equivalent by the Food and Drug Administration.[44] The content of sevoflurane from either of these sources is greater than 99.99%. The minor products include hexafluoroisopropanol, and a dehydrofluorination product **29**. Wake-up times following sedation by sevoflurane after cardiac surgery in intensive care units are comparable and relatively shorter than for the commonly used intravenous anesthetic, propofol.[45]

Sevoflurane is synthesized by the fluoroalkylation of hexafluoroisopropyl alcohol using formaldehyde and HF in the presence of fuming sulfuric acid (Abbott's procedure), or from the halide exchange reaction of the hexafluoroisopropyloxymethyl chloride (**28**) using bromine trifluoride or KF (Baxter) (Figure 15).[44] Compound **28** is synthesized in two steps from hexafluoroisopropanol, involving O-alkylation and free radical chlorination. Several other routes for the preparation of sevoflurane are also reported.[25]

5.5 Desflurane

Desflurane (2-(difluoromethoxy)-1,1,1,2-tetrafluoroethane, Suprane; former trade name: I-653), is a general inhalation anesthetic, and has approximately five times lower potency than isoflurane, although they have similar pharmacodynamic characteristics. Desflurane is marked by Anaquest since 1993. Desflurane, similar to other inhalation anesthetics, shows dose-dependent depression of cardiorespiratory and central nervous systems.[46] It is used as a racemic mixture as an

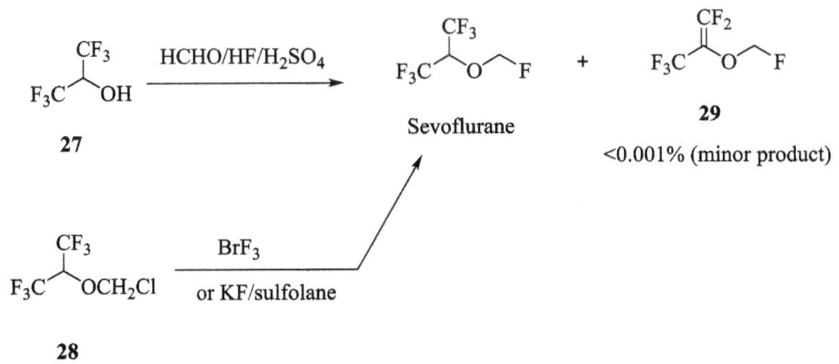

FIGURE 15 Industrial-scale synthesis of sevoflurane and formation of minor dehydrofluorination product.

anesthetic agent, and has the most rapid onset and offset of its anesthetic effects among the other fluorinated inhalation anesthetics. It is chemically relatively stable, but prolonged direct contact with soda lime, used in anesthetic circuits, gives low concentrations of fluoroform (CHF_3).[47] It is not recommended for general anesthesia via mask in children due to the high incidences of side effects such as oxyhemoglobin desaturation and severe laryngospasm.[47]

Among the currently used fluorinated inhalation anesthetics, desflurane has the lowest metabolism (<0.02%) and blood gas partition coefficient (0.42). The relatively lower metabolism of this drug is advantageous because relatively low levels of the toxic metabolites are formed. The recovery time after surgery depends on the blood gas partition coefficient; the lower the blood gas partition coefficient, the lower the recovery time, and typically the recovery of muscle coordination after the use of this anesthetic is about 5 min. This 5-min recovery time is shortest for desflurane, as compared to about 50 min for halothane, 23 min for isoflurane, and 14 min for sevoflurane.[48] These recovery times are correlated with their blood gas partition coefficient values (2.3, 1.4, and 0.7, respectively, for halothane, isoflurane and sevoflurane).[5,24,48–51]

Desflurane is currently manufactured from the commercially available isoflurane using HF and a variety of Lewis acid catalysts, including bromine trifluoride (BrF_3). The synthetic processes continue to be improved, and a vapor-phase process for the nucleophilic substitution of chloride by fluoride using antimony pentafluoride (SbF_5) supported on activated carbon is reported (Figure 16).[52,53]

Chirally pure (S)-(+)-desflurane can be synthesized from the reaction of (R)-(−)-isoflurane with BrF_3, in >96% enantiomeric excess.[26,27] Chiral syntheses of these anesthetics, combined with practical methods of chromatographic resolution, are expected to enable future technologies for the preparation of bulk amounts of anesthetics for clinical trials.

X-ray structures of desflurane bound to a pLGIC from a bacterial (*Gloeobacter violaceus*) homolog shows that it is buried deep inside the cavity and

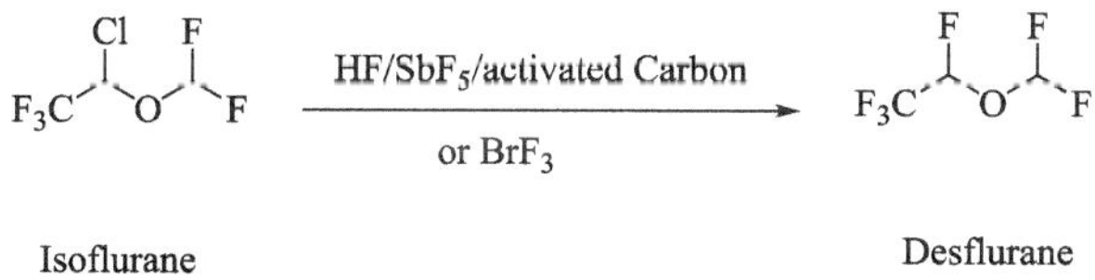

FIGURE 16 Synthesis of desflurane.

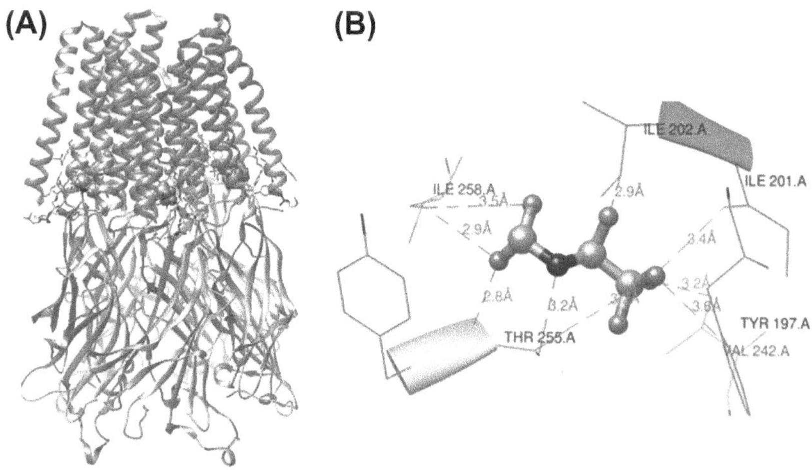

FIGURE 17 X-ray structure of desflurane (space-filling representation) bound to the LGIC from a bacterial (*Gloeobacter violaceus*) homolog (A), and an expanded view of at the binding region (B); desflurane is buried deep inside the cavity and has hydrophobic interactions with Ile201 and Ile202 (M1); Thr255 and Ile258 (M3); and Val242 (M2); the Structure was generated using UCSF Chimera software; PDB ID: 3P4W.

has hydrophobic interactions with Ile201 and Ile202 (M1); Thr255 and Ile258 (M3); and Val242 (M2).[54] The desflurane oxygen is within the hydrogen bond distance of Thr255 hydroxyl group (Figure 17). The cavity size of the active site was determined as 238 Å, which is 2.5 times larger than the surface volume of desflurane (94 Å). A single-site mutation of the F14' residue by alanine in the LGIC transforms desflurane (or chloroform) from an inhibitor to a potentiator. The F14'A mutation opens an intersubunit allosteric site, and the competitive binding of the desflurane at this site over an intrasubunit allosteric site results in its potentiating effect as opposed to its inhibitory effect in the wild-type LGIC.[55]

5.5.1 Toxicity of Desflurane

Desflurane exerts hepatotoxicity, although at somewhat lower level than that for enflurane or isoflurane. The mechanism of hepatotoxicity of desflurane is similar to that of enflurane or isoflurane; CYP2E1 oxidation of desflurane results in the

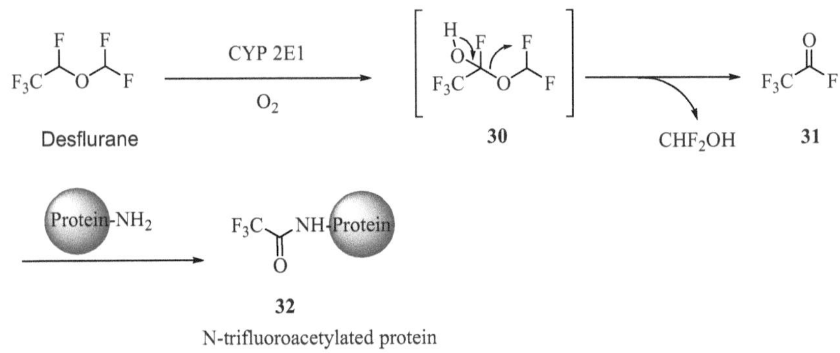

FIGURE 18 Oxidative degradation of desflurane and formation of its protein adduct.

formation of an unstable intermediate **30**, which upon elimination of difluoroethanol gives trifluoroacetyl fluoride **31** that rapidly reacts with protein side chain amino groups to give the protein adducts (**32**) (Figure 18). It is likely that the CYP oxidation of desflurane is relatively less favored as compared to that of enflurane as the free radical intermediate formed from desflurane is relatively more destabilized due to the strong electron-withdrawing CF_3 group, reflecting its relatively lower metabolic oxidation. In vitro studies in primary rat hippocampal neurons show that desflurane accelerates the neuronal cytotoxicity of amyloid-β peptide.[56] However, it was reported that a single or repeated exposure to desflurane or isoflurane does not impair acquisition learning and retention memory in young adult mice, while other studies showed that these anesthetics impair spatial memory in aged rodents.[57] In human patients, the Aβ levels in the cerebrospinal fluid (CSF) of the patients are decreased from 0.39 to 0.35 ng/mL, 24 h after surgery under desflurane anesthesia, while the Aβ levels increased from 10.9 to 12.41 ng/mL, when the surgery was performed using isoflurane as the inhalation anesthetic.[58] Thus isoflurane and desflurane have distinct effects on the CSF Aβ levels.

6. SUMMARY AND OUTLOOK

Organofluorine inhalation anesthetics are relatively safer alternatives compared to other intravenous anesthetics, from the point of view of anesthetic administration and due to their nonflammability characteristics. Inhalation anesthetics, including isoflurane, desflurane, enflurane, and sevoflurane, are widely used in general anesthesia. These anesthetics are administered as racemic mixtures due to the nonavailability of sufficient quantities of the chiral anesthetics, which are either prepared through multistep synthesis or purified through resolution of the racemic mixtures. Possible hepatotoxicity, arising from the oxidative metabolism of these compounds, is a primary concern in the use of these inhalation anesthetics, especially in case of halothane. Future studies may lead to the large-scale synthesis of chirally pure inhalation anesthetics and to the development of inhalation anesthetics that have improved metabolic stabilities.

REFERENCES

1. Kenna, J. G. Mechanism, Pathology, and Clinical Presentation of Hepatotoxicity of Anesthetic Agents. In *Drug-Induced Liver Disease,* 3rd ed.; Kaplowitz, N., DeLeve, L. D., Eds. Elsevier Ltd.: London, UK, 2013; pp 403–422.
2. Endo, S.; Yano, A.; Fukami, T.; Nakajima, M.; Yokoi, T. Involvement of miRNAs in the Early Phase of Halothane-induced Liver Injury. *Toxicology* **2014,** *319,* 75–84.
3. Schifilliti, D.; Grasso, G.; Conti, A.; Fodale, V. Anaesthetic-Related Neuroprotection: Intravenous or Inhalational Agents? *CNS Drugs* **2010,** *24,* 893–907.
4. Kalliardou, E.; Tsiotou Adelais, G.; Velegrakis, D.; Avgerinopoulou, A.; Poulakou, E.; Papadimitriou, L. Historical Aspects of Inhalation Anesthesia in Children: Ether and Chloroform. *Paediatr. Anaesth.* **2006,** *16,* 3–10.
5. Halpern, D. F. Recent Developments in Fluorine Substituted Volatile Anesthetics. *Stud. Org. Chem. (Amsterdam)* **1993,** *48,* 101–133.
6. Anders, M. W. Formation and Toxicity of Anesthetic Degradation Products. *Annu. Rev. Pharmacol. Toxicol.* **2005,** *45,* 147–176.
7. Fee, J. P.; Murray, J. M.; Luney, S. R. Molecular Sieves: an Alternative Method of Carbon Dioxide Removal Which Does Not Generate Compound a during Simulated Low-flow Sevoflurane Anaesthesia. *Anaesthesia* **1995,** *50,* 841–845.
8. Zuo, Z. Are Volatile Anesthetics Neuroprotective or Neurotoxic? *Med. Gas Res.* **2012,** *2,* 10.
9. Wei, H.; Xie, Z. Anesthesia, Calcium Homeostasis and Alzheimer's Disease. *Curr. Alzheimer Res.* **2009,** *6,* 30–35.
10. Polak, P. E.; Dull, R. O.; Kalinin, S.; Sharp, A. J.; Ripper, R.; Weinberg, G.; Schwartz, D. E.; Rubinstein, I.; Feinstein, D. L. Sevoflurane Reduces Clinical Disease in a Mouse Model of Multiple Sclerosis. *J. Neuroinflammation* **2012,** *9,* 272.
11. Lin, J.; Moore, D.; Hockey, B.; Di Lernia, R.; Gorelik, A.; Liew, D.; Nicoll, A. Drug-Induced Hepatotoxicity: Incidence of Abnormal Liver Function Tests Consistent with Volatile Anaesthetic Hepatitis in Trauma Patients. *Liver Int.* **2014,** *34,* 576–582.
12. Slater, S. J.; Cox, K. J. A.; Lombardi, J. V.; Ho, C.; Kelly, M. B.; Rubin, E.; Stubbs, C. D. Inhibition of Protein Kinase C by Alcohols and Anesthetics. *Nature* **1993,** *364,* 82–84.
13. Barber, A. F.; Liang, Q.; Covarrubias, M. Novel Activation of Voltage-gated K^+ Channels by Sevoflurane. *J. Biol. Chem.* **2012,** *287,* 40425–40432.
14. Zhang, T.; Reddy, K. S.; Johansson, J. S. Recent Advances in Understanding Fundamental Mechanisms of Volatile General Anesthetic Action. *Curr. Chem. Biol.* **2007,** *1,* 296–302.
15. Evers, A. S.; Dubois, B. W.; Burris, K. E. Saturable Binding of Volatile Anesthetics to Proteins Studied by ^{19}F-NMR Spectroscopy and Photoaffinity Labeling. *Progr. Anesth. Mech.* **1995,** *3,* 151–157.
16. Bhattacharya, A. A.; Curry, S.; Franks, N. P. Binding of the General Anesthetics Propofol and Halothane to Human Serum Albumin: High Resolution Crystal Structures. *J. Biol. Chem.* **2000,** *275,* 38731–38738.
17. Ishizawa, Y.; Pidikiti, R.; Liebman, P. A.; Eckenhoff, R. G. G Protein-coupled Receptors as Direct Targets of Inhaled Anesthetics. *Mol. Pharmacol.* **2002,** *61,* 945–952.
18. Vemparala, S.; Domene, C.; Klein, M. L. Computational Studies on the Interactions of Inhalational Anesthetics with Proteins. *Acc. Chem. Res.* **2010,** *43,* 103–110.
19. Garcao, P.; Oliveira, C. R.; Cunha, R. A.; Agostinho, P. Subsynaptic Localization of Nicotinic Acetylcholine Receptor Subunits: A Comparative Study in the Mouse and Rat Striatum. *Neurosci. Lett.* **2014,** *566,* 106–110.
20. Zhang, J.; Xue, F.; Liu, Y.; Yang, H.; Wang, X. The Structural Mechanism of the Cys-loop Receptor Desensitization. *Mol. Neurobiol.* **2013,** *48,* 97–108.

21. Spurny, R.; Billen, B.; Howard, R. J.; Brams, M.; Debaveye, S.; Price, K. L.; Weston, D. A.; Strelkov, S. V.; Tytgat, J.; Bertrand, S.; Bertrand, D.; Lummis, S. C. R.; Ulens, C. Multisite Binding of a General Anesthetic to the Prokaryotic Pentameric *Erwinia chrysanthemi* Ligand-gated Ion Channel (ELIC). *J. Biol. Chem.* **2013,** *288,* 8355–8364.
22. Krasowski, M. D.; Koltchine, V. V.; Rick, C. E.; Ye, Q.; Finn, S. E.; Harrison, N. L. Propofol and Other Intravenous Anesthetics Have Sites of Action on the Gamma-aminobutyric Acid Type a Receptor Distinct from that for Isoflurane. *Mol. Pharmacol.* **1998,** *53,* 530–538.
23. Zhang, H.; Astrof, N. S.; Liu, J.-H.; Wang, J.-h.; Shimaoka, M. Crystal structure of Isoflurane Bound to Integrin LFA-1 Supports a Unified Mechanism of Volatile Anesthetic Action in the Immune and Central Nervous Systems. *FASEB J.* **2009,** *23,* 2735–2740.
24. Halpern, D. Fluorinated Anesthetics Show How Advances in Technology Control Pharmaceutical Development. *J. Fluorine. Chem.* **2002,** *118,* 47–53.
25. Ramig, K.; Halpern, D. F. Asymmetric Syntheses and Resolutions of Fluorinated Volatile Anesthetics. *Enantiocontrolled Synth. Fluoro-Org. Compd.* **1999,** 451–468.
26. Rozov, L. A.; Huang, C.; Halpern, D. F.; Vernice, G. G. *Preparation of Purified Optical Isomers of Desflurane*; US Patent, US5283372A. Anaquest, Inc.: USA, 1994.
27. Rozov, L. A.; Huang, C. G.; Halpern, D. F.; Vernice, G. G.; Ramig, K. Enantioselective Synthesis of the Volatile Anesthetic Desflurane. *Tetrahedron: Asymmetry* **1997,** *8,* 3023–3025.
28. Yasny, J. S.; White, J. Environmental Implications of Anesthetic Gases. *Anesth. Prog.* **2012,** *59,* 154–158.
29. Takei, R.; Ito, M. *Preparation of 1-bromo-1-chloro-2,2,2-trifluoroethane from 1,2-dibromo-1,1,2-trichloroethane;* JP patent: 1989-320671; 03181430, (Asahi Glass Co., Ltd., Japan), 1991.
30. Ramig, K. Chiral Fluorinated Anesthetics; In Ramachandran, P. V., Ed.; *ACS Symp. Ser.*, Vol. 746; . American Chemical Society, 2000; pp 282–292.
31. Ramig, K.; Halpern, D. F. Asymmetric Syntheses and Resolutions of Fluorinated Volatile Anesthetics. In *Enantiocontrolled Synthesis of Fluoro-organic Compounds;* Soloshonok, V. A., Ed.; John Wiley & Sons Ltd.: Chichester, UK, 1999; pp 451–468.
32. Njoku, D.; Laster, M. J.; Gong, D. H.; Eger, E. I., II; Reed, G. F.; Martin, J. L. Biotransformation of Halothane, Enflurane, Isoflurane, and Desflurane to Trifluoroacetylated Liver Proteins: Association between Protein Acylation and Hepatic Injury. *Anesth. Analg. (Baltimore)* **1997,** *84,* 173–178.
33. Kharasch, E. D.; Hankins, D. C.; Fenstamaker, K.; Cox, K. Human Halothane Metabolism, Lipid Peroxidation, and Cytochromes P(450)2A6 and P(450)3A4. *Eur. J. Clin. Pharmacol.* **2000,** *55,* 853–859.
34. You, Q.; Cheng, L.; Ju, C. Generation of T Cell Responses Targeting the Reactive Metabolite of Halothane in Mice. *Toxicol. Lett.* **2010,** *194,* 79–85.
35. Spracklin, D. K.; Emery, M. E.; Thummel, K. E.; Kharasch, E. D. Concordance between Trifluoroacetic Acid and Hepatic Protein Trifluoroacetylation after Disulfiram Inhibition of Halothane Metabolism in Rats. *Acta Anaesthesiol. Scand.* **2003,** *47,* 765–770.
36. Minoda, Y.; Kharasch, E. D. Halothane-dependent Lipid Peroxidation in Human Liver Microsomes Is Catalyzed by Cytochrome P4502A6 (CYP2A6). *Anesthesiology* **2001,** *95,* 509–514.
37. Moody, E. J.; Harris, B. D.; Skolnick, P. The Potential for Safer Anesthesia Using Stereoselective Anesthetics. *Trends Pharmacol. Sci.* **1994,** *15,* 387–391.
38. O'Donnell, W. J.; Mazzell, P., Jr; Sprague, L. G., Jr; Elliott, A. J. Preparation of Isoflurane by the Photochemical Chlorination of 2,2,2-trifluoroethyl Difluoromethyl Ether and Chlorine in Aqueous Solutions at Low Temperatures and/or at Higher Conversions. U.S. Patent 1998–US21997 9920588, (Halocarbon Products Corporation, USA), 1999.

39. Wilhelm, W.; Hartung, A.; Larsen, R. Three Isoflurane Preparations from Various Manufacturers. A Comparison of Isoflurane Concentrations in Fresh Gas after Vaporization with a Drager Vaporizer Type 19.3. *Anaesthesist* **1995,** *44,* 284–287.
40. Biressi, G.; Rajendran, A.; Mazzotti, M.; Morbidelli, M. The GC-SMB Separation of the Enantiomers of Isoflurane. *Sep. Sci. Technol.* **2002,** *37,* 2529–2543.
41. Juza, M.; Braun, E.; Schurig, V. Preparative Enantiomer Separation of the Inhalation Anesthetics Enflurane, Isoflurane and Desflurane by Gas Chromatography on a Derivatized Γ-cyclodextrin Stationary Phase. *J. Chromatogr. A* **1997,** *769,* 119–127.
42. Huang, C. G.; Rozov, L. A.; Halpern, D. F.; Vernice, G. G. Preparation of the Isoflurane Enantiomers. *J. Org. Chem.* **1993,** *58,* 7382–7387.
43. Terrell, R. C. *1,1,2-Trifluoro-2-chloroethyl Difluoromethyl Ether as an Anesthetic Agents,* U.S. Patent 1966-583938, 3469011, (Air Reduction Co., Inc.), 1969.
44. Baker, M. T. Sevoflurane: Are There Differences in Products? *Anesth. Analg.* **2007,** *104,* 1447–1451 (Hagerstown, MD, United States).
45. Hellstroem, J.; Oewall, A.; Sackey, P. V. Wake-up Times Following Sedation with Sevoflurane versus Propofol after Cardiac Surgery. *Scand. Cardiovasc. J.* **2012,** *46,* 262–268.
46. Patel, S. S.; Goa, K. L. Desflurane: A Review of its Pharmacodynamic and Pharmacokinetic Properties and its Efficacy in General Anesthesia. *Drugs* **1995,** *50,* 742–767.
47. www.drugs.com.
48. Ebert, T. J.; Robinson, B. J.; Uhrich, T. D.; Mackenthun, A.; Pichotta, P. J. Recovery from Sevoflurane Anesthesia: a Comparison to Isoflurane and Propofol Anesthesia. *Anesthesiology* **1998,** *89,* 1524–1531.
49. Eger, E. I., II Partition Coefficients of I-653 in Human Blood, Saline, and Olive Oil. *Anesth. Analg.* **1987,** *66,* 971–973 (Baltimore, MD, United States).
50. Soares, J. H. N.; Brosnan, R. J.; Fukushima, F. B.; Hodges, J.; Liu, H. Solubility of Haloether Anesthetics in Human and Animal Blood. *Anesthesiology* **2012,** *117,* 48–55.
51. Yasuda, N.; Targ, A. G.; Eger, E. I. II Solubility of I-653, Sevoflurane, Isoflurane, and Halothane in Human Tissues. *Anesth. Analg.* **1989,** *69,* 370–373 (Baltimore, MD, United States).
52. Sivaramakrishnan, H.; Upare, A. A.; Satagopan, D.; Chambers, O. R. The Preparation of Desflurane by the Vapor-phase Fluorination of Isoflurane. *Org. Process Res. Dev.* **2011,** *15,* 585–592.
53. Rozov, L. A.; Lessor, R. A. Catalytic Fluorination Process for the Manufacture of Desflurane from Isoflurane and Hydrogen Fluoride, U.S. Patent 2003–645884 6800786, (Baxter International, Inc., USA), 2004.
54. Nury, H.; Van Renterghem, C.; Weng, Y.; Tran, A.; Baaden, M.; Dufresne, V.; Changeux, J.-P.; Sonner, J. M.; Delarue, M.; Corringer, P.-J. X-ray Structures of General Anesthetics Bound to a Pentameric Ligand-gated Ion Channel. *Nature* **2011,** *469,* 428–431.
55. Broemstrup, T.; Howard, R. J.; Trudell, J. R.; Harris, R. A.; Lindahl, E. Inhibition versus Potentiation of Ligand-gated Ion Channels Can Be Altered by a Single Mutation that Moves Ligands between Intra- and Intersubunit Sites. *Structure* **2013,** *21,* 1307–1316 (Oxford, U. K.).
56. Yu, Y.; Zhang, Y. Desflurane Accelerates Neuronal Cytotoxicity of Aβ by Downregulating MiR-214. *Neurosci. Lett.* **2013,** *554,* 28–33.
57. Kilicaslan, A.; Belviranli, M.; Okudan, N.; Nurullahoglu Atalik, E. Single and Repeated Sevoflurane or Desflurane Exposure Does Not Impair Spatial Memory Performance of Young Adult Mice. *Fundam. Clin. Pharmacol.* **2013,** *27,* 641–649.
58. Zhang, B.; Tian, M.; Zheng, H.; Zhen, Y.; Yue, Y.; Li, T.; Li, S.; Marcantonio, E. R.; Xie, Z. Effects of Anesthetic Isoflurane and Desflurane on Human Cerebrospinal Fluid Aβ and τ Level. *Anesthesiology* **2013,** *119,* 52–60.

Chapter 7

Organofluorine Compounds as Positron Emission Tomography Tracers

Chapter Outline

Organofluorine Compounds in Biology and Medicine. http://dx.doi.org/10.1016/B978-0-444-53748-5.00007-1

1. INTRODUCTION

Positron emission tomography (PET) is one of the minimally invasive medical diagnostic imaging techniques, which involves monitoring the spatial and temporal distribution of the positron-emitting isotopes in vivo, as a real-time variable. Various positron-emitting isotopes, including ^{18}F, ^{11}C, ^{13}N, and ^{15}O, have been used as medical diagnostics in health and disease. The PET is widely used in medical practice, especially as diagnostic tool for monitoring the glucose metabolism and accumulation of amyloid plaques in the brains of patients with Alzheimer disease (AD), epilepsy, and Parkinson disease (PD), and for monitoring the glucose metabolism, blood flow, and drug action in tumor and heart tissues. PET imaging is especially useful in the early diagnosis of dementia, and for the evaluation of therapies.[1] PET imaging, in addition to its wide use in personalized medicine in neurology, oncology, and cardiology, is also useful in drug discovery, for example, in monitoring the drug action in patients and in evaluating receptor–drug interactions. As most of the hospitals in the United States are currently equipped or have access to PET scanners, there is a growing trend in the discovery of convenient and safe PET imaging agents.

The half-life of ^{18}F is the highest of all other positron-emitting isotopes and therefore ^{18}F-labeled compounds are better suited for PET imaging studies as compared to others; the half-lives of ^{18}F, ^{11}C, ^{13}N, and ^{15}O radioisotopes are 110, 20, 10, and 6 min, respectively.[2] Further, in some cases the fluorinated compounds have relatively higher tumor uptake than the corresponding nonfluorinated compounds. For example, ^{18}F-fluoroacetate has about five times higher prostate tumor uptake than ^{11}C-fluoroacetate in mice as revealed by a whole body PET study.[3] Further, ^{18}F isotope has low positron energy (0.64 MeV for ^{18}F as compared to 0.96 MeV for ^{11}C isotope), thereby resulting in a short range of positron emission in tissues, and high-resolution PET images can be obtained at relatively low dose rates.[4] Due to this low positron energy and small positron range (~2 mm) the radiation toxicity of ^{18}F-labeled compounds is negligible to patients. Thus, PET imaging, for example, using ^{18}F-fluoroacetate has some unique advantages over using ^{11}C-acetate-based PET imaging, even though the latter method has relatively higher sensitivity.[3] ^{18}F-labeled fluoride ([^{18}F]F^-) is usually produced in cyclotrons through proton bombardment of $H_2{}^{18}O$, involving ^{18}O(p,n)^{18}F nuclear reaction.[5] The only limiting requirement for the synthesis of the ^{18}F-labeled compounds is the availability of cyclotron or linear particle accelerator for bombarding pure or enriched $H_2{}^{18}O$ with high-energy protons (~18 MeV). The ^{18}F-radiolabeled F_2 gas is produced by deuteron irradiation of neon, ^{20}Ne(d,α)^{18}F, with the addition of about 0.2% [^{19}F]F_2, or by proton irradiation of [^{18}O]O_2 (^{18}O(p,n)^{18}F). The theoretical maximum radiochemical yield (RCY) for [^{18}F]F_2 in the latter reaction is 50% because the remaining activity is lost in the form of [^{18}F]F^-. ^{18}F-labeled F_2 gas is used as the electrophilic fluorinating agent by itself, or is transformed into relatively milder electrophilic fluorinating agents such as acetyl hypofluorite ($CH_3CO_2{}^{18}F$), [^{18}F]XeF_2, [^{18}F]N-fluoropyridinium triflate, [^{18}F]N-fluorobenzenesulfonimide ([^{18}F]NFSI), and [^{18}F]Selectfluor (Figure 1).[6] The ^{18}F-labeled molecular fluorine is being used in the preparation of (S)-6-[^{18}F]fluoro-3,4-dihydroxyphenylalanine (6-[^{18}F]-L-DOPA, which is used in the PET imaging of parkinsonian dopaminergic neurons.[7,8]

The relatively long half-life of the ^{18}F isotope makes it possible to monitor the real-time metabolic processes and disease progression using suitable ^{18}F-labeled pharmaceuticals. The ^{18}F-PET also complements other diagnostic techniques, including X-ray scans, nuclear magnetic resonance imaging, and ^{11}C-PET.[9] Morris and coworkers have combined the ^{11}C- and ^{18}F-PET imaging techniques with magnetic resonance imaging (MRI) for correlating the rate of progression of the AD in individuals with autosomal dominant AD, towards the goal of developing a predictive tool for the onset of the symptomatic phase of the disease.[9] In these studies, [^{11}C]Pittsburgh compound B (^{11}C-PiB) was used for monitoring the amyloid-β (Aβ) deposition, and 2-[^{18}F]fluoro-2-deoxy-D-glucose (^{18}F-FDG) was used for monitoring the altered glucose metabolism.

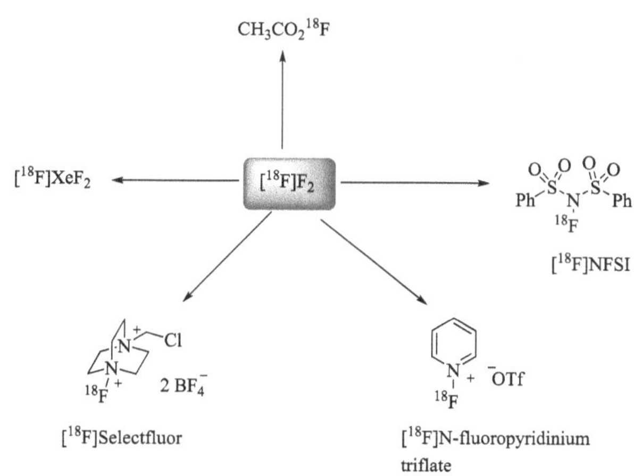

FIGURE 1 Structures of some ^{18}F-labeled electrophilic fluorinating reagents derived from the ^{18}F-labeled elemental fluorine.

MRI was used simultaneously with these two PET imaging techniques to reveal the structural atrophy of the affected brain regions. The ^{11}C-PiB imaging shows accumulation of ^{11}C-PiB in nearly every cortical region 15 years prior to the onset of the disease symptoms, while cortical glucose metabolism and cortical thinning appeared 10 and 5 years, respectively (as revealed by the ^{18}F-FDG and MRI techniques, respectively), prior to the onset of the AD symptoms. The decreased glucose metabolism is detected primarily in the precuneus, posterior cingulate, and lateral parietal lobes, while the cortical thinning on MRI was detected initially in the precuneus and posterior cingulate as well as portions of the occipital lobe and anterior temporal lobe. Most gray matter structures with Aβ plaques eventually develop hypometabolism followed by atrophy (Figure 2).[9]

Usually the radiolabeled compounds are synthesized and used for PET monitoring on the same day of their synthesis. Due to the short life time of the ^{18}F isotope, fluorination should be ideally the late stage synthetic step for the preparation of the ^{18}F-radiolabeled compounds. Numerous synthetic methodologies have been developed towards this goal, including nucleophilic aromatic substitution,[10] electrophilic fluorination of aromatics,[10] aliphatic nucleophilic substitution,[11] and click chemistry.[12] Despite these advances in the synthetic methodologies, 2-[^{18}F]fluoro-2-deoxy-D-glucose (^{18}F-FDG) had been the only ^{18}F tracer (used for monitoring the glucose hyper- or hypometabolism in tissues) approved by Food and Drug Administration (FDA) until 2012, and since then three other ^{18}F-labeled organofluorine compounds were approved for PET monitoring the Aβ plaques in the brains of AD cases: florbetapir-^{18}F ([^{18}F]AV-45; Amyvid Pharmaceuticals), florbetaben-^{18}F (Neuraceq; BAY 94-9172; Bayer

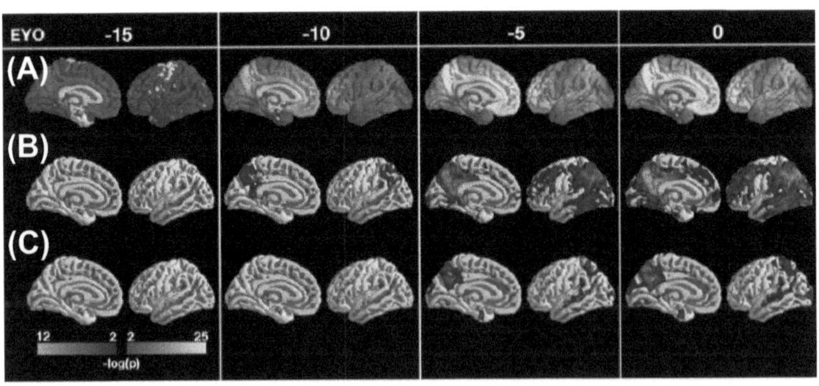

FIGURE 2 Statistical significance (*P* value) maps on medial and lateral left cortical gray surface showing differences between carriers (i.e., the individuals with autosomal dominant AD) and noncarriers in ^{11}C–PiB (A) and FDG (B), and cortical thickness (C) at −15, −10, −5, and 0 years before predicted symptom onset. Regions with significant ($P<0.01$ after correction for multiple comparisons) increases are shown in red/yellow and those with significant deceases in blue/cyan. *Adapted with permission from Reference 9.*

Schering Pharma AG), and flutemetamol-^{18}F (Vizamyl; GE Healthcare) (vide infra).[13] At least seven peptide-based ^{18}F radiopharmaceuticals are currently in clinical trials for diagnostic applications, three of them being the glycosylated peptides.[14] This chapter highlights significant aspects of the ^{18}F-PET tracers that are being targeted to monitor various pathologies.

2. ^{18}F-LABELED PHARMACEUTICALS

2.1 Antipsychotic Drugs

^{18}F-labeled version of the antipsyhotic drug haloperidol ([^{18}F]haloperidol) has been used to monitor the drug distribution in brain tissues; this compound has high affinity to basal ganglia, thalamus, and cerebellum of schizophrenic patients as compared to normal control subjects.[15] A PET imaging study showed that [^{18}F]BMY 14802, a Bristol-Myers Squibb antipsychotic drug, has relatively shorter residence time in the brain tissues, as compared to Haloperidol (Figure 3).[16] Thus subtle structural variations of the therapeutically useful drugs give rise to improved drug candidates and PET imaging agents.

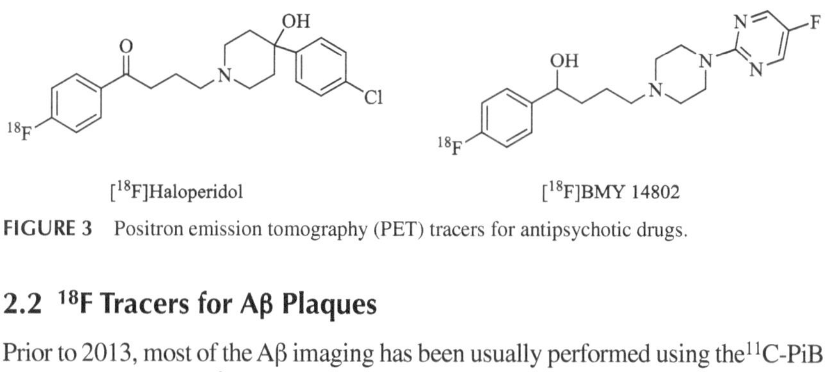

[^{18}F]Haloperidol [^{18}F]BMY 14802

FIGURE 3 Positron emission tomography (PET) tracers for antipsychotic drugs.

2.2 ^{18}F Tracers for Aβ Plaques

Prior to 2013, most of the Aβ imaging has been usually performed using the ^{11}C-PiB due to the lack of the ^{18}F-labeled PET tracers of comparable binding efficiency to Aβ plaques. However, due to the extremely short half-life of the ^{11}C isotope (20 min), on-site synthesis of this compound requires the availability of an in-house cyclotron. Therefore, the use of ^{11}C-PiB for PET imaging is limited to a few clinical facilities that are equipped with a cyclotron, and that have technical expertise for the synthesis and handling of this compound. It is expected that more and more hospitals will be using the ^{18}F-based PET tracers for the clinical diagnosis of Aβ plaques in AD, before the onset of the symptoms. In 2013 and 2014, the FDA has approved ^{18}F tracers florbetapir-^{18}F (Amyvid), flutemetamol-^{18}F (Vizamyl), and florbetaben-^{18}F (Neuraceq) that complement ^{11}C-PiB in the PET imaging of Aβ plaques (Figure 4).

People with the apolipoprotein E ε4 (ApoE4) genetic variant have triple the risk factor for developing mild cognitive impairment (MCI) leading to AD. So, early monitoring of ApoE4 patients for Aβ using the convenient ^{18}F-PET scanning would facilitate diagnosing the onset of AD before the actual symptoms of dementia

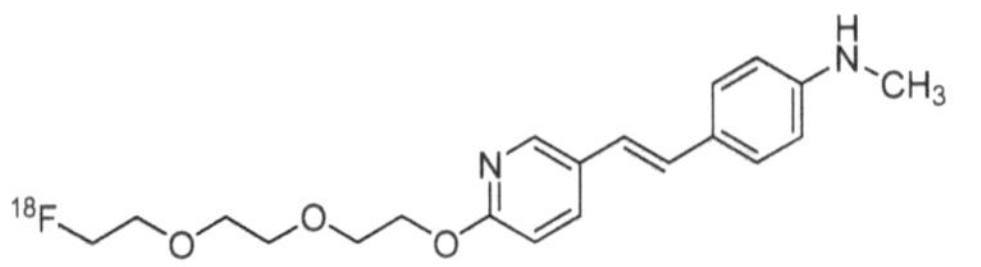

Florbetapir-^{18}F; (^{18}F-AV45)

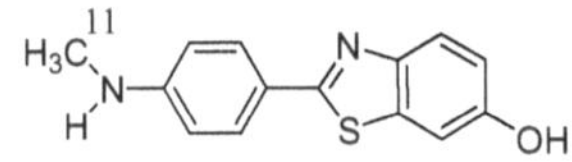

PiB-^{11}C

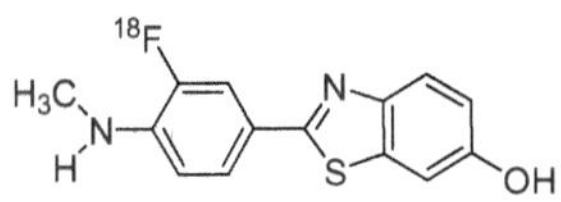

Flutemetamol-^{18}F (GE067)

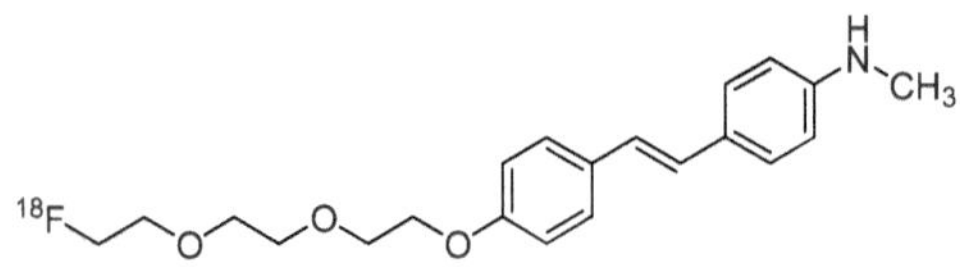

Florbetaben-^{18}F (^{18}FAV-1)

FIGURE 4 Structures of the FDA-approved ^{18}F tracers for Aβ plaques.

appear. PET scan using these radiolabeled drugs would also be expected to be a valuable tool for monitoring the effectiveness of many experimental drugs that are currently under development for AD, in animal models as well as in clinical trials.

Florbetapir-^{18}F ([^{18}F]AV-45; Amyvid), a ^{18}F-PET tracer, manufactured by the Philadelphia-based Avid Radiopharmaceuticals, has been approved by the FDA in the early 2012 to detect and quantify the Aβ plaque in vivo in the human brain.[17] The positive tests with the latter compound, however, neither indicate the development of AD-related symptoms nor can they be used for monitoring the response of the AD therapeutics. Also, it has few minor side effects in patients and binds selectively to Aβ (K_i = 2.9 nM) but not to neurofibrillary tangles (NFTs).[18,19] Florbetapir-^{18}F-Aβ imaging can be used to predict the effect of amyloid burden on cognition in healthy older individuals; lower memory performance is associated with higher amyloid burden.[20]

Florbetaben-^{18}F (Neuraceq; Piramal Imaging) has been approved as an Aβ imaging agent by both the FDA and the European commission in 2014. In clinical trials using elderly healthy individuals and AD patients, it was shown that the standardized uptake value ratios (SUVRs) for the cerebellar cortex region are comparable for the ^{11}C-PiB and the florbetaben-^{18}F radiotracers (Figure 5). These results indicate that the florbetaben-^{18}F can be used as an alternative to the ^{11}C-PiB in clinical diagnosis of the amyloid burden.[21] The SUVRs in all neocortical gray matter regions in AD were significantly higher compared with those of the healthy individuals.[22] The florbetaben-^{18}F PET scans showed a sensitivity of 80% and a specificity of 91% for discriminating AD patients from healthy individuals.[22]

Flutemetamol (Vizamyl; GE Healthcare), the ^{18}F-labeled version of the Pittsburgh Compound-B (^{11}C–PiB), was approved by FDA in 2013 for detecting the Aβ burden in AD cases. Florbetaben-^{18}F and Flutematemol-^{18}F have relatively high binding affinities for Aβ aggregates (plaques): K_i = 2.2 nM

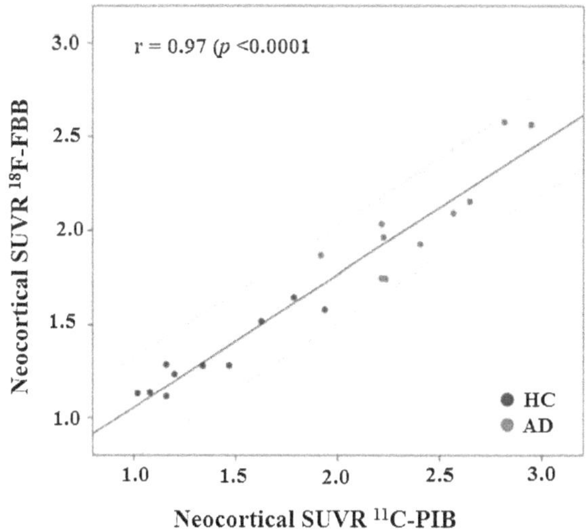

FIGURE 5 Correlation between global Aβ burdens as assessed with ^{11}C–PiB and Florbetaben-^{18}F; there is a high correlation between the ^{11}C–PiB and Florbetaben-^{18}F global Aβ burdens. The slope of 0.71 indicates that the Flrobetaben-^{18}F values tend to be lower than the ^{11}C–PiB values (a dramatic increase of the Aβ burden in the AD patients as compared to the healthy elderly controls (HC) is revealed from either of these PET tracers). *Adapted from Reference 21.*

(florbetaben-^{18}F) and 0.7 nM (flutemetamol-^{18}F).[19] Figure 6 shows Flutemetamol (Vizamyl) PET images of normal and AD brain, the red spots corresponding to the binding of the flutemetamol to the Aβ aggregates.[23] The Aβ binding affinities for these compounds are comparable to that of ^{11}C–PiB, and moreover, due to the relatively long half-life for the ^{18}F-isotope (110 min) and high sensitivity in their ^{18}F-PET imaging as compared to that of ^{11}C–PiB (cf. half-life 20 min), these compounds are expected to be used in the routine clinical diagnosis of Aβ plaques, and thereby for monitoring the progress of AD. However, flutemetamol (Vizamyl; GE Healthcare), Florbetapir-^{18}F (Amyvid; Eli Lilly Pharmaceuticals), and Florbetaben (Neuraceq; Piramal Imaging) have FDA approval not for the definitive diagnosis of AD or to gauge the AD treatment effectiveness, but for supporting other diagnostic criteria.[23,24] That is, a positive scan with these PET imaging agents does not necessarily establish AD diagnosis, but negative scan reduces the likelihood that the patient has AD at the time of the PET scan.[24]

The ^{18}F-PET agents are useful not only for the diagnosis of the Aβ plaques in the brains of individuals for monitoring the disease progression but also for evaluating the drug efficiency during treatment. Merck has included the Vizamyl scan in their 1500-person phase 3 trial involving a β-secretase inhibitor. Amyvid is similarly being used by Eli Lilly for monitoring the efficiency of the amyloid-targeting antibody Solanezumab or placebo in a 2100-people with early stage AD.[34]

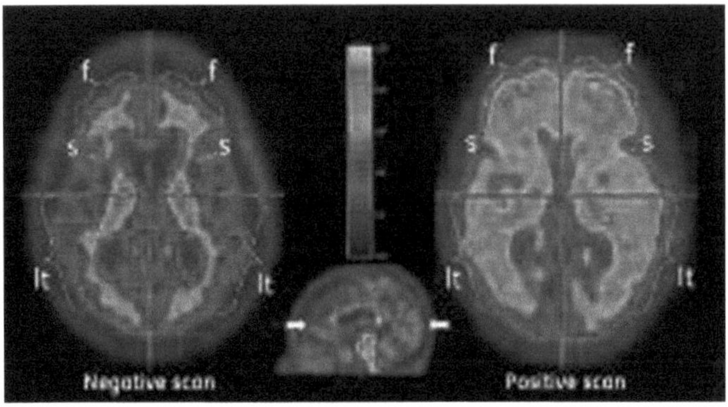

FIGURE 6 Flutemetamol (Vizamyl) PET images of a healthy brain (left) and AD brain (right). *Adapted from Reference 23.*

2.2.1 [¹⁸F]FDDNP

[^{18}F]FDDNP (2-[1-[6-[[2-[^{18}F]Fluoroethyl](methyl)amino]-2-naphthyl]ethylidene] malononitrile) is the first PET imaging probe to visualize the Aβ aggregates and neurofibrillary tangles (NFT) in living humans. So far, it has been the only tracer available for imaging the NFTs in the hippocampal region of the living humans.[25] [^{18}F]FDDNP was synthesized by the nucleophilic substitution of the corresponding tosylate using no-carrier added (i.e., in the absence of the K^{19}F) K^{18}F and the phase transfer catalyst Kryptofix-2.2.2 in acetonitrile solvent, and the process has also been fully automated.[26] [^{18}F]FDDNP binds to both NFTs and Aβ plaques and has relatively high affinity for hippocampus and frontal cortex regions of AD patients as compared to the control (healthy) people.[27] PET imaging showed that the FDDNP binding is lower in control group than in a group with mild cognitive impairment, which, in turn, is lower than a group with AD. This non-invasive PET assay is potentially useful to determine the regional cerebral patterns of Aβ aggregates and NFTs. The relatively lower accumulation of FDDNP in the control subjects, as expected, is also associated with high FDG uptake (determined by FDG PET assay), and minimal atrophy (determined by MRI analysis). However, it suffers from its high retention times in the brains and its non-specificity to Aβ plaques.[27–29]

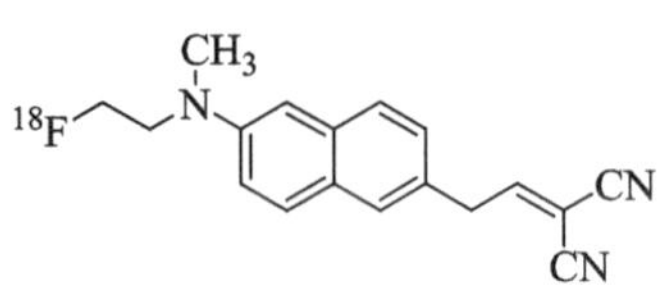

[^{18}F]FDDNP

Approximately 50% of the Parkinson's disease patients suffering from dementia (PDD) have secondary symptoms of AD arising from accumulation of Aβ aggregates and NFTs in their brains.[30] PET studies using [^{18}F]FDDNP showed extensive accumulation of Aβ aggregates and NFTs in the brains of

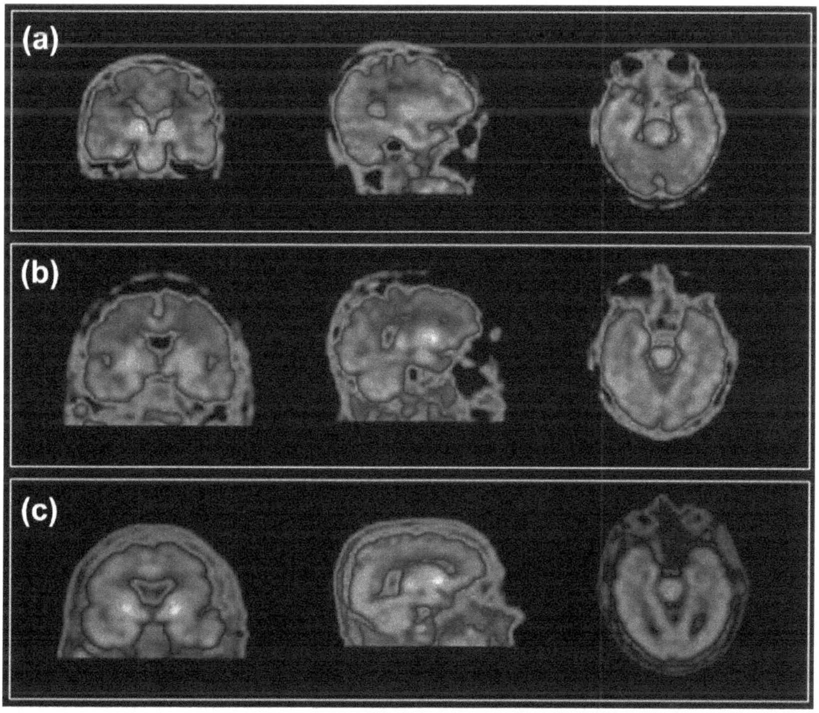

FIGURE 7 Visual assessment of [^{18}F]FDDNP PET scans (warm red color = FDDNP retention): (a) control subjects with modest uptake in the striatum (coronal section); (b) Non-dementia Parkinson's disease (PDND) subject with visually greater uptake in the striatum and modest uptake in the area corresponding to the amygdale of the hippocampus, small foci within the lateral temporo-parietal cortex, and the midbrain (coronal, sagittal and axial sections); (c) Parkinson's disease with dementia (PDD) subject with visually much greater uptake in the same areas as above and further expanding to the lateral and medial temporal and posterior cortices (coronal, sagittal and axial sections). *Reproduced from Reference 31.*

Parkinson's disease patients suffering with dementia (PDD) (Figure 7).[31] In mild cognitive impairment (MCI) patients, both ^{11}C-PIB and ^{18}F-FDDNP PET scans were able to detect increased amyloid load with time.[32] The relatively increased cortical binding of ^{11}C-PIB or ^{18}F-FDDNP in MCI patients in the follow-up time, however is not of diagnostic value for monitoring the disease progression. In AD patients, as expected based on the decrease in the glucose metabolism in the affected regions, a decrease in the ^{18}F-FDG uptake was observed.[32]

2.2.2 ^{18}F-AZD4694 (NAV4694)

Many other ^{18}F-radiolabeled compounds are being developed or in clinical trials for their PET imaging applications. Ohio-based Navidea Biopharmaceuticals' ^{18}F-PET imaging agent ^{18}F-AZD4694 (NAV4694; Figure 8) strongly binds to Aβ plaques, and is under phase 3 clinical trials.[33]

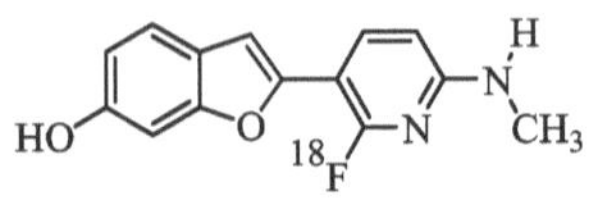

^{18}F-AZD4694 (NAV4694)

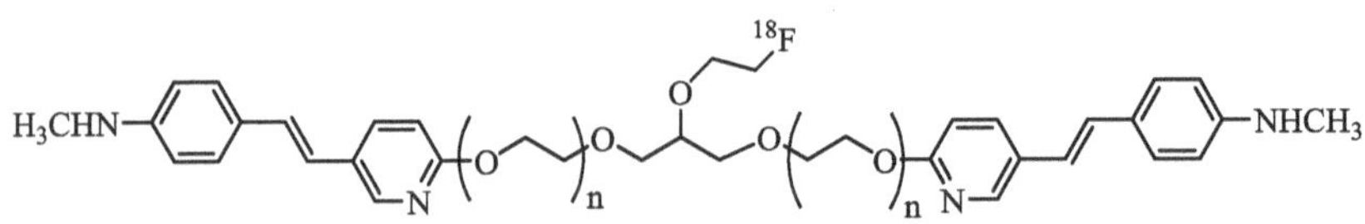

^{18}F-labeled styrylpyridne-polyethylene glycol

FIGURE 8 Structures of ^{18}F-AZD4694 and ^{18}F labeled styrylpyridine-polyethylene glycol conjugate

The cortical distribution of the NAV4694 (^{18}F-AZD4694) is almost identical with that of the ^{11}C–PiB, displaying similar dynamic range of SUVR. This compound has relatively low degree of nonspecific binding to white matter as compared to the other ^{18}F amyloid PET tracers.[33] The slope of the SUVR correlation for ^{11}C–PiB and NAV4694 (slope of 0.95) is significantly higher than that for the ^{18}F-Florbetaben (slope of 0.71), which translates to greater image contrast between AD and healthy individuals.[33] The NAV4694 PET images are similar to those of the ^{11}C–PiB so that it could be used, as in the case of the three FDA approved ^{18}F-PET tracers discussed above, as an alternative to ^{11}C–PiB whose relatively short half-life sometimes precludes clinical applications.

2.2.3 ^{18}F-Labeled Styrylpyridine-Polyethylene Glycol Conjugates

Multidentate ^{18}F-labeled styrylpyridine-polyethylene glycol conjugates (Figure 8) bind to Aβ plaques with high binding affinity and specificity (K_i = 2.9 to 7.7 nM), and therefore these compounds may be useful PET imaging agents for in vivo monitoring of Aβ in AD patients.[35]

3. SYNTHESIS OF ^{18}F-LABELED COMPOUNDS

3.1 General Synthetic Methods

Due to the short life time of the ^{18}F radio-isotope, it is necessary to shorten the reaction time for the incorporation of the ^{18}F label into the target compound, and to maximize the radiochemical yields. The yields of the products are typically measured as radiochemical yield (sometimes radioactive decay corrected), rather than the chemical yield due to the extremely small sample sizes synthesized. The synthesis of ^{18}F-labeled compounds can be achieved by electrophilic or nucleophilic fluorinations. Electrophilic fluorinations using $^{18}F(F_2)$, generated by $^{20}Ne(d,\alpha)^{18}F$ nuclear reaction, is not suitable for the preparation of PET imaging agents with high specific activity, since the ^{18}F-labeled F_2 is diluted with $^{19}F(F_2)$ gas. Nucleophilic reagents

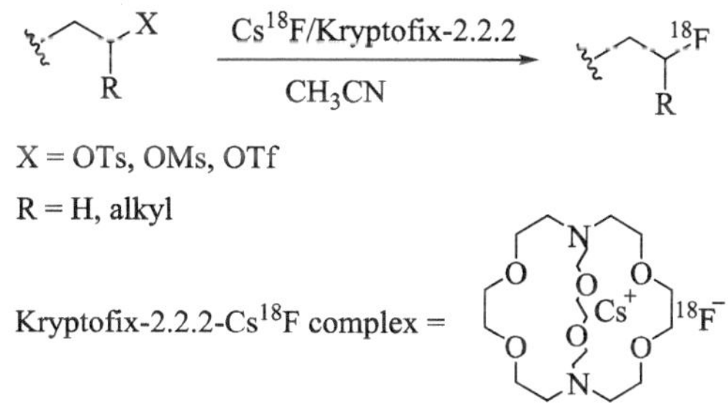

FIGURE 9 Synthesis of ^{18}F-labeled compounds by S_N2 reactions

such as $K^{18}F$ or $Bu_4N^{18}F$, prepared by the nuclear reaction of $^{18}O(p,n)^{18}F$, on the other hand, can be free of added ^{19}F fluoride anions, and afford high specific activity to the radiopharmaceuticals. The former process involving ^{19}F-diluted reagents is called carrier-added synthesis, while the latter one without any added ^{19}F-isotopic "reagents is called as carrier-free synthesis. The ideal PET imaging probes should have high specific activity so that minimal amounts of the samples can be used for PET analysis. The use of ^{18}F-labeled compounds with low specific activity necessitates their use in relatively high concentrations for PET imaging, which would potentially interfere with other metabolic pathways. In order to obtain high specific activity, and also to avoid experimental inconvenience associated with gaseous $^{18}F_2$, nucleophilic fluorination involving carrier-free reagents is widely used for the preparation of PET compounds.[36]

3.2 Aliphatic Nucleophilic Substitution Reactions

Most of the methods for the synthesis of ^{18}F-labeling rely on the aliphatic bimolecular nucleophilic substitution (S_N2) of primary or secondary tosylates, mesylates, or triflates by "naked" fluoride anion, which is formed upon mixing of the $K^{18}F$ (or $Cs^{18}F$) with a [2.2.2]-cryptand (commercially labeled as Kryptofix-2.2.2, Kryptofix 222, or K222) (Figure 9). Typically, as in the other S_N2 reactions, polar aprotic solvents, such as DMSO, DMF, and acetonitrile, are used as solvents in these S_N2 reactions. The strong basicity of the free fluoride anion, in some cases, also results in unwanted elimination products, which have to be purified by chromatography.

3.2.1 2-[^{18}F]Fluoro-2-Deoxy-D-Glucose (FDG)

2-[^{18}F]Fluoro-2-deoxy-D-glucose (^{18}F-FDG) was synthesized by the reaction of the 1,3,4,6-tetra-O-acetyl-2-(trifluoromethylsufonyl)-β-D-mannopyranose with CsF in the presence of the Kryptofix-2.2.2 in acetonitrile, followed by *in situ* hydrolysis and purification by eluting through a disposable chromatographic column and sterilization (Figure 10). The [^{18}F]-FDG was obtained in over 60% radiochemical yields, with an overall reaction time for the synthesis of about 30 min, and this synthetic method has been automated for routine synthesis in the hospital rooms.[37] The simplicity of preparation and purification of the ^{18}F-FDG is primarily responsible for its wide use in PET imaging to monitor the glucose metabolism.

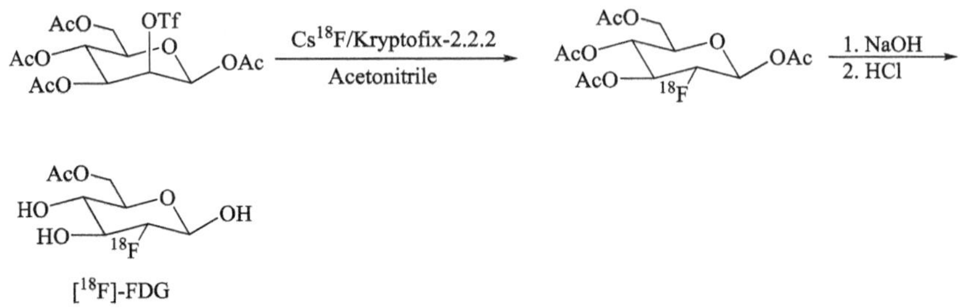

FIGURE 10 Synthesis of 2-[^{18}F]fluoro-2-deoxy-D-glucose (FDG)

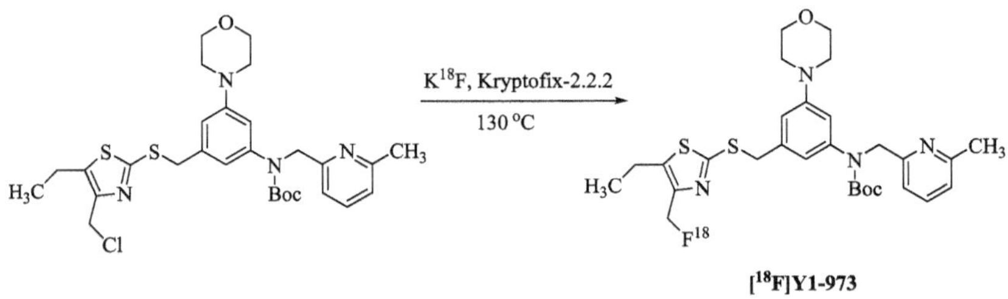

FIGURE 11 Synthesis of NPY Y1 PET tracer, [^{18}F]Y1-973

3.2.2 [^{18}F]Y1-973

$K^{18}F$ was used as the source of ^{18}F-labeled fluoride anion for the preparation of [^{18}F]Y1-973, a newly developed PET tracer for neuropeptide Y receptor subtype 1 (NPYY1). The compound was synthesized in high radiochemical purity (>98 %) and with high specific activity (>1000 Ci/mmol) through nucleophilic substitution reaction of the corresponding alkyl chloride using $K^{18}F$-Kryptofix-2.2.2 complex (Figure 11). PET studies on rhesus monkeys show the distribution of this compound in the striatum and cortical regions; and in cerebellum nuclei and brain stem, consistent with the known NPY Y1 distribution, showing that this PET imaging agent has a high potential for human PET applications.[38]

3.2.3 [^{18}F]-Fluorocaptopril

PET imaging using ^{18}F-labeled enzyme inhibitors provide convenient means for evaluating spatio- and temporal bio-distribution of the enzymes involved in biological activity. For example, ^{18}F-labeled version of the angiotensin converting enzyme (ACE) inhibitor Fluorocaptopril can be used to follow the bio-distribution of the enzyme.[39] In vivo bio-distribution of fluorocaptopril in humans is comparable with that of Captopril in the lung and kidney, although it is not currently used as a PET tracer for clinical studies. The radiolabeled [^{18}F]-fluorocaptopril can be synthesized by the S_N2 substitution of the thiol-protected (*E*)-captopril-4-trifluoromethane sulfonate using no-carrier added $K^{18}F$/Kryptofix-[2.2.2] in acetonitrile, followed by hydrolysis (Figure 12).[39]

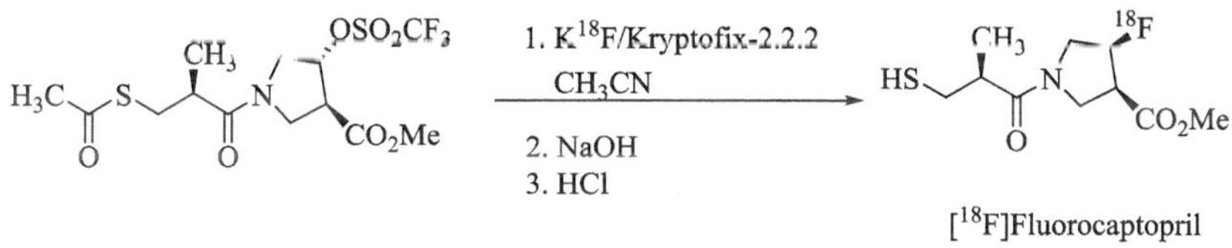

FIGURE 12 Synthesis of [^{18}F]fluorocaptopril.

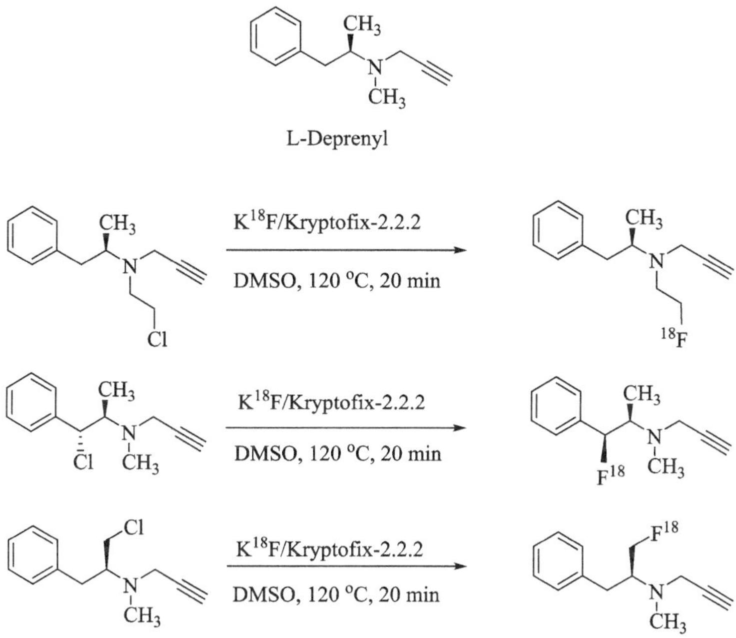

FIGURE 13 Synthesis of [^{18}F]-L-deprenyl analogs.

3.2.4 L-Deprenyl

Nucleophilic substitution reaction of the L-deprenyl derived alkyl chloride using ^{18}F-labeled KF–Kryptofix-2.2.2 complex is generally used for the preparation of the ^{18}F-labeled analogs of L-deprenyl. The latter compound is a prescribed drug for PD, and exerts its pharmacological activity by selectively binding to the monoamine oxidase B (MAO-B).[40] MAO-A and MAO-B enzymes regulate the levels of monoaminergic neurotransmitters by catalyzing their deamination, and therefore specific inhibitors of these enzymes are required for the treatment of depression (inhibition of MAO-A) and PD (inhibition of MAO-B). L-deprenyl selectively inhibits MAO-B and thus the ^{18}F-labeled analogs of the drug candidates are promising as PET tracers for studying the drug interactions. Compounds closely related to L-deprenyl are also synthesized for PET assays. The L-deprenyl derivatives are synthesized through S_N2 reactions of the corresponding halogenated compounds using $K^{18}F$ complexed to Kryptofix-2.2.2 (Figure 13).

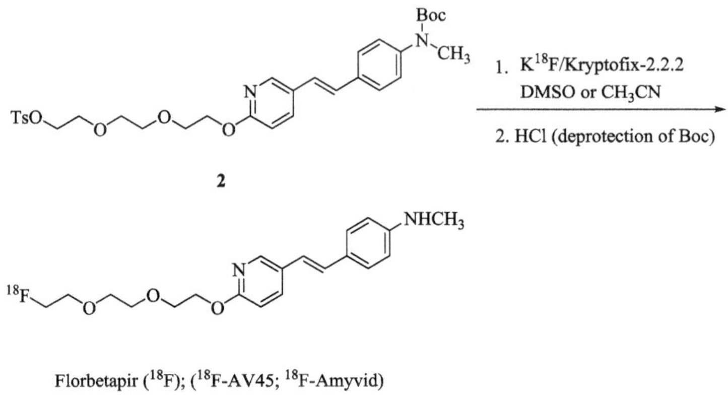

FIGURE 14 Synthesis of [^{18}F]florbetapir, a PET tracer for Aβ plaques.

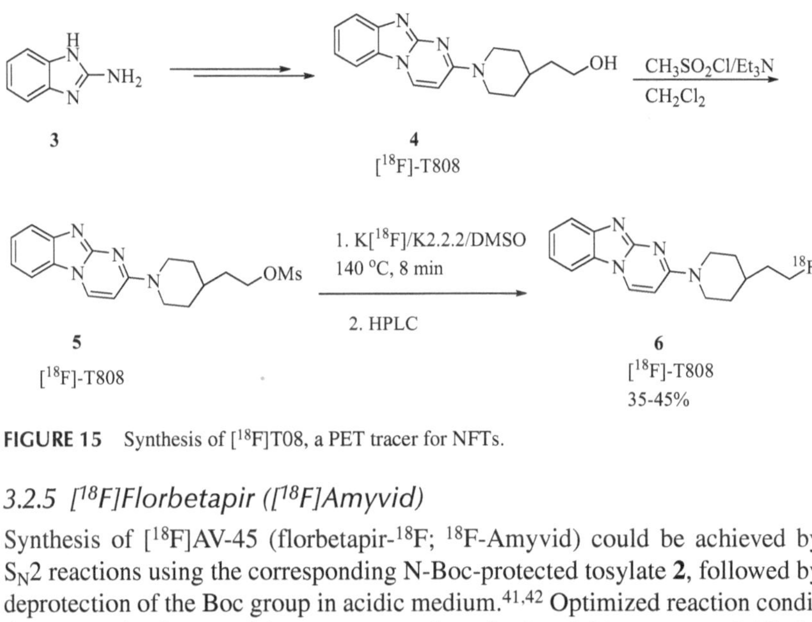

FIGURE 15 Synthesis of [^{18}F]T08, a PET tracer for NFTs.

3.2.5 [^{18}F]Florbetapir ([^{18}F]Amyvid)

Synthesis of [^{18}F]AV-45 (florbetapir-^{18}F; ^{18}F-Amyvid) could be achieved by S_N2 reactions using the corresponding N-Boc-protected tosylate **2**, followed by deprotection of the Boc group in acidic medium.[41,42] Optimized reaction conditions were implemented in an automated synthesis of this compound. Under these optimized reaction conditions, florbetapir was obtained with a radiochemical purity of >95% in an overall synthesis time of 50 min (Figure 14).

3.2.6 [^{18}F]T08

The ^{18}F-labeled compound [^{18}F]T08 has been developed for selective PET imaging of NFT by Siemens, and is in clinical trials.[43] A laboratory-scale multistep synthesis of this radiolabeled compound has been developed, from 1H-benzo[3]-imidazol-2-amine **3**. The final step of the synthesis involves the nucleophilic substitution of the mesylate **5** with K[^{18}F]/ Kryptofix-2.2.2 complex (Figure 15).[44]

3.2.7 [^{18}F]Fluorination by S_N2 Reactions in Protic Solvents

Tetrabutylammonium [^{18}F]fluoride ([^{18}F]TBAF) in protic solvents can be used as an alternative to CsF, without the need of using the aza-crown ether as an additive. Use of *tert*-butyl alcohol as the reaction solvent in these reactions (using TBAF or CsF), without the cryptand additive, was found to give up to 82% yield of the products.[37] This unexpected rate-enhancing effect of the protic solvent was rationalized as due to the hydrogen-bond-stabilization of the fluoride anion by the alcohol, resulting in the decreased strength of the electrostatic interactions in CsF and TBAF. At the same time, the use of the sterically crowded alcohols minimizes the extensive solvation of the fluoride anion so that its nucleophilicity is not significantly decreased.

Various ^{18}F-labeled PET tracers, including ^{18}F-FDG, 3′-[^{18}F]fluoro-2′,3′-dideoxythymidine (^{18}F-FLT), N-2-[^{18}F]fluoropropyl-2β-carbomethoxypropyl-3β-(4-iodophenyl)nortropane, and 1-[^{18}F]fluoro-3-(2-nitroimidazol-1-yl)propan-2-ol have been synthesized in improved RCYs using sterically crowded alcohols as solvents, such as *tert*-butyl alcohol, without having to use the cryptand complexing agent (Figure 16).[37]

3.3 Fluorination via Aromatic Nucleophilic Substitution Reactions

ipso-Nucleophilic aromatic substitution of the nitro group by fluoride anions can be used for the synthesis of the ^{18}F-labeled compounds. This reaction requires the presence of electron-withdrawing groups (e.g., carbonyl) at *ortho* or *para* positions to the nitro group to stabilize the intermediate Meisenheimer complex. ^{18}F-labeled 4-(2′-methoxyphenyl)-1-[2′-(N-2″-pyridinyl)-4-fluorobenzamido]ethylpiperazine (*p*-MPPF), a PET tracer for selective imaging of a serotonin receptor, 5-hydroxytryptamine-1A (5-HT_{1A}) receptor, is prepared by this method in a fully automated way[45]; The carbonyl ^{11}C-labeled version of the *p*-MPPF, WAY-100635, has been widely used for selective PET imaging of 5-HT_{1A} receptors.[46]

ipso-Nucleophilic aromatic substitution of the nitro group by carrier-free [^{18}F] fluoride ions similarly gives the ^{18}F-fluoroflumazenil, which is a potential PET tracer for the visualization and quantification of the central benzodiazepine receptor and γ-aminobutyric acid-A ($GABA_A$) receptor complex in the human brain (Figure 17).[47]

3.3.1 Aromatic Nucleophilic Substitution Reactions of Quaternary Ammonium Salts

Nucleophilic aromatic substitutions of tetramethylammonium group, *ortho* or *para* to the carbonyl group, by ^{18}F-fluoride ion gives *ortho*- or *para*-[^{18}F]fluorobenzaldehydes in high RCYs; these compounds are useful starting materials for the preparation of various ^{18}F-labeled amino acids (Figure 18).[36]

3.3.2 Aromatic Nucleophilic Substitution Reactions of Diaryliodonium Salts

Diaryliodonium salts upon reaction with carrier-free [^{18}F]fluoride anion give the corresponding ^{18}F-labeled aromatic compounds (Figure 19).[48–51] In this

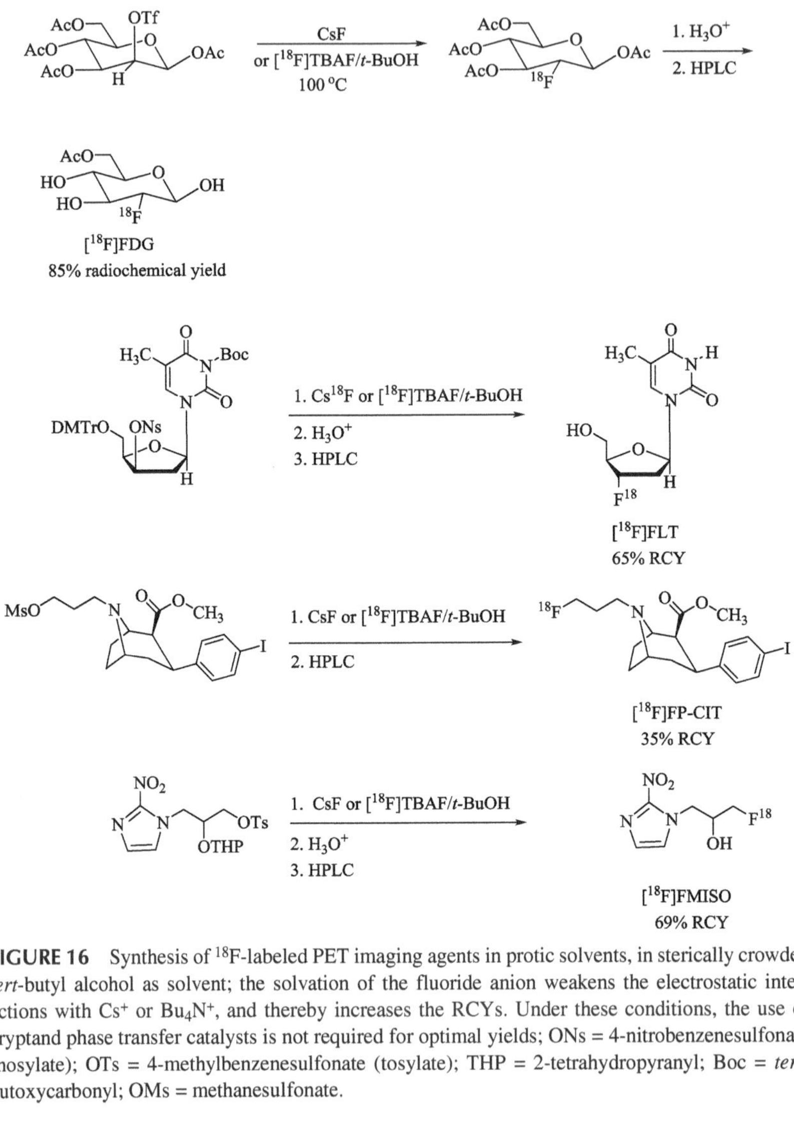

FIGURE 16 Synthesis of ^{18}F-labeled PET imaging agents in protic solvents, in sterically crowded *tert*-butyl alcohol as solvent; the solvation of the fluoride anion weakens the electrostatic interactions with Cs^+ or Bu_4N^+, and thereby increases the RCYs. Under these conditions, the use of cryptand phase transfer catalysts is not required for optimal yields; ONs = 4-nitrobenzenesulfonate (nosylate); OTs = 4-methylbenzenesulfonate (tosylate); THP = 2-tetrahydropyranyl; Boc = *tert*-butoxycarbonyl; OMs = methanesulfonate.

reaction, one of the aryl groups is chosen such that it is relatively more electron rich so that the nucleophilic substitution reaction does not occur on this ring; typically, 2,4,6-trimethoxyphenyl or 2-thenyl rings serve this purpose. These radiolabeled aryl fluorides serve as precursors to various radiopharmaceuticals that are useful as PET tracers.

3.4 Pd-Catalyzed ^{18}F Labeling of Proteins

A modified Pd-catalyzed Suzuki–Miyaura coupling of ^{18}F-labeled arylboronic acids with aryl iodides affords the corresponding ^{18}F-labeled compounds; the use

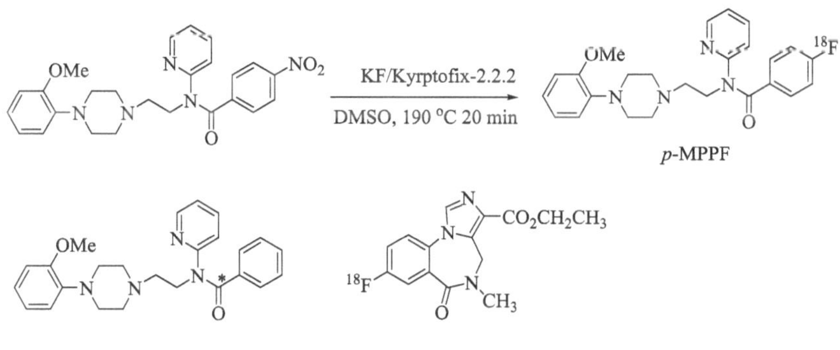

WAY-100635; * = ^{11}C 2'-[^{18}F]fluoroflumazenil

FIGURE 17 Synthesis of ^{18}F-labeled compounds by aromatic nucleophilic substitution reactions.

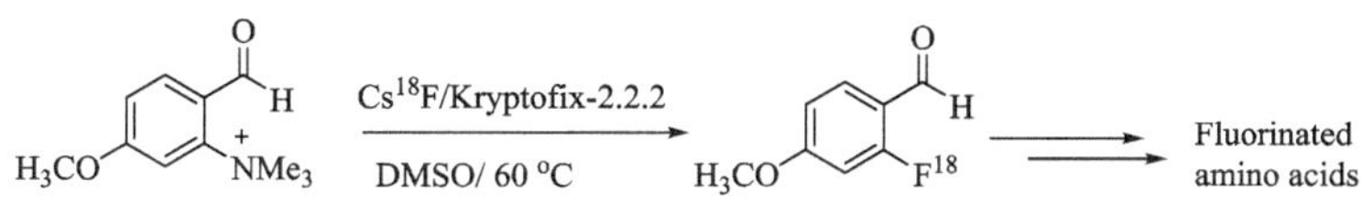

52% RCY

FIGURE 18 [^{18}F]fluorination by aromatic nucleophilic substitution reactions of aryl quaternary ammonium salts (DMSO = dimethylsulfoxide).

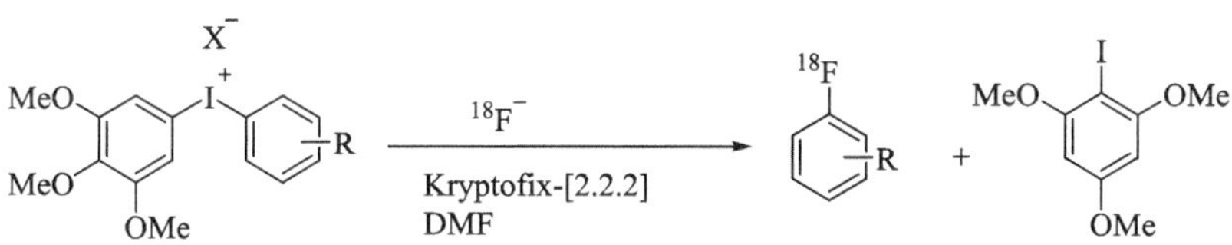

R = e.g., 4-Me, 4-OMe, H, 4-Cl,
4-MeOCO, 4-Bz, 3-EtOCO

FIGURE 19 [^{18}F]fluorination by aromatic nucleophilic substitution reactions of diaryliodonium salts.

of N,N-dimethylguanidine as the ligand allows the reaction to proceed under mild conditions (Figure 20). This modified method, using the [^{18}F]-4-fluorophenylboronic acid as the prosthetic group, allows the synthesis of ^{18}F-labeled small molecules, peptides, and proteins, under biologically compatible conditions.[52]

The ^{18}F-prosthetic arylating agent **9** was synthesized in two steps from the di(4-iodophenyl)iodonium triflate **7**; the aromatic nucleophilic substitution reaction of the triflate **7** with fluoride anion using K^{18}F-Kryptofix-2.2.2 complex gave 4-[^{18}F]fluoroiodobenzene **8**, which was transformed into the 4-[^{18}F]fluoroboronic acid **9** in a single step using tetrahydroxydiborane under Pd catalysis. The direct attachment of this ^{18}F-labeled prosthetic group to the haloarylamino acid residue (Pic156) of the protein subtilisin from *Bacillus lentus* (SBL) (**10**) was carried out in phosphate buffer (pH 8.0) and $(L)_2Pd(OAc)_2$ at 37 °C. At the end of the reaction (30 min), 3-mercaptopropanoic acid was

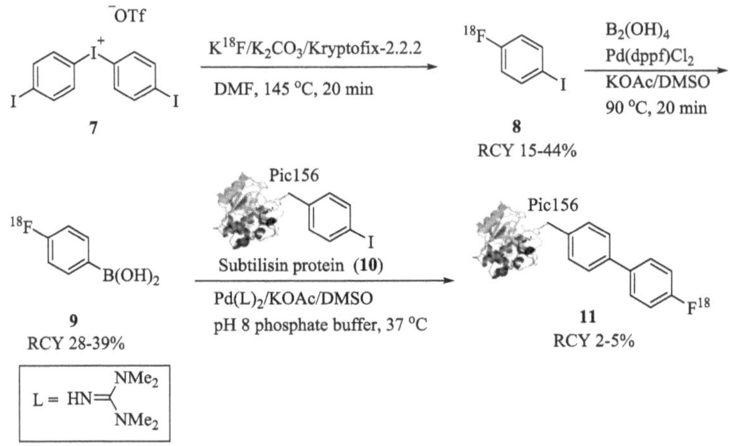

FIGURE 20 Pd(0)-catalyzed arylation of proteins using ^{18}F-labeled arylboronic acids (dppf = 1,2-bis(diphenylphosphino)ethane; DMF = N,N-dimethylformamide; DMSO = dimethylsulfoxide).

added to scavenge the palladium catalyst, and the radiolabeled protein **11** was purified by size exclusion chromatography; RCYs of about 2–5% were obtained for this conversion (Figure 20).[52]

3.4.1 *Pd-Catalyzed Fluorination of Arylboronic Acids*

Ritter and coworkers have pioneered Pd-catalyzed transformation of aryl boronic acids or boronates to [^{18}F]-labeled arenes using the ^{18}F-labeled fluoride source.[53,54] In this process the palladium-based electrophilic fluorinating agent **16** was obtained from the reaction of the radiolabeled [^{18}F]F^- with the Pd(IV) complex **15.** This synthetic method is conceptually interesting in that an appropriately ligated Pd(IV) species allows conversion of nucleophilic fluoride anion into an electrophilic fluorinating agent. The oxidative transfer of "F^+" from Pd(IV)F (**16**) to ArPd(II) (**14**) species results in the formation of a high-valency ArPd(IV)F species (**18**), which undergoes reductive elimination to give the corresponding aryl fluorides. Pharmaceutically interesting compounds such as [^{18}F]fluorodeoxyestrone (**20**) and a serotonin receptor agonist **21** could be synthesized starting from the corresponding arylboronic acid in RCYs of about 33% (Figure 21). Although the synthesis of the specific arylboronates involves multistep process, radiofluorination is the last step of the synthesis so that high RCYs could be achieved. ^{18}F-labeled derivatives of selective serotonin reuptake inhibitors, such as [^{18}F]paroxetine, have been synthesized using this procedure.[55]

3.5 [^{18}F]Trifluoromethylation

3.5.1 *Cu(I)-Catalyzed [^{18}F]Trifluoromethylation of Aryl and Heteroaryl Iodides*

Aryl and heteroaryl iodides can be transformed into the corresponding [^{18}F]-labeled trifluoromethylated derivatives through the in situ generated

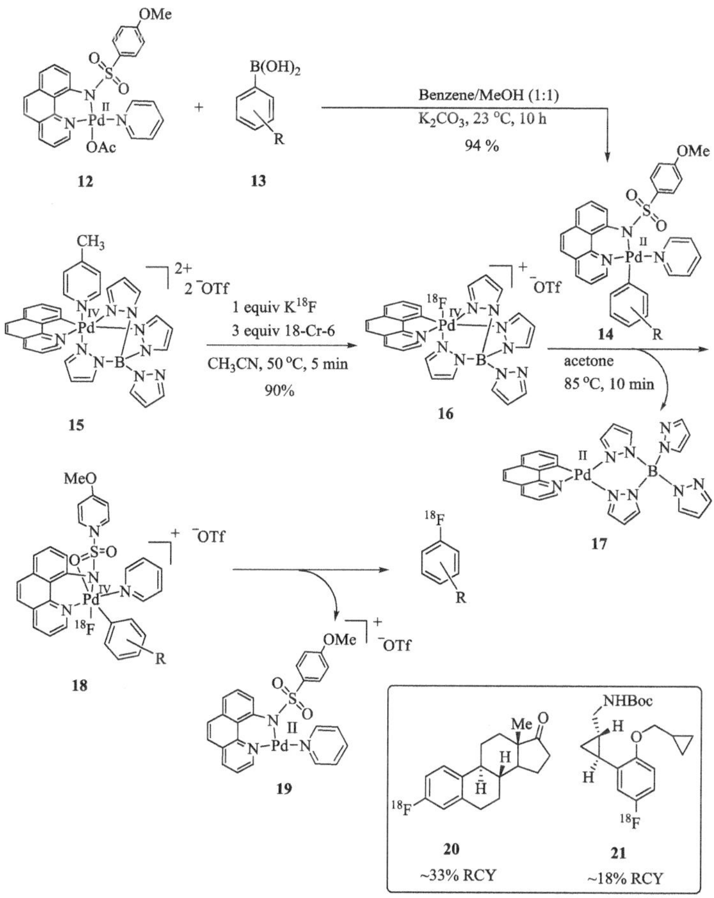

FIGURE 21 Pd(II)-catalyzed [^{18}F]fluorination of arylboronic acids (18-Cr-6 = 18-crown-6).

[^{18}F]-$CuCF_3$.[56] Reaction of the methyl chlorodifluoromethylacetate (**22**) with Cu(I) salts in the presence of a ^{18}F-labeled fluoride source (K^{18}F) forms ^{18}F-labeled trifluoromethylcopper through the intermediate difluorocarbene. Thus generated [^{18}F]CF_3Cu **23** reacts with aryl or heteroaryl iodides to give the corresponding ^{18}F-labeled trifluoromethylarenes, **24**. The reaction proceeds at 150 °C in DMF solvent to give the ^{18}F-labeled trifluoromethylarenes in moderate RCYs. This aryl-trifluoromethylation was demonstrated to be broadly applicable for the preparation of a variety of aromatics including those derived from carbohydrates, dipeptides, and a derivative of uracil (Figure 22).[56]

3.5.2 ^{18}F-Fluorination of Enolsilyl Ethers

Direct fluorination of α,α-difluoroenolsilyl ethers using ^{18}F-labeled molecular fluorine gives the corresponding ^{18}F-labeled trifluoromethyl ketones in RCYs of 45–55% (Figure 23). The synthesis time corresponded to about 40 min from

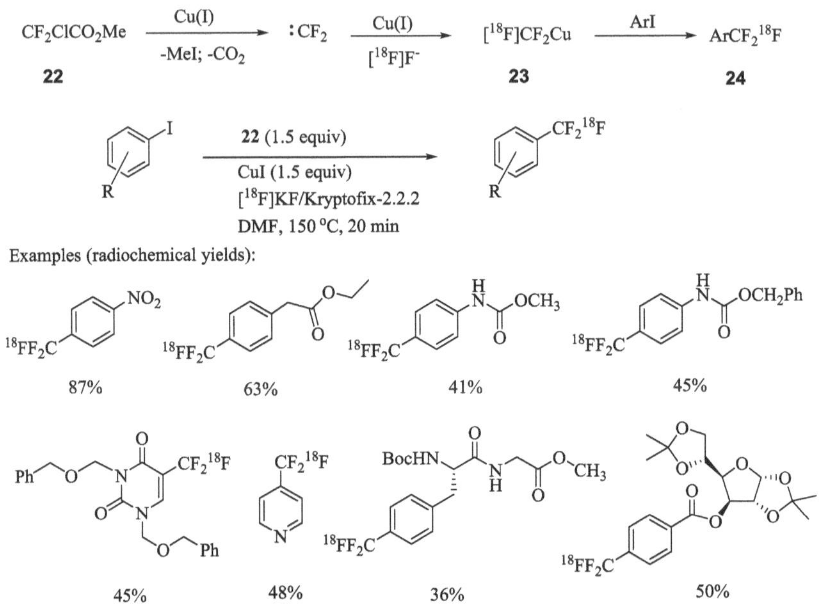

FIGURE 22 Cu(I)-catalyzed trifluoromethylation of aryl and heteroaryl iodides (DMF = N,N-dimethylformamide).

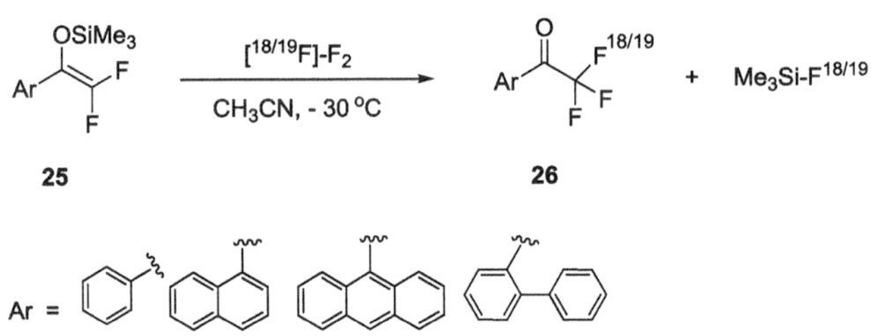

FIGURE 23 Direct fluorination of enolsilyl ethers with ^{18}F-labeled elemental fluorine.

the time of end of the bombardment.[57,58] This late-stage fluorination involving only a single step is potentially useful for the preparation of biologically active ^{18}F-labeled organic compounds.

4. ^{18}F-LABELED NEUROTRANSMITTERS

The neurotransmitters, 6-[^{18}F]-L-dopamine, 6[^{18}F]fluoronorepinephrine, 6-[^{18}F]-L-DOPA, and their structural analogs (e.g., 6-[^{18}F]fluoro-*m*-tyrosine and 6-[^{18}F]fluoro-β-fluoromethylene-*m*-tyrosine have been used to monitor cerebral blood flow, cerebral glucose metabolism, and functioning of the neurotransmitter systems—including cholinergic, dopaminergic, and serotonergic systems—in neurological disorders (Figure 24).[59–61]

The design and use of PET tracers depends on specific applications.[7] Generally, the synthesis of these compounds could be achieved by fluorodestannylation

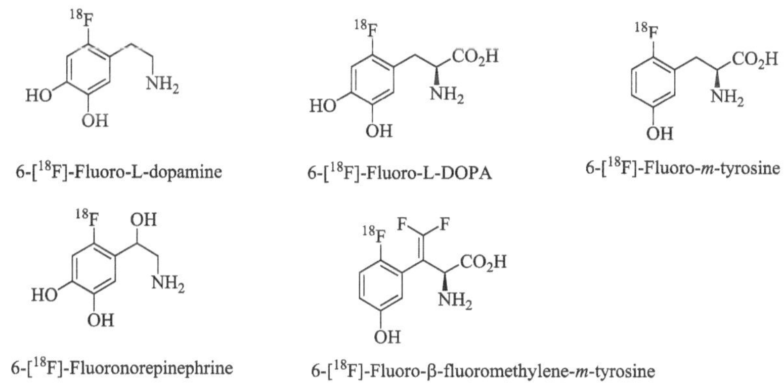

FIGURE 24 Structures of some ^{18}F-labeled neurotransmitters.

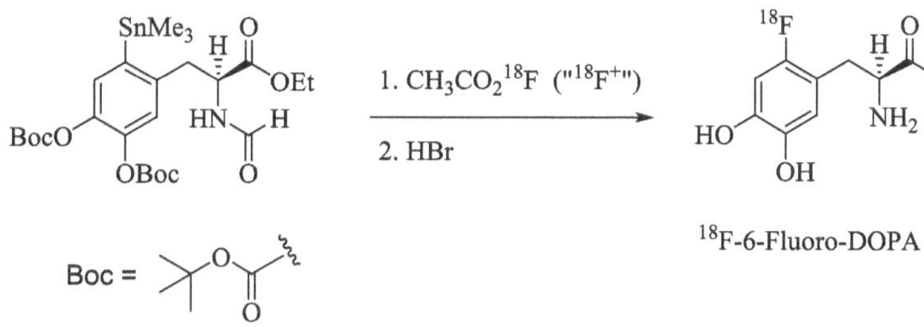

27

FIGURE 25 Synthesis of ^{18}F-6-fluoro-L-DOPA using $CH_3CO_2{}^{18}F$.

or fluorodemercuriation of the corresponding compounds using ^{18}F-labeled acetylhypofluorite ($CH_3CO_2{}^{18}F$; synthesized from the ^{18}F-labeled elemental fluorine), or the ^{18}F-labeled elemental fluorine as the electrophilic fluorinating agents.[62] ^{18}F-labeled electrophilic fluorinating reagent Selectfluor ditriflate has also been used for similar transformations.[63]

4.1 [^{18}F]-L-DOPA

The ^{18}F-labeled acetylhypofluorite ($CH_3CO_2{}^{18}F$) (synthesized from the ^{18}F-labeled elemental fluorine ([^{18}F]F_2), and the adiolabeled elemental fluorine gas ([^{18}F]F_2)) have been used as the electrophilic fluorinating agents for the fluorodestannylation of the appropriately protected DOPA ethyl ester, **27**. After deprotection using HBr followed by high-performance liquid chromatography (HPLC) purification, the radiolabeled [^{18}F]L-DOPA was obtained in RCYs of up to 25% (Figure 25).[62] In the brain, the ^{18}F-labeled DOPA is localized to serotonin, noradrenaline, and dopamine neurons. It has also been used to monitor the regional changes in the monoamine neurons in PD patients.[64]

[^{18}F]Selectfluor-mediated transformation of the aryl boronic esters into the ^{18}F-labeled aromatics has been developed, and using this method

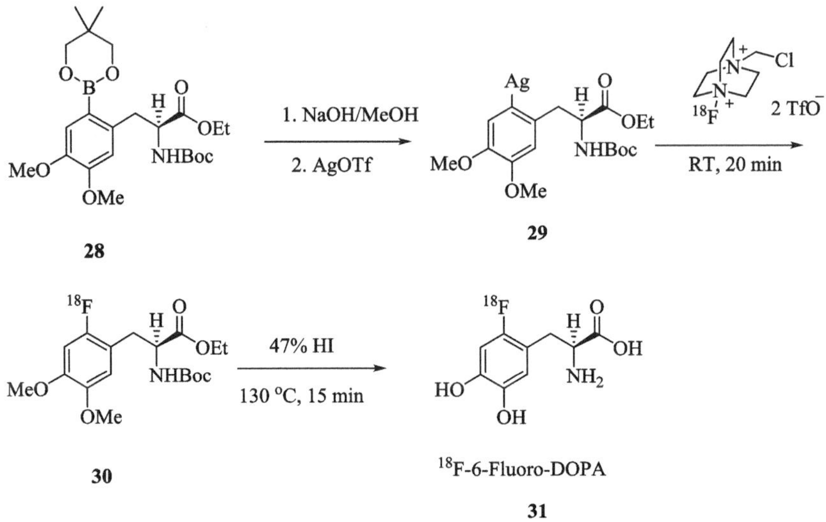

FIGURE 26 Synthesis of ^{18}F-6-fluoro-L-DOPA using ^{18}F-labeled Selectfluor.

synthesis of ^{18}F-6-fluoro-L-DOPA has been achieved in RCYs of 19 ± 12% (Figure 26).[8] This synthetic strategy involves conversion of the aryl boronate esters, **28**, into the corresponding aryl silver, followed by electrophilic fluorination by the [^{18}F]-Selectfluor ditriflate, and a final deprotection using aqueous HI.

4.2 6-[^{18}F]Fluoronorepinephrine

Synthesis of the 6-[^{18}F]fluoronorepinephrine has been achieved under carrier-free conditions, that is, with exclusion of the stable ^{19}F isotope, by aromatic nucleophilic substitution reaction of the nitrobenzaldehyde derivatives with $K^{18}F$. The carrier-free $K^{18}F$ is produced by neutron irradiation of $Li_2{}^{18}O$, followed by cation exchange purification, by eluting with acetonitrile solution of Kryptofix-2.2.2 and K_2CO_3.[65] The cryptand Kryptofix-2.2.2 is conventionally used in the preparation of the radiolabeled fluorinated PET agents, as it selectively solvates the K^+ ions to release the fluoride anion.

The aromatic nucleophilic substitution reaction of the *o*-nitrobenzaldehyde derivative **32**, using Kryptofix-2.2.2-activated $K^{18}F$ in DMSO (10 min at 120 °C) gave the ^{18}F-labeled compound **33** in RCYs of 40–45% (Figure 27). Reaction of the crude product **33** with trimethylsilyl cyanide gave the corresponding cyanohydrin trimethylsilyl ether. In situ reduction of the latter compound using $LiAlH_4$ to the primary amine, followed by hydrolysis and purification (semipreparative liquid chromatography followed by chiral HPLC) gave the enantiomerically pure 6-[^{18}F]fluoronorepinephrine in an overall 6% RCY for each of the chirally pure compounds and 20% RCY for the racemic mixture (total synthesis time = ~93 min for the racemic compound and ~128 min for the chirally pure material; Figure 27).[65]

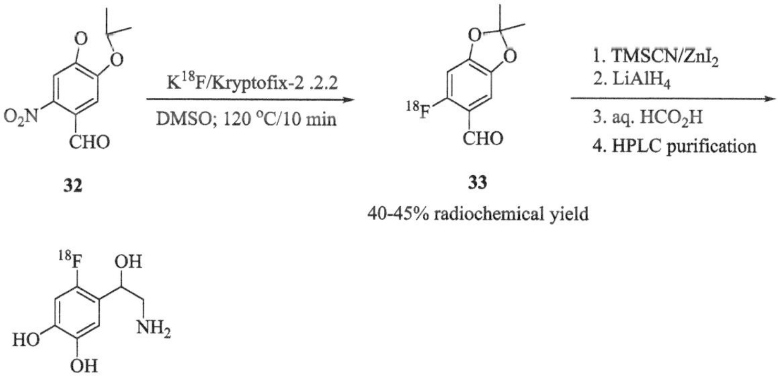

32

33

40-45% radiochemical yield

6-[^{18}F]-Fluoronorepinephrine
~6% overall radiochemical yield

FIGURE 27 Synthesis of 6-[^{18}F]fluoronorepinephrine (DMSO = dimethylsulfoxide).

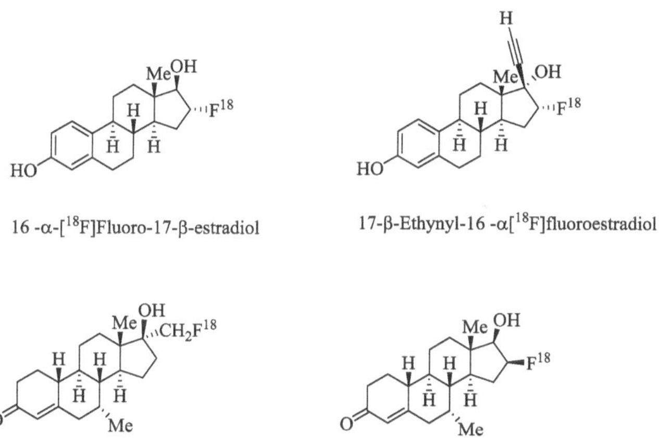

16 -α-[^{18}F]Fluoro-17-β-estradiol

17-β-Ethynyl-16 -α[^{18}F]fluoroestradiol

20-[^{18}F]-Fluoromibolerone

16-β-[^{18}F]Fluoro-(4,5α)-dihydrotestosterone (^{18}F-FDHT)

FIGURE 28 Structures of some ^{18}F-labeled corticosteroids.

5. [^{18}F]CORTICOSTEROIDS

Steroidal hormones (corticosteroids), 16α-[^{18}F]fluoro-17β-estradiol, 17β-ethynyl-16α-[^{18}F]fluoroestradiol, ^{18}F-FDHT, and 20-[^{18}F]fluoromibolerone are used for the diagnosis of breast tumors (Figure 28).[66–68] These PET tracers provide a noninvasive means for measuring the estrogen receptor expression in tumors and thereby provide a basis for the selection of appropriate endocrine therapeutic agents.

Prostate cancer is the second leading cause of mortality among men in the United States. PET using ^{18}F-labeled radioisotopes provides a noninvasive and quantitative imaging technique for the detection of, and for monitoring, the

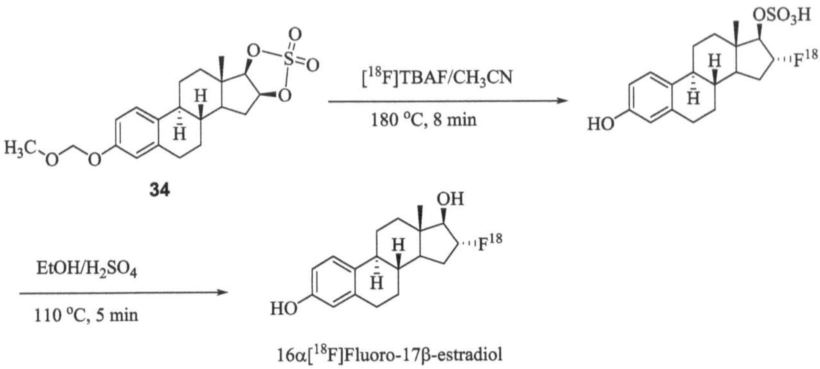

FIGURE 29 Synthesis of 16α-[18F]fluoroestradiol (TBAF = tetrabutylammonium fluoride; EtOH = ethanol).

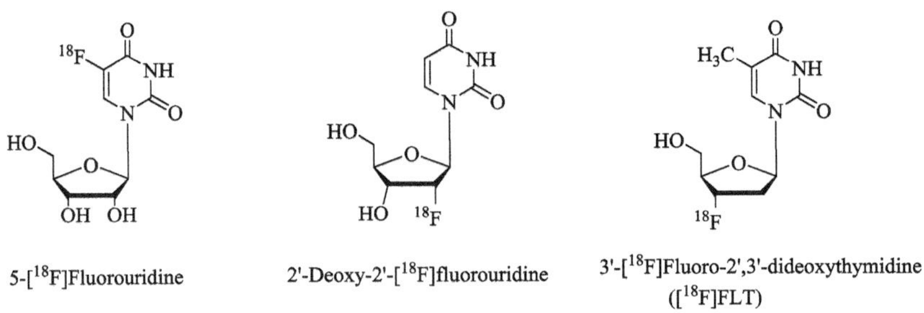

FIGURE 30 Structures of 5-[18F]fluorouridine, 2′-deoxy-2′-[18F]fluorouridine, and [18F]FLT.

progress of prostate cancer.[69] 2-[18F]Fluoro-2-deoxy-D-glucose (18F-FDG) is not suitable for monitoring prostate cancer progression because the glucose utilization in prostate cancer cells is relatively low and, in addition, intense accumulation of FDG in the proximally located urinary bladder overshadows the tumor uptake.[69] 18F-labeled fluoromibolerone has been used for detecting and quantifying androgen receptors in metastatic prostatic cancers,[70] and 16β-[18F]fluoro-(4, 5α)-dihydrotestosterone has been used as a PET tracer to monitor the metastatic prostate cancer.[71]

The 16α-[18F]fluoro-17β-estradiol is readily synthesized from the corresponding cyclic sulfone **34** by nucleophilic substitution with [18F]F⁻ using [18F]-labeled TBAF, followed by acid-catalyzed hydrolysis (Figure 29).[71]

6. 18F-LABELED NUCLEOSIDES

The 18F-labeled anticancer drug 5-[18F]fluorouridine is used for monitoring tumor metabolism and proliferation.[73] A thymidine analog, 3′-[18F]fluoro-2′,3′-dideoxythymidine (18F-FLT) has been used as PET tracer in cancer therapy for monitoring the levels of thymidine kinase-1, which is responsible for thymidylate synthase activity.[74,75] 2′-Deoxy-2′-[18F]fluorouridine is also useful as a PET tracer for studying tumor proliferation (Figure 30).[73]

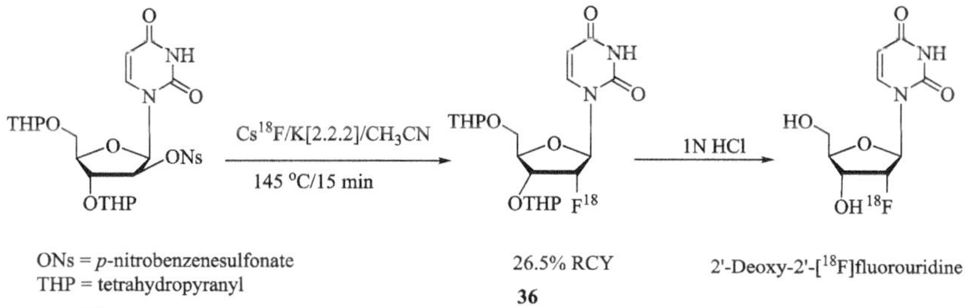

35

36

FIGURE 31 Synthesis of 2′-deoxy-2′-[^{18}F]fluorouridine.

Synthesis of 2′-deoxy-2-[^{18}F]fluorouridine is achieved through S_N2 substitution reaction of the bis(3′,5′-O-tetrahydropyranyl)-D-arabinofuranosyluracil-2′-*p*-nitrobenzenesulfonate **35** using ^{18}F-labeled fluoride ion complexed to Kryptofix-2.2.2, followed by deprotection of the tetrahydropyranyl group of **36** under acidic aqueous conditions.[73] After a final purification by HPLC, the radiolabeled fluorouridine was obtained in 26.5% RCY and 98% purity. Overall this synthesis involves four steps starting from uridine, with an overall yield of 9.4% (Figure 31).

High RCYs (~65%) of this compound were obtained using an improved method, in which the late-stage nucleophilic substitution reaction was carried out in a polar, protic solvent using [^{18}F]CsF and tetrabutylammonium hydroxide as the phase transfer catalyst; the use of the Kryptofix-2.2.2 is not needed under these reaction conditions so that additional purification steps are avoided.[37,76–79] Various other radiopharmaceuticals, including [^{18}F]FLT could also be synthesized using this procedure.

7. ^{18}F-RADIOLABELING OF PEPTIDES AND PROTEINS

7.1 3-[^{18}F]Fluorosilylbenzamide Derivatives

A promising technique for the incorporation of ^{18}F label into proteins uses N-succinimidyl 3-(di-*tert*-butyl[^{18}F]fluorosilyl)benzoate ([^{18}F]SiFB) as the synthon for reaction with terminal amino groups of proteins to derivatize them as (3-(di-*tert*-butyl[^{18}F]fluorosilyl)benzamides) (Figure 32).[80,81] The [^{18}F]SiFB synthon is conveniently synthesized by fluoride anion exchange of [^{19}F]SiFB **37** using ^{18}F-labeled fluoride anion in the presence of oxalic acid. After purification by solid-phase extraction, [^{18}F]SiFB could be obtained with high specific activities (~525 Ci/mmol). Various proteins, including rat serum albumin, apotransferrin, and erythropoietin could be conveniently labeled with [^{18}F]SiFB in RCYs of 19–36% (Figure 32).

7.2 ^{18}F-Labeled 2-Cyanobenzothiazole for Radiofluorination of Peptides and Proteins

Condensation of appropriately ^{18}F-labeled 2-cyanobenzothiazole **38** and terminal cysteine groups of proteins or peptides could be achieved under relatively mild reaction conditions (RT, 20–30 min) to give their corresponding dithiazole

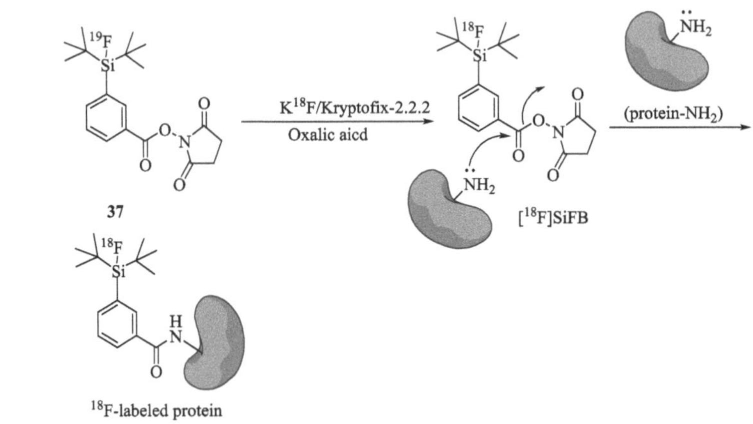

FIGURE 32 Synthesis of 3-[^{18}F]fluorosilylbenzamide conjugates of proteins.

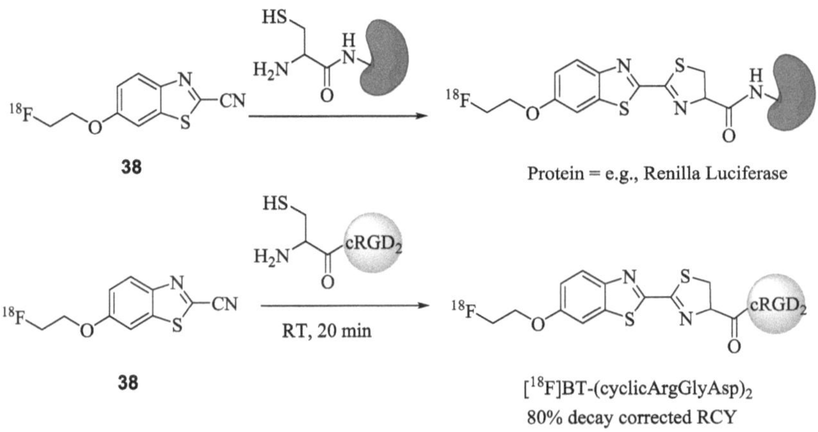

FIGURE 33 Peptide and protein conjugates of ^{18}F-labeled 2-cyanobenzothiazole.

adducts (Figure 33).[82] RCYs range from 80% for small peptides to 12% for proteins. Thus, using this procedure, site-specific modification of proteins could potentially be achieved under physiological conditions for PET studies. The incorporation of the cysteine moiety into peptides or proteins could be achieved by the standard peptide coupling routes using the N-Boc-Cys(Trt) succinimidyl ester (Trt = trityl; used for protecting the sulfhydryl group of the cysteine).[82] Similarly, click reactions (vide infra) using strained aza-dibenzocyclooctyne-substituted peptides with appropriately ^{18}F-labeled azides results in the radiolabeling of the proteins under physiological conditions.[83]

7.3 ^{18}F-Labeled Peptide Derivatives

Peptides with terminal amino groups functionalized by 4-nitro-3-trifluoromethylbenzoyl groups (**39**) react with K^{18}F/ Kryptofix-2.2.2 by aromatic nucleophilic

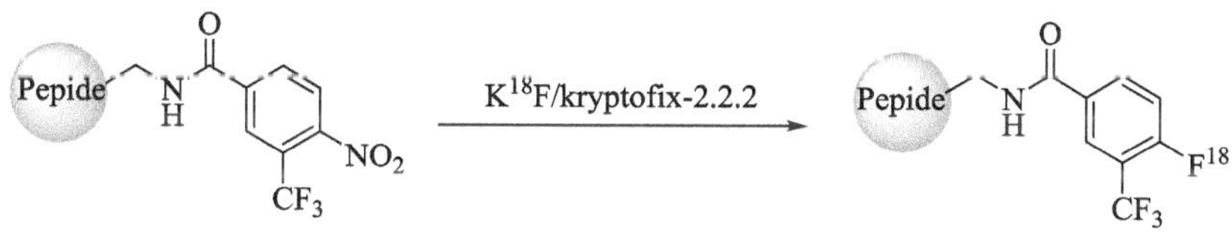

39

FIGURE 34 Radiofluorination of 4-nitro-3-trifluoromethylbenzamide derivatives of peptides.

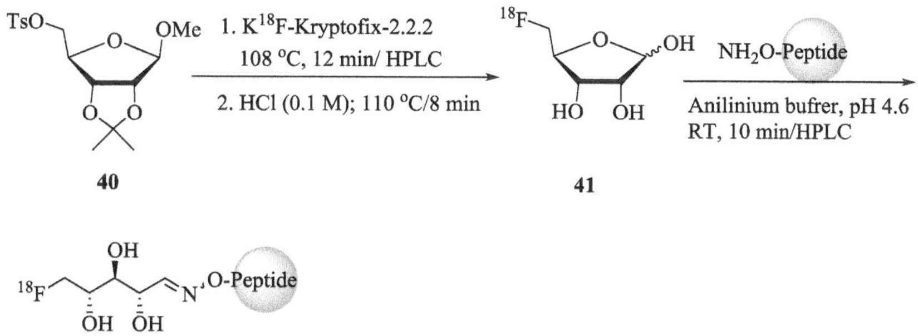

40 **41** **42**

FIGURE 35 Glycosylation of peptides using 5-deoxy-5-[^{18}F]fluororibose ([^{18}F]FDR).

substitution of the nitro group by $^{18}F^-$ anion. In this way, dimeric cyclic arginylglycylaspartic acid (cRGD) peptide was synthesized with high specific activity (79 GBq/μmol) and evaluated in vivo in an animal tumor xenograft model (Figure 34).[84]

Peptides have been modified by glycosylation using 5-deoxy-5-[^{18}F]fluororibose ([^{18}F]FDR; **41**) as a prosthetic group for potential applications in PET imaging. The synthesis of [^{18}F]FDR was achieved by a nucleophilic substitution reaction of **40** using K^{18}F-Kryptofix-2.2.2 complex as the fluorinating agent, followed by acid-catalyzed hydrolysis. The [^{18}F]FDR is conjugated with terminal aminoxy (NH_2O)-functionalized peptides in anilinium buffer (overall reaction time = 10 min), and after HPLC purification the ^{18}F-labeled peptides could be obtained in sufficient purity for their formulations.[14] The final products in this synthesis were obtained in >98% radiochemical purity and in RCYs of 27–37% (Figure 35).

7.3.1 Synthesis of ^{18}F-Labeled Peptides Through Click Chemistry

Sharpless and coworkers introduced a modified version of Huisgen 1,3-dipolar cycloaddition reactions of terminal alkynes and azides using Cu(I) catalysis to give 1,4-dialkl(aryl)-1,2,3-triazoles as the exclusive product under mild conditions. In the absence of Cu(I) catalysis, the original Huisgen reaction required high temperature and longer reaction times, and usually form mixtures of 1,4-dialkyl(aryl) and 1,5-dialkyl(aryl) triazoles.

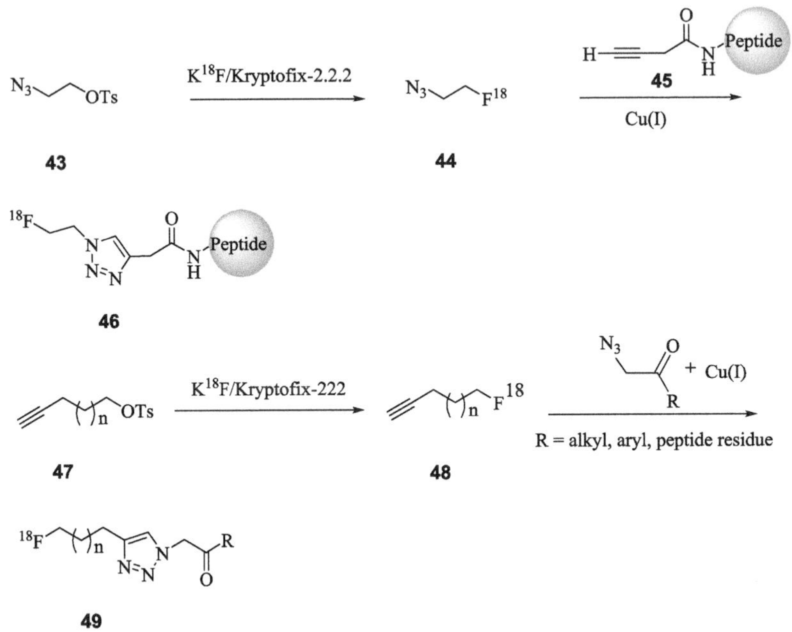

FIGURE 36 Synthesis of ^{18}F-labeled peptides through click chemistry.

There are several reports of adapting Sharpless click reactions to synthesize ^{18}F-labeled small molecule analogs of drug candidates,[85–87] and ^{18}F-incorporated peptides.[85,88–91] ^{18}F label could be incorporated into alkyl azides by S_N2 reaction of azidoalkyl tosylates, **43**, with K^{18}F under the usual reaction conditions (carrier-free [^{18}F]F$^-$, Kryptofix-2.2.2, and polar solvents such as acetonitrile or DMSO) in high RCYs. These ^{18}F-labeled azides react with compounds of interest, which are suitably modified by terminal alkyne moieties, to give the 1,4-disubstituted 1,2,3-triazoles in short reaction times (typically 10 min at about 80 °C). The terminal amino groups of peptides or other small molecules are easily functionalized using the alkynyl acid chlorides under mild reaction conditions at ambient temperatures (Figure 36).

Alternatively the ^{18}F label could be incorporated into the terminal alkynes by nucleophilic substitution reaction of the corresponding terminal-alkynyl-substituted alkyl tosylates, **47**, using ^{18}F-labeled fluoride ions, and these ^{18}F-labeled alkynes then reacted with terminal azido-substituted peptides (or various small molecule analogs of drug candidates) to give the corresponding triazole derivatives **49**. A relatively large, 37-residue low-pH insertion peptide (pHLIP) was labeled with ^{18}F isotope using the click protocol. Nucleophilic displacement of 6-bromo-2-ethynylpyridine **50** by K^{18}F gives the radiolabeled alkyne **51**, which reacts readily with azido-functionalized pHLIP peptide to give the corresponding triazole. In vitro assays revealed high stability of this peptide in human and mouse plasma after 120 min with the parent tracer remaining intact at 65% and 85%, respectively (Figure 37).[85]

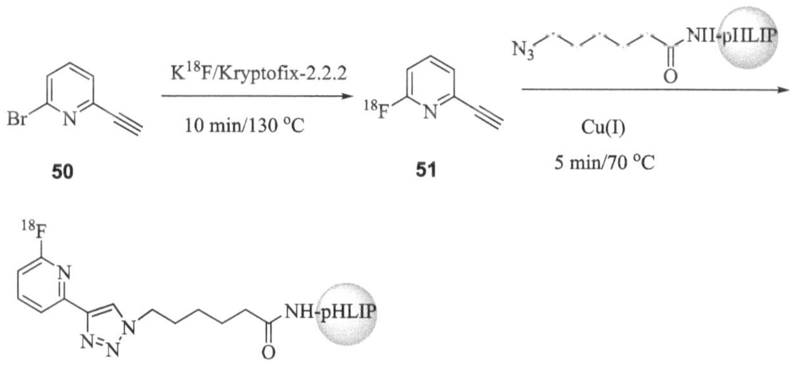

Radiochemical yield: 5-20%

FIGURE 37 ^{18}F labeling of a 37-residue low-pH insertion peptide (pHLIP) using the click chemistry.

Alternatively, synthesis of the radiolabeled peptides such as **54** could be achieved by the click reactions of ^{18}F-labeled 2-deoxy-2-fluoro-β-(D)-glucopyranosyl azide **52** with terminal-alkynyl-substituted peptides **53** (Figure38).[89] Similar triazole complexes derived from cRGD peptide and [^{18}F]-2-deoxy-2-fluoro-D-glucose (and those derived from other radiolabeled carbohydrates, such as [^{18}F]-6-deoxy-2-fluoro-D-glucose and [^{18}F] maltose) were prepared using the Cu(I)-catalyzed click chemistry from the corresponding azides and propargyl-tethered cRGD peptides. The latter arginylglycylaspartic acid (ArgGlyAsp; RGD)-based triazoles (**57**) have high specificity to the integrin receptor and favorable uptake and biodistribution in the U87MG tumor, and have favorable tissue clearance in vivo.

An RGD-based ^{18}F-PET tracer, Al[^{18}F]F-NOTA-PRGD2, in which the Al^{18}F is chelated to a macrocyclic chelator-conjugated dimeric RGD peptide, is currently in clinical studies to image tumor angiogenesis by targeting αvβ3 integrin receptors in tumors (Figure 39).[92] This complex is synthesized by a single-step halide exchange of Al-Cl bond by [^{18}F]F$^-$, and this synthetic method is amenable for the preparation of a library of analogous compounds.[93] It shows high tumor uptake and fast clearance from the body, and has good tumor to normal organ ratios.

7.3.2 Bioorthogonal Click Chemistry

Bioorthogonal coupling of strained dibenzocyclooctyne derivatives, **60**, with ^{18}F-labeled polyethyleneoxide-derived azides **59** gives the ^{18}F-labeled triazoles, **61**, under physiological conditions. The strained dibenzocyclooctyne derivatives react rapidly and with high chemoselectivity with azides, even in the presence of a plethora of functionality under physiological conditions. The unreacted cyclooctyne **60** is removed by chemoorthogonal purification reaction with an azide resin under these conditions. The synthesis of radiolabeled fluoroalkyl polyethyleneoxide azide

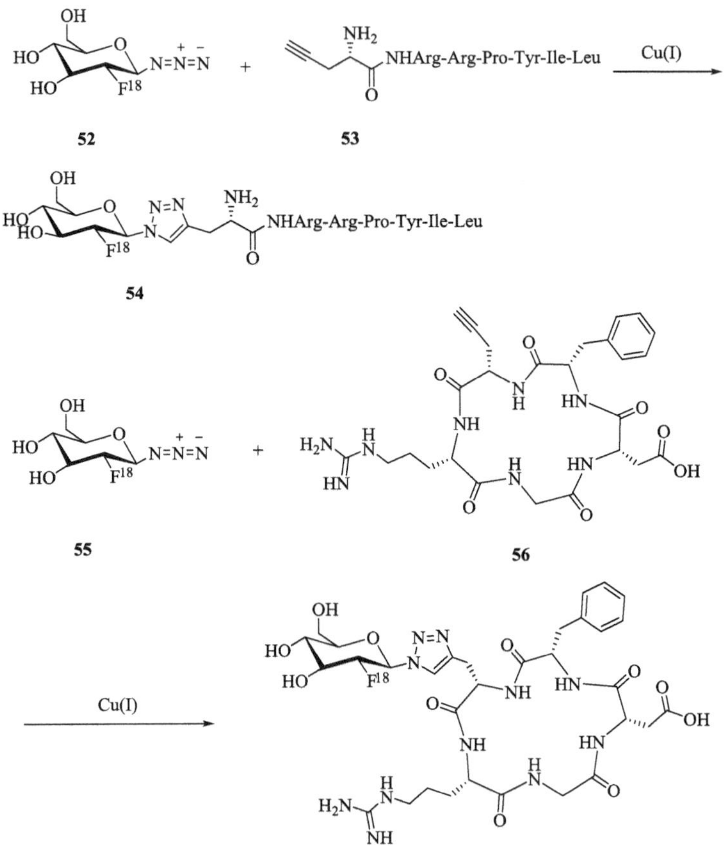

FIGURE 38 Click chemistry of the ^{18}F-labeled 2-deoxy-2-fluoro-β-(D)-glucopyranosyl azide with terminal-alkynyl-substituted peptides.

could be achieved in 63% decay-corrected RCYs from the corresponding mesylates using tetrabutylammonium fluoride(^{18}F).[83] ^{18}F-labeled bioactive peptides such as cRGD, bombesin, and apoptosis-targeting peptides were synthesized using this protocol (Figure 40).

A variety of ^{18}F-labeled azides such as 2-fluoroethylazide,[94] *p*-fluorobenzyl azide,[95] and 1-(3-azidopropyl)-4-(3-fluoropropyl)piperazine,[96] and alkynes such as 4-^{18}F-fluorobutyne and 1-(but-3-ynyl)-4-(3-fluoropropyl)piperazine suitable for the bioorthogonal click chemistry have been synthesized and used as PET tracers (Figure 41).[96]

8. ENZYMATIC SYNTHESIS OF ^{18}F-PET TRACERS

Fluorinase (5′-fluoro-5′-deoxyadenosine synthase, EC 2.5.1.63) enzyme-mediated conversion of S-adenosyl-L-methionine (SAM) to 5′-fluoro-5′-deoxyadenosine ([^{18}F]-5′-FDA) and L-methionine has attracted considerable attention, as in this

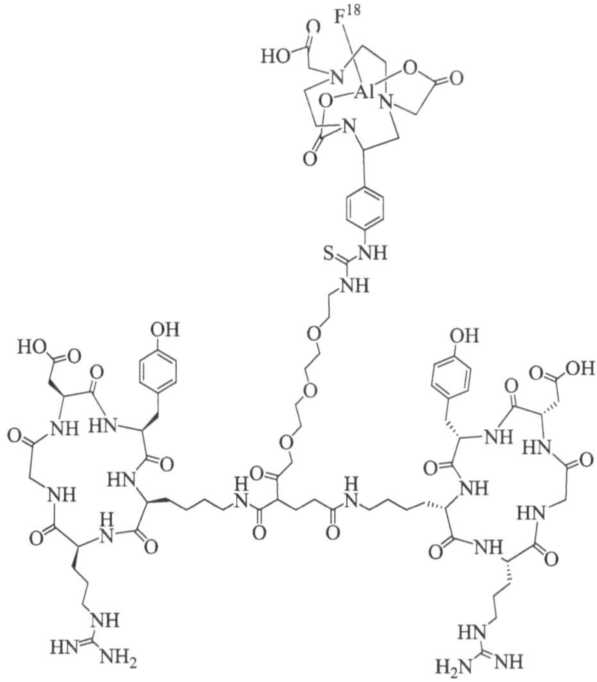

Al[18F]F-NOTA-PRGD2

FIGURE 39 Structure of Al[^{18}F]F-NOTA-PRGD2, used for PET imaging of tumor angiogenesis.

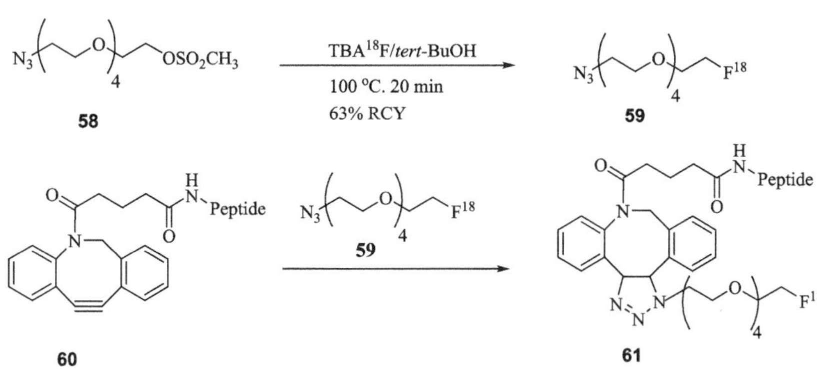

Peptide = e.g., cyclicRGD

FIGURE 40 Synthesis of ^{18}F-labeled bioactive peptides through bioorthogonal click chemistry (TBAF = tetrabutylammonium fluoride; *tert*-BuOH = *tert*-butyl alcohol; RCY = radiochemical yield).

process, the enzyme uses inorganic fluoride ion as the nucleophilic reactant. Moreover, the fluorinase enzyme can be produced in significant quantities (~40 mg/L) by fermentation of an engineered, overproducing *Escherichia coli* strain.[97] Enzymatic reaction of a mixture of fluorinase, immobilized purine nucleotide

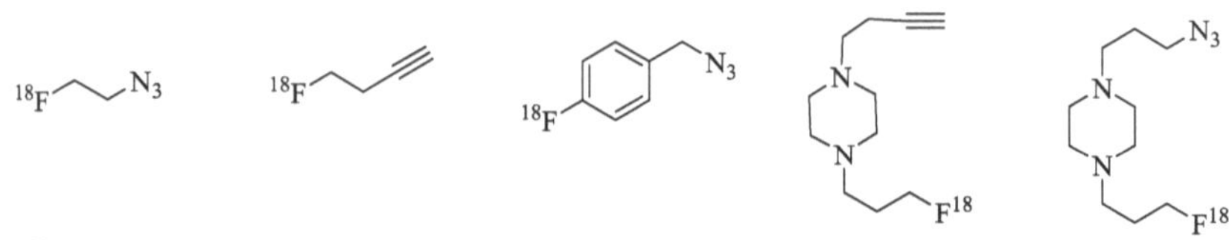

FIGURE 41 Structures of ^{18}F-labeled azides and alkynes useful for bioorthogonal click chemistry. [^{18}F]FEA, 2-fluoroethylazide; [^{18}F]FBnA, *p*-fluorobenzyl azide; [^{18}F]AFP, 1-(3-azidopropyl)-4-(3-fluoropropyl)piperazine; [^{18}F]BFP, 1-(but-3-ynyl)-4-(3-fluoropropyl)piperazine.

phosphorylase, and a phytase at 35 °C gives [^{18}F]-5′-FDR, a ^{18}F-labeled starting material for the synthesis of other carbohydrate derivatives.[98] An RCY of 45% was obtained after 4 h under these reaction conditions. On the other hand, enzymatic deamination of the radiolabeled ([^{18}F]-5′-FDA) using adenylic acid deaminase gives 5′-fluoro-5′-deoxyinosine ([^{18}F]-5′-FDI).[98]

Solid-phase fluorination using polymer-immobilized fluorinase has been developed for the synthesis of [^{18}F]-5′-FDA from SAM and ^{18}F-labeled fluoride ions. The fluorinase enzyme immobilized on the polymeric support, poly(glycidyl methacrylate-*co*-ethylene glycol dimethacrylate) is reusable up to four cycles without loss of appreciable enzyme activity, and the [^{18}F]-5′-FDA was synthesized with 68% fluorination efficiency (Figure 42).[99]

An X-ray crystal structural data of the substrate-bound fluorinase enzyme is in accordance with the S_N2 type of mechanism for this reaction. The C-F bond in the newly formed product is stabilized by hydrogen bonds with Ser 158-amide and OH groups, and the leaving group methionine and the CF bond are in antiperiplanar orientation (structural details of the enzyme-substrate complexes are provided in chapter 2).[100]

A ^{18}F-labeled 5-deoxy-5-fluoroadenosine can be conveniently transformed into fluoroacetic acid, a starting material for many other ^{18}F-labeled compounds. Thus, chromic acid oxidation of 5′-fluoro-5′-deoxyadenosine, derived from the fluorinase-catalyzed reaction, gives ^{18}F-fluoroacetic acid, isolated as the sodium salt in about 36% RCYs and over 96% radiochemical purity.[97] The RCYs were improved by adding a coenzyme, L-amino acid oxidase, to the fluorinating medium so that the byproduct methionine is oxidized, and thereby is incapable of reversing the equilibrium. Typically the SAM is incubated with the enzyme mixture and the no-carrier-added inorganic fluoride (e.g., $Na^{18}F$) for 30 min at 37 °C to obtain the 5′-fluoroadenosine in nearly quantitative RCYs.

9. SUMMARY AND OUTLOOK

The half-life of the ^{18}F isotope is the highest among all the positron-emitting isotopes, and therefore PET using ^{18}F-labeled pharmaceuticals, has practical advantage in the clinical settings. Until 2013, 2-[^{18}F]fluoro-2-deoxy-D-glucose (^{18}F-FDG) was the only FDA-approved compound for ^{18}F PET imaging. Three

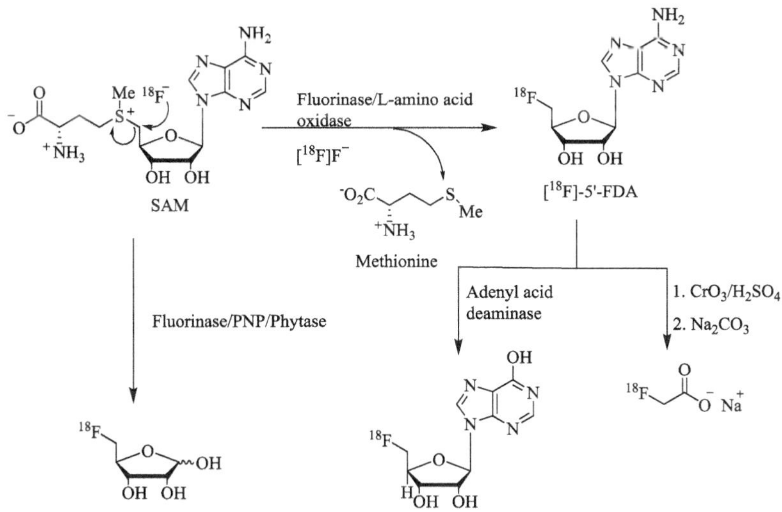

FIGURE 42 Biosynthesis of [^{18}F]-5′-FDI and [^{18}F]-5′-FDR.

other ^{18}F-labeled compounds have been approved in 2013 and 2014 for monitoring the Aβ plaques in the brains of AD patients: florbetapir-^{18}F ([^{18}F]AV-45; Amyvid Pharmaceuticals), florbetaben-^{18}F (Neuraceq; BAY 94-9172; Bayer Schering Pharma AG), and flutemetamol-^{18}F (Vizamyl; GE Healthcare) (vide supra). Many more ^{18}F-labeled pharmaceuticals are in clinical trials for human PET imaging.

Due to the relatively short half-life of the ^{18}F isotope, it is necessary to shorten the synthesis time for the ^{18}F-radiolabeled compound for PET imaging. Many efficient synthetic methods have been developed through either electrophilic or nucleophilic fluorinations. The electrophilic fluorination is achieved through using either the ^{18}F-labeled elemental F_2, or milder fluorinating agents, such as acetyl hypofluorite ($CH_3CO_2{}^{18}F$), [^{18}F]XeF_2, [^{18}F]N-fluoropyridinium triflate, [^{18}F]NFSI, or ([^{18}F] Selectfluor). The nucleophilic fluorination is generally achieved using ^{18}F-labeled reagents, $K^{18}F$ or $Bu_4N^{18}F$, in the presence of the aza-crown ether Kryptofix-2.2.2 as the phase transfer catalyst. ^{18}F-labeled peptides and proteins could also be synthesized using the latter method, for PET applications. For example, the ^{18}F-labeled cRGD peptide has been shown to be a viable PET tracer using an animal tumor xenograft model (vide supra).

Synthetic methodologies and automated synthesis for the ^{18}F labeling are being continuously improved so that minute samples of high specific activity ^{18}F-PET tracers are conveniently prepared in relatively short reaction time, onsite or in the vicinity of the PET-administering hospitals. Practically every drug candidate or proteins/peptides can be modified by introduction of the ^{18}F label, and thus many future PET tracers are expected to be those derived from ^{18}F-labeling.

REFERENCES

1. Vlassenko, A. G.; Benzinger, T. L. S.; Morris, J. C. PET Amyloid-beta Imaging in Pre-clinical Alzheimer's Disease. *Biochimica et Biophysica Acta - Mol. Basis Dis.* **2012,** *1822,* 370–379.
2. Coleman, R. E. Single Photon Emission Computed Tomography and Positron Emission Tomography in Cancer Imaging. *Cancer* **1991,** *67,* 1261–1270.
3. Ponde, D. E.; Dence, C. S.; Oyama, N.; Kim, J.; Tai, Y.-C.; Laforest, R.; Siegel, B. A.; Welch, M. J. ^{18}F-fluoroacetate: a Potential Acetate Analog for Prostate Tumor Imaging-in Vivo Evaluation of ^{18}F-fluoroacetate ^{11}C-acetate. *J. Nucl. Med.* **2007,** *48,* 420–428.
4. Couturier, O.; Luxen, A.; Chatal, J.-F.; Vuillez, J.-P.; Rigo, P.; Hustinx, R. Fluorinated Tracers for Imaging Cancer with Positron Emission Tomography. *Eur. J. Nucl. Med. Mol. Imaging* **2004,** *31,* 1182–1206.
5. Lasne, M.-C.; Perrio, C.; Rouden, J.; Barre, L.; Roeda, D.; Dolle, F.; Crouzel, C. Chemistry of β^+-emitting Compounds Based on Fluorine-18. *Top. Curr. Chem.* **2002,** *222,* 201–258.
6. Teare, H.; Robins, E. G.; Kirjavainen, A.; Forsback, S.; Sandford, G.; Solin, O.; Luthra, S. K.; Gouverneur, V. Radiosynthesis and Evaluation of [^{18}F]Selectfluor bis(triflate). *Angew. Chem. Int. Ed.* **2010,** *49,* 6821–6824.
7. Chang, C. W.; Wang, H. E.; Lin, H. M.; Chtsai, C. S.; Chen, J. B.; Liu, R. S. Robotic Synthesis of 6-[^{18}F]fluoro-L-dopa. *Nucl. Med. Commun.* **2000,** *21,* 799–802.
8. Stenhagen, I. S. R.; Kirjavainen, A. K.; Forsback, S. J.; Jorgensen, C. G.; Robins, E. G.; Luthra, S. K.; Solin, O.; Gouverneur, V. [^{18}F]Fluorination of an arylboronic ester using [^{18}F] selectfluor bis(triflate): Application to 6-[^{18}F]fluoro-L-DOPA. *Chem. Commun.* **2013,** *49,* 1386–1388.
9. Benzinger, T. L. S.; Blazey, T.; Jack, C. R., Jr; Koeppe, R. A.; Su, Y.; Xiong, C.; Raichle, M. E.; Snyder, A. Z.; Ances, B. M.; Bateman, R. J.; Cairns, N. J.; Fagan, A. M.; Goate, A.; Marcus, D. S.; Aisen, P. S.; Christensen, J. J.; Ercole, L.; Hornbeck, R. C.; Farrar, A. M.; Aldea, P.; Jasielec, M. S.; Owen, C. J.; Xie, X.; Mayeux, R.; Brickman, A.; McDade, E.; Klunk, W.; Mathis, C. A.; Ringman, J.; Thompson, P. M.; Ghetti, B.; Saykin, A. J.; Sperling, R. A.; Johnson, K. A.; Salloway, S.; Correia, S.; Schofield, P. R.; Masters, C. L.; Rowe, C.; Villemagne, V. L.; Martins, R.; Ourselin, S.; Rossor, M. N.; Fox, N. C.; Cash, D. M.; Weiner, M. W.; Holtzman, D. M.; Buckles, V. D.; Moulder, K.; Morris, J. C. Regional Variability of Imaging Biomarkers in Autosomal Dominant Alzheimer's Disease. *Proc. Natl. Acad. Sci. U.S.A.* **2013,** *110,* E4502–E4509.
10. Tredwell, M.; Gouverneur, V. ^{18}F Labeling of Arenes. *Angew. Chem. Int. Ed.* **2012,** *51,* 11426–11437.
11. Sachinidis, J. I.; Poniger, S.; Tochon-Danguy, H. J. Automation for Optimised Production of Fluorine-18-labelled Radiopharmaceuticals. *Curr. Radiopharm.* **2010,** *3,* 248–253.
12. Glaser, M.; Robins, E. G. Click Labelling' in PET Radiochemistry. *J. Label. Compd. Radiopharm.* **2009,** *52,* 407–414.
13. Zeng, F.; Goodman, M. M. Fluorine-18 Radiolabeled Heterocycles as PET Tracers for Imaging β-amyloid Plaques in Alzheimer's Disease. *Curr. Top. Med. Chem.* **2013,** *13,* 909–919.
14. Li, X.-G.; Helariutta, K.; Roivainen, A.; Jalkanen, S.; Knuuti, J.; Airaksinen, A. J. Using 5-deoxy-5-[^{18}F]fluororibose to Glycosylate Peptides for Positron Emission Tomography. *Nat. Protoc.* **2014,** *9,* 138–145.
15. Schlyer, D. J.; Volkow, N. D.; Fowler, J. S.; Wolf, A. P.; Shiue, C. Y.; Dewey, S. L.; Bendriem, B.; Logan, J.; Raulli, R.; et al. Regional Distribution and Kinetics of Haloperidol Binding in Human Brain: a PET Study with [^{18}F]haloperidol. *Synapse (N. Y.)* **1992,** *11,* 10–19.

16. Ding, Y. S.; Fowler, J. S.; Dewey, S. L.; Wolf, A. P.; Logan, J.; Gatley, S. J.; Volkow, N. D.; Shea, C.; Taylor, D. P. Synthesis and PET Studies of Fluorine-18-BMY 14802: a Potential Antipsychotic Drug. *J. Nucl. Med.* **1993,** *34,* 246–254.
17. http://www.Drugs.com/stats/top100.
18. Kingwell, K. Alzheimer Disease: Florbetapir-a Useful Tool to Image Amyloid Load and Predict Cognitive Decline in Alzheimer Disease. *Nat. Rev. Neurol* **2012,** *8(9),* 471.
19. Kung, H. F. The β-amyloid Hypothesis in Alzheimer's Disease: Seeing Is Believing. *ACS Med. Chem. Lett.* **2012,** *3,* 265–267.
20. Sperling, R. A.; Johnson, K. A.; Doraiswamy, P. M.; Reiman, E. M.; Fleisher, A. S.; Sabbagh, M. N.; Sadowsky, C. H.; Carpenter, A.; Davis, M. D.; Lu, M.; Flitter, M.; Joshi, A. D.; Clark, C. M.; Grundman, M.; Mintun, M. A.; Skovronsky, D. M.; Pontecorvo, M. J. Amyloid Deposition Detected with Florbetapir F-18 (^{18}F-AV-45) Is Related to Lower Episodic Memory Performance in Clinically Normal Older Individuals. *Neurobiol. Aging* **2013,** *34,* 822–831.
21. Villemagne, V. L.; Mulligan, R. S.; Pejoska, S.; Ong, K.; Jones, G.; O'Keefe, G.; Chan, J. G.; Young, K.; Tochon-Danguy, H.; Masters, C. L.; Rowe, C. C. Comparison of ^{11}C-PiB and ^{18}F-Florbetaben for Aβ Imaging in Ageing and Alzheimer's Disease. *Eur. J. Nucl. Med. Mol. Imaging* **2012,** *39,* 983–989.
22. Barthel, H.; Gertz, H.-J.; Dresel, S.; Peters, O.; Bartenstein, P.; Buerger, K.; Hiemeyer, F.; Wittemer-Rump, S. M.; Seibyl, J.; Reininger, C.; Sabri, O. Cerebral Amyloid-β PET with Florbetaben (^{18}F) in Patients with Alzheimer's Disease and Healthy Controls: a Multicentre Phase 2 Diagnostic Study. *Lancet Neurol.* **2011,** *10,* 424–435.
23. Anon Alzheimer's imaging agents struggle to find a market outside trials. *Nat. Med. (N. Y., NY, U. S.)* **2013,** *19,* 1551.
24. www.drugs.com.
25. Ashford, J. W.; Rosen, A.; Adamson, M.; Bayley, P.; Sabri, O.; Furst, A.; Black, S. E.; Weiner, M.; Shin, J.; Kepe, V.; Barrio, J. R.; Small, G. W. The Merits of FDDNP-PET Imaging in Alzheimer's Disease. *J. Alzheimer's Dis.* **2011,** *26,* 135–145.
26. Chang, K. W.; Lee, S. Y.; Lin, W. J.; Chen, C. C.; Chen, J. T.; Wang, H. E. Auto-synthesis of Fluoro-18-FDDNP on Alzheimer's Research. *J. Label. Compd. Radiopharm.* **2010,** *53,* 414–417.
27. Small, G. W.; Kepe, V.; Ercoli, L. M.; Siddarth, P.; Bookheimer, S. Y.; Miller, K. J.; Lavretsky, H.; Burggren, A. C.; Cole, G. M.; Vinters, H. V.; Thompson, P. M.; Huang, S. C.; Satyamurthy, N.; Phelps, M. E.; Barrio, J. R. PET of Brain Amyloid and Tau in Mild Cognitive Impairment. *N. Engl. J. Med.* **2006,** *355,* 2652–2663.
28. Agdeppa, E. D.; Kepe, V.; Liu, J.; Flores-Torres, S.; Satyamurthy, N.; Petric, A.; Cole, G. M.; Small, G. W.; Huang, S. C.; Barrio, J. R. Binding Characteristics of Radiofluorinated 6-dialkylamino-2-naphthylethylidene Derivatives as Positron Emission Tomography Imaging Probes for Beta-amyloid Plaques in Alzheimer's Disease. *J. Neurosci.* **2001,** *21.*
29. Nordberg, A. PET Imaging of Amyloid in Alzheimer's Disease. *Lancet Neurol.* **2004,** *3,* 519–527.
30. Irwin, D. J.; Lee, V. M. Y.; Trojanowski, J. Q. Parkinson's disease dementia: convergence of α-synuclein, tau and amyloid-β pathologies. *Nat. Rev. Neurosci* **2013,** *14,* 626–636.
31. Buongiorno, M.; Compta, Y.; Marti, M. J. Amyloid-β and τ biomarkers in Parkinson's disease-dementia. *J. Neurol. Sci* **2011,** *310,* 25–30.
32. Ossenkoppele, R.; Tolboom, N.; Foster-Dingley, J. C.; Adriaanse, S. F.; Boellaard, R.; Yaqub, M.; Windhorst, A. D.; Barkhof, F.; Lammertsma, A. A.; Scheltens, P.; Flier, W. M.; Berckel, B. N. M. Longitudinal imaging of Alzheimer pathology using [^{11}C]PIB, [^{18}F]FDDNP and [^{18}F]FDG PET. *Eur. J. Nucl. Med. Mol. Imaging* **2012,** *39,* 990–1000.

33. Rowe, C. C.; Pejoska, S.; Mulligan, R. S.; Jones, G.; Chan, J. G.; Svensson, S.; Cselenyi, Z.; Masters, C. L.; Villemagne, V. L. Head-to-head Comparison of ^{11}C-PiB and ^{18}F-AZD4694 (NAV4694) for β-amyloid Imaging in Aging and Dementia. *J. Nucl. Med.* **2013,** *54,* 880–886.
34. Alzheimer's Imaging Agents Struggle to Find a Market outside Trials. *Nat. Med.* **2013,** *19,* 1551.
35. Zha, Z.; Choi, S. R.; Ploessl, K.; Lieberman, B. P.; Qu, W.; Hefti, F.; Mintun, M.; Skovronsky, D.; Kung, H. F. Multidentate ^{18}F-Polypegylated Styrylpyridines as Imaging Agents for Aβ Plaques in Cerebral Amyloid Angiopathy (CAA). *J. Med. Chem.* **2011,** *54,* 8085–8098.
36. Lemaire, C.; Libert, L.; Plenevaux, A.; Aerts, J.; Franci, X.; Luxen, A. Fast and Reliable Method for the Preparation of *ortho- and para-*[^{18}F]fluorobenzyl Halide Derivatives: Key Intermediates for the Preparation of No-Carrier-Added PET Aromatic Radiopharmaceuticals. *J. Fluorine Chem.* **2012,** *138,* 48–55.
37. Kim, D. W.; Ahn, D.-S.; Oh, Y.-H.; Lee, S.; Kil, H. S.; Oh, S. J.; Lee, S. J.; Kim, J. S.; Ryu, J. S.; Moon, D. H.; Chi, D. Y. A New Class of S_N2 Reactions Catalyzed by Protic Solvents: Facile Fluorination for Isotopic Labeling of Diagnostic Molecules. *J. Am. Chem. Soc.* **2006,** *128,* 16394–16397.
38. Hostetler, E. D.; Sanabria-Bohorquez, S.; Fan, H.; Zeng, Z.; Gantert, L.; Williams, M.; Miller, P.; O'Malley, S.; Kameda, M.; Ando, M.; Sato, N.; Ozaki, S.; Tokita, S.; Ohta, H.; Williams, D.; Sur, C.; Cook, J. J.; Burns, H. D.; Hargreaves, R. Synthesis, Characterization, and Monkey Positron Emission Tomography (PET) Studies of [^{18}F]Y1-973, a PET Tracer for the Neuropeptide Y Y1 Receptor. *Neuroimage* **2011,** *54,* 2635–2642.
39. Hwang, D. R.; Eckelman, W. C.; Mathias, C. J.; Petrillo, E. W., Jr; Lloyd, L.; Welch, M. J. Positron-labeled Angiotensin-converting Enzyme (ACE) Inhibitor: Fluorine-18-fluorocaptopril. Probing the ACE Activity In Vivo by Positron Emission Tomography. *J. Nucl. Med.* **1991,** *32,* 1730–1737.
40. Nag, S.; Lehmann, L.; Heinrich, T.; Thiele, A.; Kettschau, G.; Nakao, R.; Gulyas, B.; Halldin, C. Synthesis of Three Novel Fluorine-18 Labeled Analogues of L-deprenyl for Positron Emission Tomography (PET) Studies of Monoamine Oxidase B (MAO-B). *J. Med. Chem.* **2011,** *54,* 7023–7029.
41. Liu, Y.; Zhu, L.; Ploessl, K.; Choi, S. R.; Qiao, H.; Sun, X.; Li, S.; Zha, Z.; Kung, H. F. Optimization of Automated Radiosynthesis of [^{18}F]AV-45: a New PET Imaging Agent for Alzheimer's Disease. *Nucl. Med. Biol.* **2010,** *37,* 917–925.
42. Moon, D. H.; Chi, D. Y.; Kim, D. W.; Oh, S. J.; Ryu, J.-S. Method for Preparation of Organofluoro Compounds in Alcohol Solvents. US patent 2009-605518; 20100113763 *Chem. Abstr.* **2010,** *152,* 525573 (Futurechem Co., Ltd., S. Korea; The Asan Foundation).
43. Zhang, W.; Arteaga, J.; Cashion, D. K.; Chen, G.; Gangadharmath, U.; Gomez, L. F.; Kasi, D.; Lam, C.; Liang, Q.; Liu, C.; Mocharla, V. P.; Mu, F.; Sinha, A.; Szardenings, A. K.; Wang, E.; Walsh, J. C.; Xia, C.; Yu, C.; Zhao, T.; Kolb, H. C. A Highly Selective and Specific PET Tracer for Imaging of Tau Pathologies. *J. Alzheimer's Dis.* **2012,** *31,* 601–612.
44. Gao, M.; Wang, M.; Zheng, Q.-H. Concise and High-yield Synthesis of T808 and T808P for Radiosynthesis of [^{18}F]-T808, a PET Tau Tracer for Alzheimer's Disease. *Bioorg. Med. Chem. Lett.* **2014,** *24,* 254–257.
45. Hayashi, K.; Furutsuka, K.; Ito, T.; Muto, M.; Aki, H.; Fukumura, T.; Suzuki, K. Fully Automated Synthesis and Purification of 4-(2'-methoxyphenyl)-1-[2'-(N-2'-pyridinyl)-*p*-[^{18}F]fluorobenzamido]ethylpiperazine. *J. Label. Compd. Radiopharm.* **2012,** *55,* 120–124.
46. Wooten, D. W.; Moraino, J. D.; Hillmer, A. T.; Engle, J. W.; Dejesus, O. J.; Murali, D.; Barnhart, T. E.; Nickles, R. J.; Davidson, R. J.; Schneider, M. L.; Mukherjee, J.; Christian, B. T. In Vivo Kinetics of [F-18]MEFWAY: a Comparison with [C-11]WAY100635 and [F-18]MPPF in the Nonhuman Primate. *Synapse* **2011,** *65,* 592–600.

47. Massaweh, G.; Schirrmacher, E.; La Fougere, C.; Kovacevic, M.; Waengler, C.; Jolly, D.; Gravel, P.; Reader, A. J.; Thiel, A.; Schirrmacher, R. Improved Work-up Procedure for the Production of [^{18}F]flumazenil and first results of its use with a high-resolution research tomograph in human stroke. *Nucl. Med. Biol.* **2009,** *36,* 721–727.
48. Chun, J.-H.; Pike, V. W. Single-step Syntheses of No-Carrier-Added Functionalized [^{18}F] fluoroarenes as Labeling Synthons from Diaryliodonium Salts. *Org. Biomol. Chem.* **2013,** *11,* 6300–6306.
49. Ross, T. L.; Ermert, J.; Hocke, C.; Coenen, H. H. Nucleophilic ^{18}F-Fluorination of Heteroaromatic Iodonium Salts with No-Carrier-Added [^{18}F]fluoride. *J. Am. Chem. Soc.* **2007,** *129,* 8018–8025.
50. Shah, A.; Pike, V. W.; Widdowson, D. A. Synthesis of [^{18}F]fluoroarenes from the Reaction of Cyclotron-Produced [^{18}F]fluoride Ion with Diaryliodonium Salts. *J. Chem. Soc., Perkin Trans.* **1998,** *1,* 2043–2046.
51. Chun, J.-H.; Lu, S.; Pike, V. W. Rapid and Efficient Radiosyntheses of *meta*-Substituted [^{18}F] fluoroarenes from [^{18}F]fluoride Ion and Diaryliodonium Tosylates within a Microreactor. *Eur. J. Org. Chem.* **2011,** *2011,* 4439–4447.
52. Gao, Z.; Gouverneur, V.; Davis, B. G. Enhanced Aqueous Suzuki-Miyaura Coupling Allows Site-Specific Polypeptide ^{18}F-labeling. *J. Am. Chem. Soc.* **2013,** *135,* 13612–13615.
53. Lee, E.; Kamlet, A. S.; Powers, D. C.; Neumann, C. N.; Boursalian, G. B.; Furuya, T.; Choi, D. C.; Hooker, J. M.; Ritter, T. A Fluoride-derived Electrophilic Late-stage Fluorination Reagent for PET Imaging. *Science* **2011,** *334,* 639–642.
54. Campbell, M. G.; Ritter, T. Late-Stage Fluorination: From Fundamentals to Application. *Org. Process Res. Dev.* **2014,** *18,* 474–480.
55. Kamlet, A. S.; Neumann, C. N.; Lee, E.; Carlin, S. M.; Moseley, C. K.; Stephenson, N.; Hooker, J. M.; Ritter, T. Application of Palladium-Mediated ^{18}F-fluorination to PET Radiotracer Development: Overcoming Hurdles to Translation. *PLoS One* **2013,** *8,* e59187.
56. Huiban, M.; Tredwell, M.; Mizuta, S.; Wan, Z.; Zhang, X.; Collier, T. L.; Gouverneur, V.; Passchier, J. A Broadly Applicable [^{18}F]trifluoromethylation of Aryl and Heteroaryl Iodides for PET Imaging. *Nat. Chem.* **2013,** *5,* 941–944.
57. Prakash, G. K. S.; Hu, J.; Alauddin, M. M.; Conti, P. S.; Olah, G. A. A General Method of Halogenation for Synthesis of α-halodifluoromethyl Ketones and [^{18}F]-Labeled Trifluoromethyl Ketones. *J. Fluorine Chem.* **2003,** *121,* 239–243.
58. Prakash, G. K. S.; Alauddin, M. M.; Hu, J.; Conti, P. S.; Olah, G. A. Expedient Synthesis of [^{18}F]-Labeled α-trifluoromethyl Ketones. *J. Labelled Compd. Radiopharm.* **2003,** *46,* 1087–1092.
59. Newberg, A. B.; Moss, A. S.; Monti, D. A.; Alavi, A. Positron Emission Tomography in Psychiatric Disorders. *Ann. N. Y. Acad. Sci.* **2011,** *1228,* E13–E25.
60. Någren, K.; Halldin, C.; Rinne, J. O. Radiopharmaceuticals for Positron Emission Tomography Investigations of Alzheimer's Disease. *Eur. J. Nucl. Med. Mol. Imaging* **2010,** *37,* 1575–1593.
61. Kadir, A.; Nordberg, A. Target-specific PET Probes for Neurodegenerative Disorders Related to Dementia. *J. Nucl. Med.* **2010,** *51,* 1418–1430.
62. Namavari, M.; Bishop, A.; Satyamurthy, N.; Bida, G.; Barrio, J. R. Regioselective Radiofluorodestannylation with Fluorine-18 and [^{18}F] Acetyl Hypofluorite: a High Yield Synthesis of 6-[^{18}F]fluoro-L-DOPA. *Appl. Radiat. Isot.* **1992,** *43,* 989–996.
63. Stenhagen, I. S. R.; Kirjavainen, A. K.; Forsback, S. J.; Jorgensen, C. G.; Robins, E. G.; Luthra, S. K.; Solin, O.; Gouverneur, V. [^{18}F]Fluorination of an Arylboronic Ester Using [^{18}F] selectfluor Bis(triflate): Application to 6-[^{18}F]fluoro-L-DOPA. *Chem. Commun.* **2013,** *49,* 1386–1388.

64. Moore, R. Y.; Whone, A. L.; McGowan, S.; Brooks, D. J. Monoamine Neuron Innervation of the Normal Human Brain: an ^{18}F-DOPA PET Study. *Brain Res.* **2003,** *982,* 137–145.
65. Ding, Y. S.; Fowler, J. S.; Gatley, S. J.; Dewey, S. L.; Wolf, A. P. Synthesis of High Specific Activity (+)- and (-)-6-[^{18}F]fluoronorepinephrine via the Nucleophilic Aromatic Substitution Reaction. *J. Med. Chem.* **1991,** *34,* 767–771.
66. Lee, J. H.; Peters, O.; Lehmann, L.; Dence, C. S.; Sharp, T. L.; Carlson, K. E.; Zhou, D.; Jeyakumar, M.; Welch, M. J.; Katzenellenbogen, J. A. Synthesis and Biological Evaluation of Two Agents for Imaging Estrogen Receptor β by Positron Emission Tomography: Challenges in PET Imaging of a Low Abundance Target, *Nucl. Med. Biol.* **2012,** *39,* 1105–1116.
67. Kurland, B. F.; Peterson, L. M.; Lee, J. H.; Linden, H. M.; Schubert, E. K.; Dunnwald, L. K.; Link, J. M.; Krohn, K. A.; Mankoff, D. A. Between-Patient and Within-Patient (Site-to-Site) Variability in Estrogen Receptor Binding, Measured In Vivo by ^{18}F-fluoroestradiol PET. *J. Nucl. Med.* **2011,** *52,* 1541–1549.
68. Fowler, A. M.; Chan, S. R.; Sharp, T. L.; Fettig, N. M.; Zhou, D.; Dence, C. S.; Carlson, K. E.; Jeyakumar, M.; Katzenellenbogen, J. A.; Schreiber, R. D.; Welch, M. J. Small-animal PET of Steroid Hormone Receptors Predicts Tumor Response to Endocrine Therapy Using a Preclinical Model of Breast Cancer. *J. Nucl. Med.* **2012,** *53,* 1119–1126.
69. Hong, H.; Zhang, Y.; Sun, J.; Cai, W. Positron Emission Tomography Imaging of Prostate Cancer. *Amino Acids* **2010,** *39,* 11–27.
70. Bonasera, T. A.; O'Neil, J. P.; Xu, M.; Dobkin, J. A.; Cutler, P. D.; Lich, L. L.; Choe, Y. S.; Katzenellenbogen, J. A.; Welch, M. J. Preclinical Evaluation of Fluorine-18-Labeled Androgen Receptor Ligands in Baboons. *J. Nucl. Med.* **1996,** *37,* 1009.
71. Larson, S. M.; Morris, M.; Gunther, I.; Beattie, B.; Humm, J. L.; Akhurst, T. A.; Finn, R. D.; Erdi, Y.; Pentlow, K.; Dyke, J.; Squire, O.; Bornmann, W.; McCarthy, T.; Welch, M.; Scher, H. Tumor Localization of 16β-^{18}F-fluoro-5α-dihydrotestosterone versus ^{18}F-FDG in Patients with Progressive, Metastatic Prostate Cancer. *J. Nucl. Med.* **2004,** *45,* 366–373.
72. Knott, K. E.; Graetz, D.; Huebner, S.; Juettler, S.; Zankl, C.; Mueller, M. Simplified and Automatic One-Pot Synthesis of 16α-[^{18}F]fluoroestradiol without High-Performance Liquid Chromatography Purification. *J. Label. Compd. Radiopharm.* **2011,** *54,* 749–753.
73. Kang, S. H.; Oh, S. J.; Yoon, M. K.; Ryu, J. S.; Lee, W. K.; Choi, S. J.; Park, K. P.; Moon, D. H. Simple and High Radiochemical Yield Synthesis of 2'-Deoxy-2'-[^{18}F]fluorouridine via a New Nosylate Precursor. *J. Labelled Compd. Radiopharm.* **2006,** *49,* 1237–1246.
74. Plotnik, D. A.; McLaughlin, L. J.; Krohn, K. A.; Schwartz, J. L. The Effects of 5-fluoruracil Treatment on 3'-fluoro-3'-deoxythymidine (FLT) Transport and Metabolism in Proliferating and Non-proliferating Cultures of Human Tumor Cells. *Nucl. Med. Biol.* **2012,** *39,* 970–976.
75. Wu, C. Y.; Wang, H. E.; Lin, M. H.; Chu, L. S.; Liu, R. S. Radiolabeled Nucleosides for Predicting and Monitoring the Cancer Therapeutic Efficacy of Chemodrugs. *Curr. Med. Chem.* **2012,** *19,* 3315–3324.
76. Shinde, S. S.; Lee, B. S.; Chi, D. Y. Synergistic Effect of Two Solvents, *tert*-Alcohol and Ionic Liquid, in One Molecule in Nucleophilic Fluorination. *Org. Lett.* **2008,** *10,* 733–735.
77. Lee, S.-J.; Oh, S.-J.; Moon, W.-Y.; Choi, M.-S.; Kim, J.-S.; Chi, D.-Y.; Moon, D.-H.; Ryu, J.-S. New Automated Synthesis of [^{18}F]FP-CIT with Base Amount Control Affording High and Stable Radiochemical Yield: a 1.5-year Production Report. *Nucl. Med. Biol.* **2011,** *38,* 593–597.
78. Lee, S. J.; Oh, S. J.; Chi, D. Y.; Kil, H. S.; Kim, E. N.; Ryu, J. S.; Moon, D. H. Simple and Highly Efficient Synthesis of 3'-deoxy-3'-[^{18}F]fluorothymidine Using Nucleophilic Fluorination Catalyzed by Protic Solvent. *Eur. J. Nucl. Med. Mol. Imaging* **2007,** *34,* 1406–1409.

79. Lee, S. J.; Oh, S. J.; Chi, D. Y.; Kang, S. H.; Kil, H. S.; Kim, J. S.; Moon, D. H. One-Step High-Radiochemical Yield Synthesis of [^{18}F]FP CIT Using a Protic Solvent System. *Nucl. Med. Biol.* **2007,** *34,* 345–351.
80. Kostikov, A. P.; Chin, J.; Orchowski, K.; Niedermoser, S.; Kovacevic, M. M.; Aliaga, A.; Jurkschat, K.; Wangler, B.; Wangler, C.; Wester, H.-J.; Schirrmacher, R. Oxalic Acid Supported Si-^{18}F-Radiofluorination: One-Step Radiosynthesis of N-Succinimidyl 3-(Di-*tert*-butyl[^{18}F]fluorosilyl)benzoate ([^{18}F]SiFB) for Protein Labeling, *Bioconjug. Chem.* **2012,** *23,* 106–114.
81. Kostikov, A. P.; Chin, J.; Orchovski, K.; Schirrmacher, E.; Niedermoser, S.; Jurkschat, K.; Iovkova-Berends, L.; Waengler, C.; Waengler, B.; Schirrmacher, R. Synthesis of [^{18}F]SiFB: a Prosthetic Group for Direct Protein Radiolabeling for Application in Positron Emission Tomography. *Nat. Protoc.* **2012,** *7,* 1956–1963.
82. Jeon, J.; Shen, B.; Xiong, L.; Miao, Z.; Lee, K. H.; Rao, J.; Chin, F. T. Efficient Method for Site-specific ^{18}F-Labeling of Biomolecules Using the Rapid Condensation Reaction between 2-Cyanobenzothiazole and Cysteine. *Bioconjug. Chem.* **2012,** *23,* 1902–1908.
83. Sachin, K.; Jadhav, V. H.; Kim, E.-M.; Kim, H. L.; Lee, S. B.; Jeong, H.-J.; Lim, S. T.; Sohn, M.-H.; Kim, D. W. F-18 Labeling Protocol of Peptides Based on Chemically Orthogonal Strain-Promoted Cycloaddition under Physiologically Friendly Reaction Conditions. *Bioconjug. Chem.* **2012,** *23,* 1680–1686.
84. Jacobson, O.; Zhu, L.; Ma, Y.; Weiss, I. D.; Sun, X.; Niu, G.; Kiesewetter, D. O.; Chen, X. Rapid and Simple One-Step F-18 Labeling of Peptides. *Bioconjug. Chem.* **2011,** *22,* 422–428.
85. Daumar, P.; Wanger-Baumann, C. A.; Pillarsetty, N.; Fabrizio, L.; Carlin, S. D.; Andreev, O. A.; Reshetnyak, Y. K.; Lewis, J. S. Efficient ^{18}F-labeling of Large 37-amino-acid PHLIP Peptide Analogues and Their Biological Evaluation. *Bioconjug. Chem.* **2012,** *23,* 1557–1566.
86. Flagothier, J.; Kaisin, G.; Mercier, F.; Thonon, D.; Teller, N.; Wouters, J.; Luxen, A. Synthesis of Two New Alkyne-Bearing Linkers Used for the Preparation of siRNA for Labeling by Click Chemistry with Fluorine-18. *Appl. Radiat. Isot.* **2012,** *70,* 1549–1557.
87. Carpenter, R. D.; Hausner, S. H.; Sutcliffe, J. L. Copper-Free Click for PET: Rapid 1,3-Dipolar Cycloadditions with a Fluorine-18 Cyclooctyne. *ACS Med. Chem. Lett.* **2011,** *2,* 885–889.
88. Hausner, S. H.; Marik, J.; Gagnon, M. K. J.; Sutcliffe, J. L. In Vivo Positron Emission Tomography (PET) Imaging with an Avβ6 Specific Peptide Radiolabeled Using ^{18}F-"Click" Chemistry: Evaluation and Comparison with the Corresponding 4-[^{18}F]fluorobenzoyl- and 2-[^{18}F]fluoropropionyl-Peptides. *J. Med. Chem.* **2008,** *51,* 5901–5904.
89. Maschauer, S.; Einsiedel, J.; Haubner, R.; Hocke, C.; Ocker, M.; Huebner, H.; Kuwert, T.; Gmeiner, P.; Prante, O. Labeling and Glycosylation of Peptides Using Click Chemistry: A General Approach to ^{18}F-Glycopeptides as Effective Imaging Probes for Positron Emission Tomography. *Angew. Chem. Int. Ed.* **2010,** *49,* 976–979. S976/971-S976/918.
90. Li, J.; Shi, L.; Jia, L.; Jiang, D.; Zhou, W.; Hu, W.; Qi, Y.; Zhang, L. Radiolabeling of RGD Peptide and Preliminary Biological Evaluation in Mice Bearing U87MG Tumors. *Bioorg. Med. Chem.* **2012,** *20,* 3850–3855.
91. Gill, H. S.; Marik, J. Preparation of ^{18}F-labeled Peptides Using the Copper(I)-catalyzed Azide-alkyne 1,3-dipolar Cycloaddition. *Nat. Protoc.* **2011,** *6,* 1718–1725.
92. Lang, L.; Li, W.; Guo, N.; Ma, Y.; Zhu, L.; Kiesewetter, D. O.; Shen, B.; Niu, G.; Chen, X. Comparison Study of [^{18}F]FAl-NOTA-PRGD2, [^{18}F]FPPRGD2, and [^{68}Ga]Ga-NOTA-PRGD2 for PET Imaging of U87MG Tumors in Mice. *Bioconjug. Chem.* **2011,** *22,* 2415–2422.
93. Liu, S.; Liu, H.; Jiang, H.; Xu, Y.; Zhang, H.; Cheng, Z. One-step Radiosynthesis of ^{18}F-AlFNOTA-RGD2 for Tumor Angiogenesis PET Imaging. *Eur. J. Nucl. Med. Mol. Imaging* **2011,** *38,* 1732–1741.

94. Antunes, I. F.; Haisma, H. J.; Elsinga, P. H.; Sijbesma, J. W. A.; Waarde, A. v.; Willemsen, A. T. M.; Dierckx, R. A.; de Vries, E. F. J. In Vivo Evaluation of [^{18}F]FEAnGA-Me: a PET Tracer for Imaging β-glucuronidase (β-GUS) Activity in a Tumor/Inflammation Rodent Model. *Nucl. Med. Biol.* **2012,** *39,* 854–863.
95. Haradahira, T.; Hasegawa, Y.; Furuta, K.; Suzuki, M.; Watanabe, Y.; Suzuki, K. Synthesis of a F-18 Labeled Analog of Antitumor Prostaglandin Δ^7-*p*GA$_1$ Methyl Ester Using *p*-[^{18}F]fluorobenzylamine. *Appl. Radiat. Isot.* **1998,** *49,* 1551–1556.
96. Pretze, M.; Mamat, C. Automated Preparation of [^{18}F]AFP and [^{18}F]BFP: Two Novel Bifunctional ^{18}F-labeling Building Blocks for Huisgen-Click. *J. Fluorine Chem.* **2013,** *150,* 25–35.
97. Li, X.-G.; Domarkas, J.; O'Hagan, D. Fluorinase Mediated Chemoenzymatic Synthesis of [^{18}F]-Fluoroacetate. *Chem. Commun.* **2010,** *46,* 7819–7821.
98. Deng, H.; Cobb, S. L.; Gee, A. D.; Lockhart, A.; Martarello, L.; McGlinchey, R. P.; O'Hagan, D.; Onega, M. Fluorinase Mediated C-^{18}F Bond Formation, an Enzymatic Tool for PET Labelling. *Chem. Commun.* **2006,** 652–654.
99. Sergeev, M. E.; Morgia, F.; Javed, M. R.; Doi, M.; Keng, P. Y. Polymer-immobilized Fluorinase: Recyclable Catalyst for Fluorination Reactions. *J. Mol. Catal. B Enzym.* **2013,** *92,* 51–56.
100. Dong, C.; Huang, F.; Deng, H.; Schaffrath, C.; Spencer, J. B.; O'Hagan, D.; Naismith, J. H. Crystal Structure and Mechanism of a Bacterial Fluorinating Enzyme. *Nature* **2004,** *427,* 561–565.

Chapter 8

Organofluorine Compounds in Neurological Disorders

Chapter Outline

1. INTRODUCTION

Oxidative stress is invariably involved in the onset of neurological disorders, including Parkinson disease (PD), Alzheimer disease (AD), amyotrophic lateral sclerosis (ALS), and Pick's disease. Among these diseases, AD is the least understood pathology in terms of its mechanism of onset, and therefore there are no viable treatment options for this disease to date. While the Centers for Disease Control and Prevention estimated that AD is the sixth leading cause of death in the United States,[1] other studies suggests that it is the third leading cause of death among people aged 75 years and older, heart disease and cancer being the first and second leading causes, respectively.[2] About 30 million people worldwide are inflicted by AD, and 30–40% of those older than 85 years suffer from this disease.[3] Alzheimer's Association estimates that nearly 5.4 million people in the United States are affected by AD, and that if the current trend continues, the AD population is expected to reach 16 million by the year 2050.[4,5] The clinical features of AD include progressive and irreversible memory loss, learning disabilities, and other psychiatric features such as dementia.

The pharmaceuticals currently available for treating AD symptoms, including the acetylcholinesterase inhibitors, N-methyl-D-aspartate (NMDA) receptor antagonists, and antiinflammatory drugs, are at best palliative, and memory

Organofluorine Compounds in Biology and Medicine. http://dx.doi.org/10.1016/B978-0-444-53748-5.00008-3

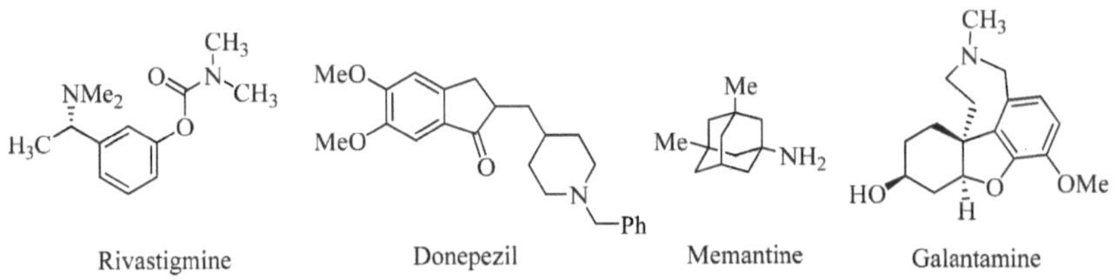

FIGURE 1 Structures of rivastigmine, donepezil, memantine, and galantamine, the palliative therapeutics for Alzheimer disease (AD).

restoration through the use of these drugs could not be achieved for any reasonable time periods. Rivastigmine, donepezil, and galantamine are the acetylcholinesterase inhibitors that are currently used in clinical practice to treat AD symptoms. Memantine is an NMDA receptor antagonist, and is used in combination with acetylcholinesterase inhibitors and antiinflammatory agents (Figure 1).[6] There is hope that with increasing sophistication in the technology for early diagnosis and with ever-increasing understanding of the molecular mechanisms involved in the onset of AD, novel and effective therapeutics that target the major causative factors of AD, including amyloid-β (Aβ) peptides, neurofibrillary tangles (NFTs), and cell cycle inhibitors (vide infra), may be able to cure or circumvent AD and related neurological disorders in the near future.

AD patients suffer from progressive decline in cognitive function that leads to the eventual loss of autonomous control. Majority of the cases of AD are sporadic, that is, the onset of AD is due to environmental and other exogenous risk factors as opposed to familial AD, which is a rare autosomal dominant disease resulting from mutations in the Aβ-precursor protein (APP) and presenilin genes. The onset of familial AD usually occurs before the age of 65 years.[7]

Because a vast majority of the currently existing AD cases are of nongenetic origin, most of the current research is focused on the mechanistic pathways involved in the pathogenesis of the sporadic AD. In both sporadic and familial cases, the progression of AD is characterized by extracellular senile plaques, which are aggregates of Aβ peptides (mostly consisting of $A\beta_{1-40}$ and/or $A\beta_{1-42}$), and also by intracellular NFTs (Figure 2). The NFTs are aggregates of hyperphosphorylated tau protein in the brain tissues, located mainly in the hippocampus and surrounding regions. The formation of the NFTs and Aβ aggregates is associated with synapse and neuronal loss, thereby causing gradual memory loss in cases of AD. The deposition of extracellular Aβ aggregates and intracellular NFTs may precede the onset of AD symptoms by many years, at which stage of disease pathogenesis, widespread neuronal loss results from the neuroinflammation. According to the amyloid cascade hypothesis, the formation of the Aβ aggregates is the early stage of AD, although there is controversy surrounding whether the deposition of Aβ is the cause or the consequence of AD. Due to the uncertainty in the mechanism of onset of AD, several approaches are being used to monitor the progression

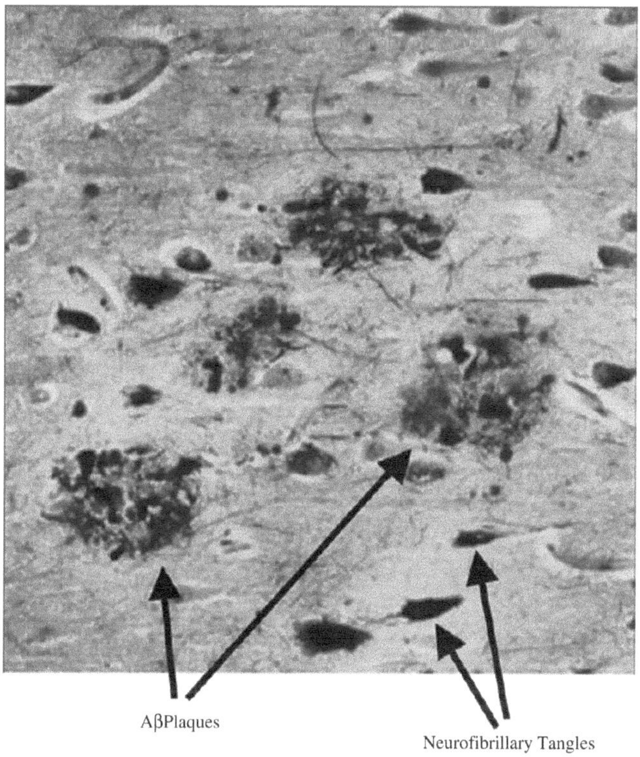

FIGURE 2 The extracellular Aβ plaques and the intracellular neurofibrillary tangle (NFT) in the cerebral cortex of Alzheimer disease (AD) brains. *Adapted with permission from Ref. 7. Copyright 2006, Elsevier Ltd.*

of the disease, and for formulating therapeutics. None of the existing therapeutics, however, provide long-lasting cure for AD, underscoring the importance of the delineations of the causative factors of AD and related neurological disorders.

Oxidative stress is probably the major pathway that leads to the formation of the peptide or protein aggregates.[8,9] These protein/peptide aggregates are the basis for the onset of not only AD but also many other neurological disorders. The self-assembly of the proteins or peptides under the influence of the excessive reactive oxygen species (ROS) and reactive nitrogen species is one of the hallmarks for AD (aggregates of Aβ peptides, and NFTs), PD (α-synuclein aggregation), Huntington disease (Huntingtin aggregation), and ALS (superoxide dismutase 1 (SOD1) aggregation). Excessive accumulation of ROS is generally implicated in neuronal cell death and thereby onset of neurodegenerative diseases.

Thus, some of the current therapeutics for treating the symptoms of these neurodegenerative diseases is directed to protein disaggregation and/or inhibitors of protein aggregation and toward antiinflammatory agents that attenuate oxidative stress.[10,11]

For over 20 years, the Aβ and NFT hypotheses were intensely scrutinized in order to understand the principal causative factors of the onset and progression of AD. Various therapies that target Aβ formation and accumulation, for example, β-secretase (β-site amyloid precursor protein cleaving enzyme-1; BACE-1) inhibitors[12–19] and γ-secretase modulators (GSMs),[16,20–22] have been the focus of the major pharmaceutical companies. However, the underlying difficulty in designing therapeutics, based on the BACE inhibitors or GSMs, is that they have to overcome the blood–brain barrier (BBB), which is a specialized brain endothelial structure of the neurovascular system that protects neurons from the toxic and unneeded components of the circulating blood. The breakdown of the BBB in certain susceptible individuals leads to the altered transport of molecules to the brain from the vasculature, or vice versa, and may initiate hypoperfusion that may eventually lead to AD and other neurological disorders, including PD, ALS, and multiple sclerosis.[23] On the other hand, the potential drug candidates have to reach neuronal cells through this barrier in order to exert their therapeutic actions. It is, therefore, a challenging task to design the drug candidates that have favorable BBB permeability characteristics, and that, at the same time, would not irreversibly break the BBB.

The native Aβ is a polypeptide and consists of 39–43 amino acid residues. Even though all of these peptides show propensity to fibril formation, $A\beta_{1-42}$ is the major constituent of the amyloid plaques. The Aβ peptides are formed by cleavage of the Aβ-precursor protein (APP) by successive proteolytic actions of β- and γ-secretases (GSs) (Figure 3).[18] APP is a glycoprotein of 695–770 amino acids and consists of three segments: the extracellular N-terminal region, a hydrophobic transmembrane region, and a cytoplasmic C-terminal domain. The precursor Aβ residues 1–28 are in the extracellular region, while the remaining residues are embedded in the transmembrane region. The APP is encoded by genes on chromosome 21. Down's syndrome patients overexpress APP and therefore have more probability for the early onset of AD. It is generally assumed that slowing the formation of Aβ peptides or their clearance from the brain tissues would slow the progression of AD. However, the neurotoxicity of the diffusible oligomeric prefibrillar components of Aβ far exceeds that of the fibrillar Aβ plaques, suggesting that slowing or prevention of the formation of the Aβ may be a more effective strategy than clearing the Aβ plaques.[24] Aβ antibodies decrease the levels of Aβ plaques (senile plaques) in animal studies,

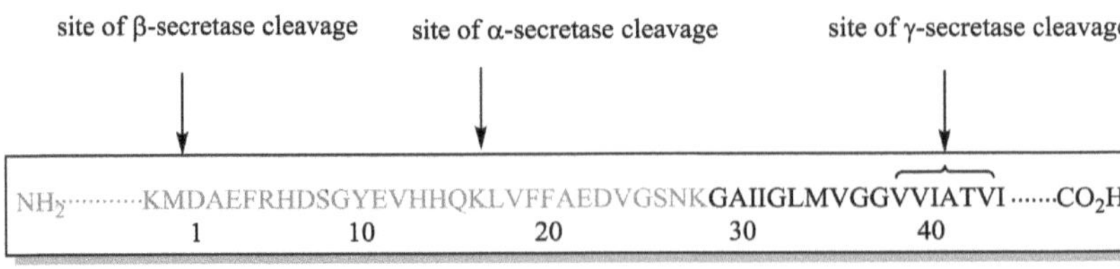

FIGURE 3 Schematic representation of the processing of APP by α, β, and γ-secretase enzymes; the extracellular region is highlighted in orange.

based on which, and toward the goal of attenuating the levels of Aβ plaques in AD patients, Eli Lilly has completed phase III clinical trials of solanezumab (LY2062430), a humanized anti-Aβ immunoglobulin G (IgG)-1 monoclonal antibody.[25–27] This humanized monoclonal antibody (mAb) binds to the central region of Aβ and promotes its clearance from the brain, and in phase II clinical trials it showed good safety profile. However, contrary to expectations, solanezumab failed to improve cognition or functional ability, in phase III clinical trials.[27] Thus, the ability of Aβ-clearing drugs to slow the rate of deterioration in the advanced stages of the AD is in question.

Positron emission tomography is currently widely used to validate the efficacy of the antiamyloid drugs by monitoring the Aβ plaques and NFTs, using ^{11}C- and ^{18}F-tracer-containing imaging agents (Chapter 7 is devoted to a detailed discussion of the ^{18}F-labeled PET tracers).[28–33]

2. BACE-1 INHIBITORS

BACE-1 is a member of the aspartic protease superfamily, which includes renin and human immunodeficiency virus (HIV) proteases. There is abundant evidence indicating that even a partial pharmacological inhibition of the BACE-1 can significantly reduce amyloidosis.[34] The active site of BACE-1, as compared with that of renin or HIV proteases, is more open and less hydrophobic, and tight binding of the inhibitor requires interaction with amino acid residues, D289, D93, G291, Y132, and F169 at the active site.[35] Fluorinated versions of the pharmaceutical candidates often have improved pharmacokinetics (PK), afford increased potency, and, in some cases, have excellent BBB penetrability; however, in spite of the surge in the advancement of the drug candidates to clinical trials, none of the BACE-1 inhibitors have so far entered into clinical practice. BACE-2, a close homolog of BACE-1 cleaves the Aβ-precursor protein (APP) at F691, which is downstream of the BACE-1 cleavage site, and the resulting Aβ peptide is nonamyloidogenic. Consequently, most of the attention in this area is focused on the selective inhibition of the BACE-1 over other aspartic proteases, and on those BACE-1 inhibitors that could pass through the BBB. Moreover, the BACE-1 inhibitor drug candidates should have good PK and pharmacodynamic profiles. Animal models, such as transgenic mice expressing human APP are usually used for the initial screening of Aβ-lowering effect and for the long-term efficacy of the BACE-1 inhibitors.

The mechanism of the hydrolysis of the peptides by BACE-1 is similar to that of the other aspartyl proteases. Two of the BACE-1 aspartyl residues at the active site (one of the aspartates acting as the general acid and the other as the general base) are involved in the hydration of the amide bond to form the intermediate **3**. The general acid-catalyzed protonation of the amine followed by expulsion of the amino group in **3** gives the fragmented peptides (Figure 4).[36a,37a]

In the early stages of the development of BACE inhibitors, a large-molecular-weight hydroxyethyl peptidomimetic OM99-2 was synthesized and

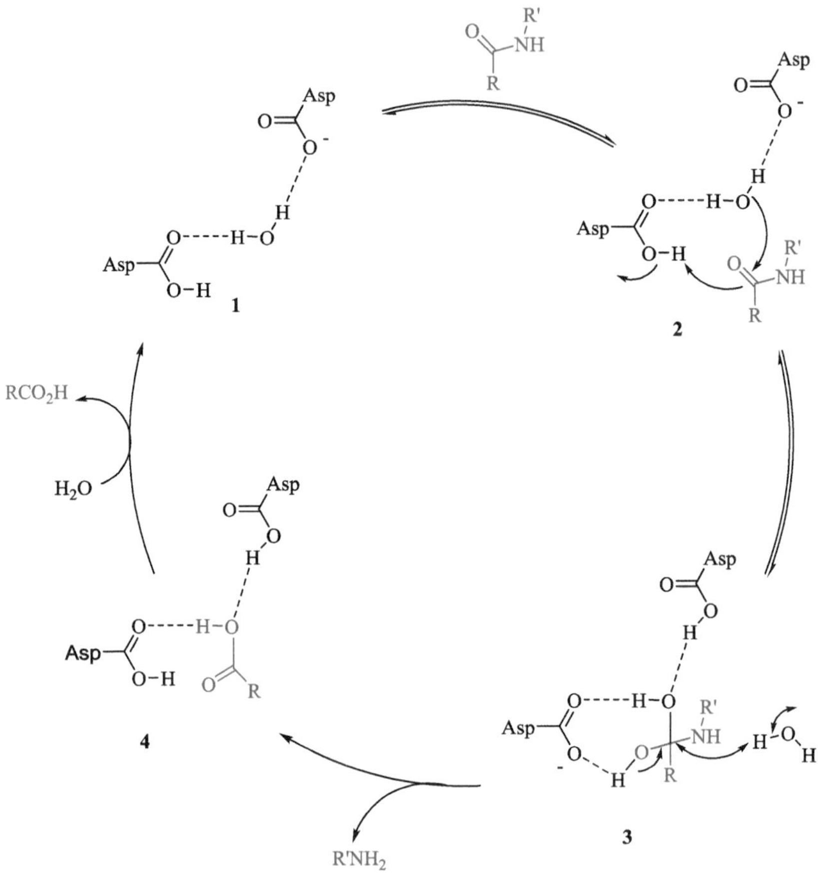

FIGURE 4 A generalized mechanism for peptide bond hydrolysis, catalyzed by BACE-1; the enzyme undergoes several conformational changes during the course of hydrolysis, and the two aspartyl residues that are involved in the hydration of the amide bond have distinct pK_a values, in accordance with the proposed general acid/general base catalysis. A general acid catalysis may be involved in the protonation of the departing amino group in intermediate **3**.

cocrystallized with BACE-1 (Figure 5).[38] Although this compound has impressively high affinity to BACE-1 active site (K_i = 1.6 nM), due to its large molecular mass and high polarity, it has unacceptable BBB permeability, and therefore it could not be used as a drug candidate for the reduction of Aβ peptides from the brain tissues. Search for small molecule peptidomimetic or nonpeptide inhibitors for BACE-1 is currently being pursued in many laboratories, partly, using this as a lead compound, and based on the X-ray structural details of the enzyme's active site when bound to this compound.

Trabanco and coworkers have synthesized a series of 3-amino-5,6-dihydro-1-methylpiperazine-2-one-derived BACE-1 inhibitors, and identified a 40 nM inhibitor **6**, which was able to reduce the Aβ levels in vivo.[39] Interestingly,

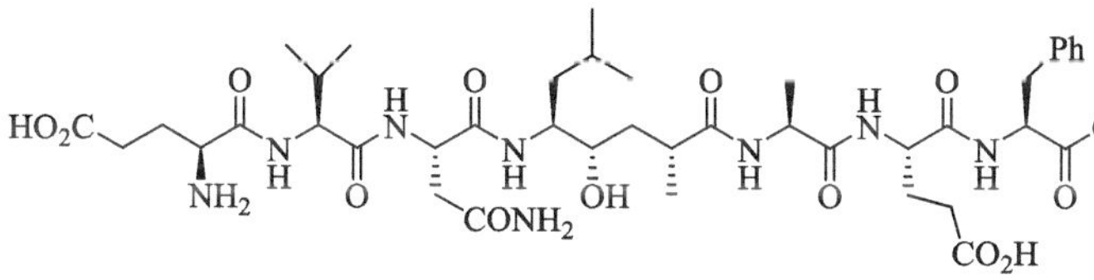

OM99-2; $K_i = 1.6$ nM

FIGURE 5 Structure of OM99-2, a BACE-1 inhibitor.

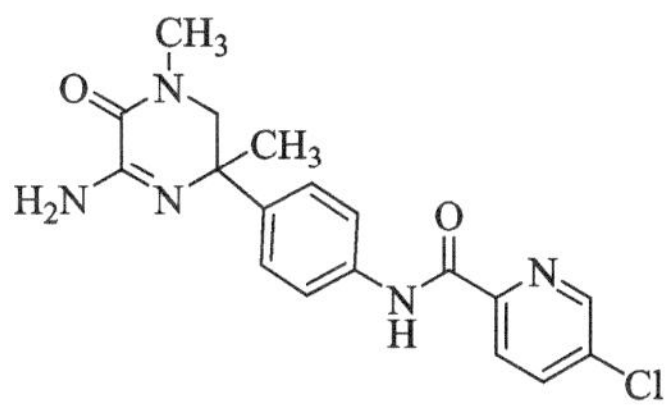

5

$K_i = 0.295\ \mu M$

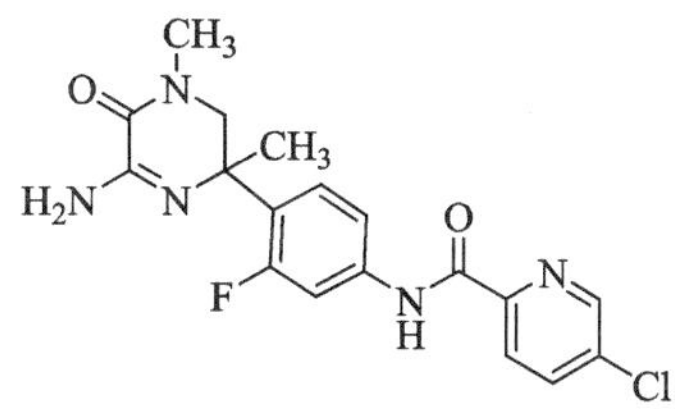

6

$K_i = 40$ nM

FIGURE 6 Structures of piperazinone-derived BACE-1 inhibitors **5** and **6.**

the piperazinone analog that lacked the aryl fluorine, **5**, shows a dramatically reduced BACE-1 inhibition (K_i of 0.295 μM) (Figure 6). The fluorinated compound **6** also exhibits relatively higher BBB permeability, and increased plasma concentration as compared with its nonfluorinated versions, when administered subcutaneously in mice. This increased BBB may be attributed to the relatively enhanced lipophilicity due to the presence of the fluorine. These compounds may serve as lead structures for developing therapeutics with improved metabolic stability, oral bioavailability, and enhanced BBB.

Hilpert and coworkers have developed small molecule BACE-1 inhibitors, based on the 1,3-oxazine structure, that possess high in vivo efficacy and that are suitable for clinical evaluation in AD.[35] Through an extensive "fluorine scan," they have identified oxazine **7** as the candidate drug molecule, with favorable BBB characteristics and optimal BACE-1 binding affinity (Figure 7). The trifluoromethyl group in the oxazine ring significantly lowers the pK_a of the amino group in oxazine **7** (pK_a 7.0) as compared with that of the oxazine **8** that lacks the CF_3 group (pK_a 9.7). Due to this lowered pK_a, **7** is relatively more lipophilic ($\log P = 2.9$ and 2.3, respectively, for oxazines **7** and **8**), and thereby is relatively more permeable through BBB.[35] Even though both compounds show comparable in vitro human BACE-1 inhibition (12 nM and 51 nM, respectively), only oxazine **7** shows significant dose-dependent decrease in the Aβ levels upon oral administration in wild-type mice (Figure 8); oxazine **8** is virtually inactive in vivo.[35] Oxazine **7** shows a reasonably good pharmacological profile, with a bioavailability of 68% and a terminal half-life time of 9–11 h.[35] Oxazine **9**, in which the electron-withdrawing CF_2 moiety is in relatively closer

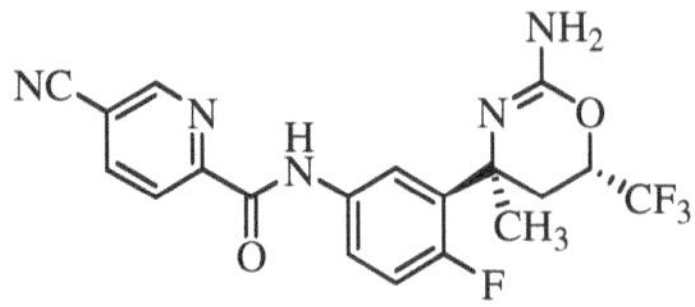

7; potent in vivo $A\beta_{1-40}$ suppression

hBACE1: IC_{50} = 12 nM
Cell ($A\beta_{1-40}$); IC_{50} = 2 nM

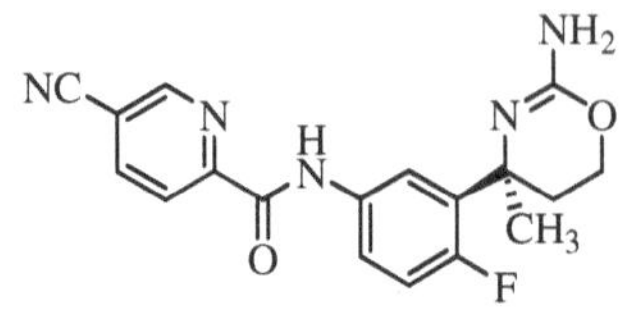

8; Marginal in vivo $A\beta_{1-40}$ suppression

hBACE1: IC_{50} = 51 nM

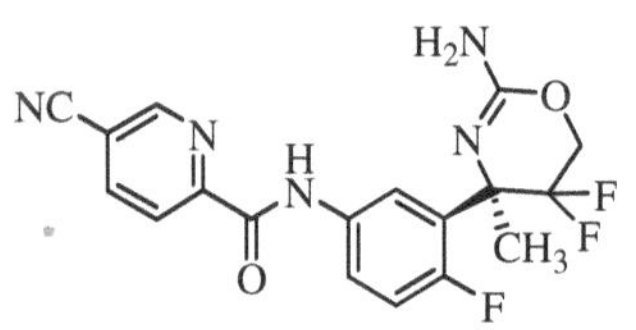

9

hBACE1: IC_{50} = 23 nM
Cell ($A\beta_{1-40}$); IC_{50} = 21 nM

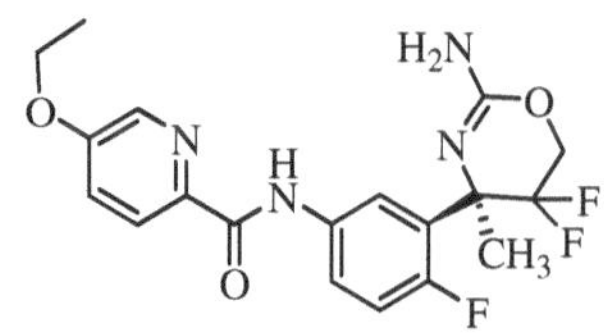

10

hBACE1: IC_{50}= 18.4 nM
Cell ($A\beta_{1-40}$); IC_{50} = 48 nM

FIGURE 7 Structures of oxazine-derived BACE-1 inhibitors, **7**, **8**, **9**, and **10.**

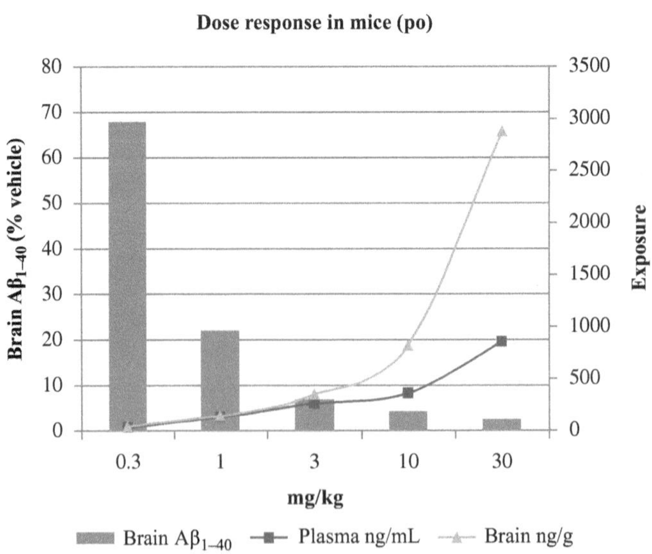

FIGURE 8 Dose-dependent suppression of brain Aβ40 formation (4 h after oral dose) in mice by BACE-1 inhibitor **7.** *Adapted with permission from the Ref. 35. Copyright 2013, American Chemical Society.*

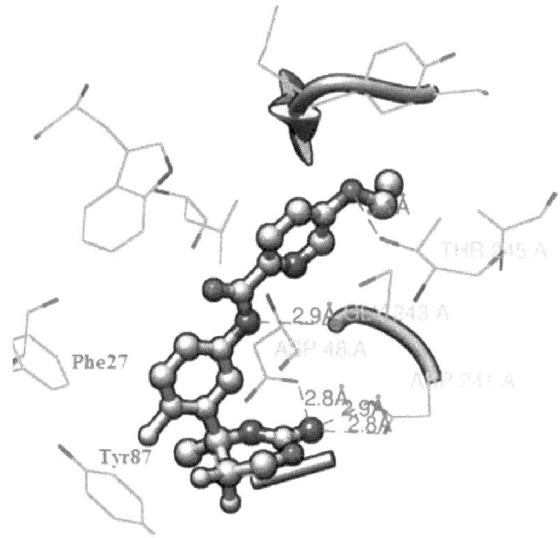

FIGURE 9 Oxazine-**10** (ball and stick model) bound to the BACE-1 active site (N, blue; F, green; O, red); the oxaamidine moiety is tightly bound to the aspartate residues D48 and D241, and the geminal fluorines in the oxazine ring and the *ortho*-fluorine in the aryl ring have van der Waals interactions to Y87and F27, respectively; the structure was created using UCSF CHIMERA software; PDB 6T9.

proximity to the amino group, the amine pK_a is markedly reduced to 5.8, but its lipophilicity (log P = 2.5) is relatively lower than that for the oxazine **9**. Thus, the relatively more hydrophobic CF_3 group imparts increased lipophilicity to the oxazine **7** as compared with that of the CF_2 group, even though compound **9** has relatively weaker basicity of the amino group. Oxazine **10** also has comparable amyloid-lowering propensity as that for the oxazine **7**.[34] These fluorinated oxazine derivatives have not entered into clinical trials as yet, but these and similar compounds may prove to have potent therapeutic effect, provided the amyloid cascade hypothesis holds for the AD pathogenesis.

The reported X-ray structure of the oxazine **10** bound to BACE-1 (Figure 9) reveals that the *geminal* fluorines and the aryl fluorine atom have van der Waals interactions with the Tyr87 and Phe27 residues, respectively.[35] Perhaps due to the strong electron-withdrawing inductive effect of the geminal difluoromethylene group, the oxaamidine moiety is tightly bound to the aspartate residues, Asp48 and Asp241 (N–O bond distances ~2.8 Å).

2.1 Synthesis of the BACE-1 Inhibitor 7

3-Methyl-5-trifluoromethyl-4,5-dihydroisoxazole (**11**), synthesized through a dipolar cycloaddition reaction of acetonitrile oxide with 3,3,3-trifluoropropene, was reacted with 2-fluorophenyllithium (generated in situ from the reaction of 2-bromofluorobenzene with *n*-BuLi) to give the arylated product **12**, with high diastereoselectivity.[35] Transfer hydrogenation using ammonium formate and Pd/C gave the ring-opened amino alcohol **13**, which was

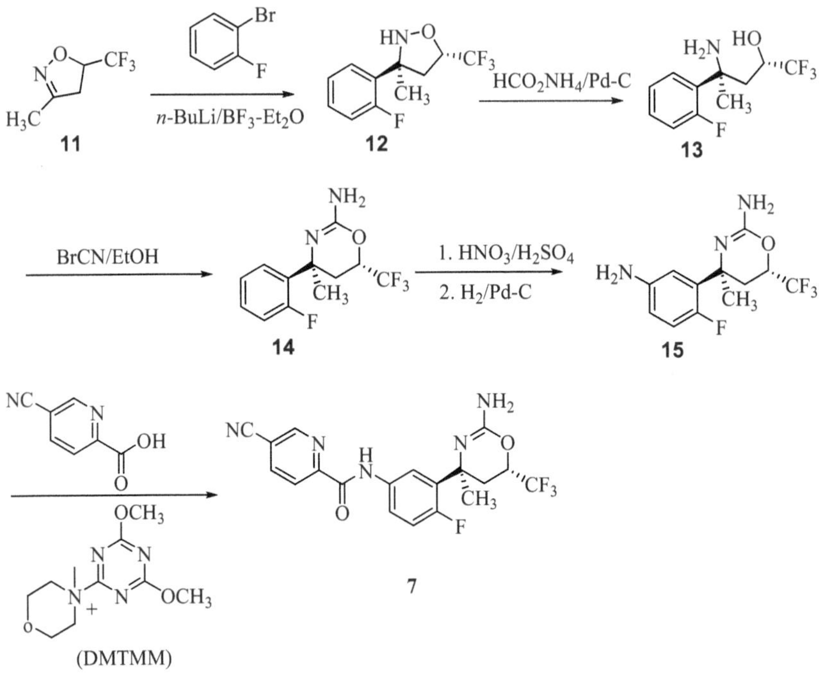

FIGURE 10 Synthesis of oxazine 7, a BACE-1 inhibitor.

converted to the oxazine derivative **14** by reaction with cyanogen bromide (CNBr). The oxazine derivative **14**, upon electrophilic nitration, followed by catalytic hydrogenation afforded compound **15**, which after coupling with 5-cyano-2-pyridinylmethanoic acid, in the presence of a peptide-coupling agent 4-(4,6-dimethoxy-1,3,5-triazin-2-yl)-4-methylmorpholinium chloride; (DMTMM), gave the oxazine **7** (Figure 10). Using similar synthetic strategies, other fluorinated oxazine derivatives were prepared and assayed for their BACE-1 inhibitory activities in vitro and in vivo.[35]

3. GSMs AND γ-SECRETASE INHIBITORS

Aβ-precursor protein (APP) is initially cleaved by β-secretase to produce the soluble N-terminal APP fragments and a C-99 fragment bound to the cell membrane. The latter is then cleaved by the GS resulting in the formation of the Aβ peptides, in which the number of amino acid residues ranges from 37 to 43.[4] A review by Gijsen and coworkers summarizes the current trends in the development of the γ-secretase modulating pharmaceuticals (GSMs) that control the relative amounts of the Aβ isoforms.[4] Of the several Aβ species, $A\beta_{1-42}$, constituting 5–10% of the total Aβ, is more prone to aggregation and senile

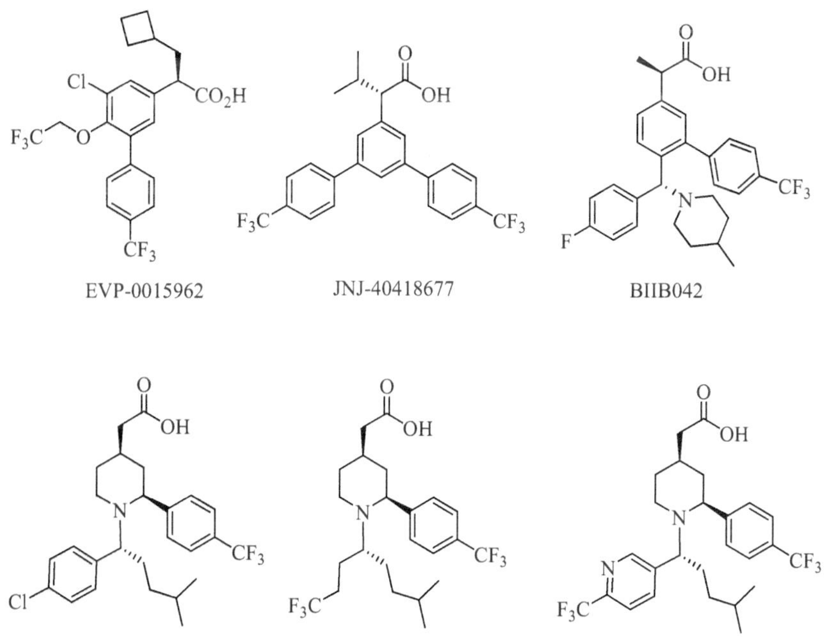

FIGURE 11 Structures of selected γ-secretase modulators (GSMs).

plaque formation. Thus the goal in the development of this class of AD therapeutics (GSM therapeutics) is to significantly reduce the relative proportion of the $A\beta_{1-42}$ peptide among the other Aβ isoforms, while maintaining normal rates of APP cleavage by GS. Most of these pharmaceuticals are derivatives of various nonsteroidal antiinflammatory drugs (NSAIDs), and a majority of these compounds incorporate either fluorinated aromatic rings or a trifluoromethyl substituent. NSAIDs, such as sulindac sulfide, indomethacin, (R)-ibuprofen, and (R)-flurbiprofen show GSM activity. However, in clinical trials they did not prove to have significant therapeutic effect. The second-generation GSM, based on the NSAIDs, such as EVP-0015962, JNJ-40418677, BIIB042, GSM-1, GSM-2, and GSM-10h have significant $A\beta_{1-42}$-lowering effect, and are potential therapeutics for treating AD (Figure 11).[40-43]

Rogers and coworkers have developed a fluorinated GSM, EVP-0015962 (Figure 11), and tested for its effect on the reduction of the $A\beta_{1-42}$ levels, and elevation of the $A\beta_{1-38}$ levels in Tg2576 mice. It decreased $A\beta_{42}$ in H4 cells with an IC_{50} of 67 nM, and increased $A\beta_{38}$ by 1.7-fold at this concentration. EVP-0015962 preserved the GS activity and did not alter the total Aβ peptides in the Tg2576 mouse brain. This compound is effective in reducing Aβ aggregates, Aβ plaques, cognitive deficits, and inflammatory markers in

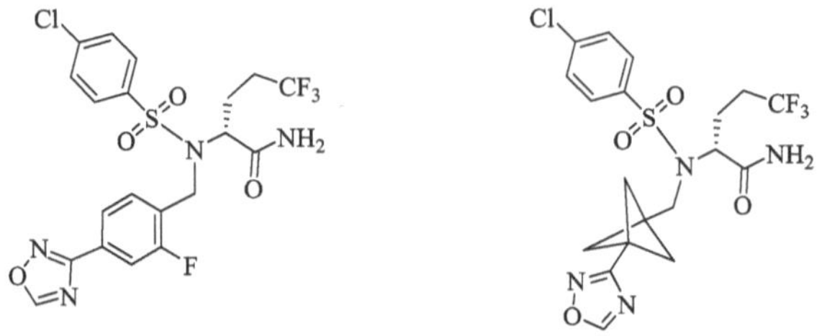

BMS-708,163 **16**

FIGURE 12 Structures of BMS-708163, a γ-secretase inhibitor (GSI) and a bioisosteric analog **16.**

Tg2576 mice, showing its potential as a therapeutic candidate to treat AD patients.[44]

It is important to note that the GSIs, although they significantly reduce the Aβ burden, showed adverse effects in clinical studies. These adverse effects of the GSIs are due to their indiscriminate inhibition of Notch signaling by binding to the Notch family of transmembrane receptors.[43] Such mechanism-based inhibition of Notch signaling results in undesired adverse effects, and therefore counterscreening of the GS substrates for Notch signaling is a required criterion in this area of drug discovery. Thus exclusive specificity for the GS and against the Notch signaling is an arduous task. Accordingly, the current trend is to develop potent GSMs, instead of the initially appealing GSI, although research in the area of GSI is ongoing. Bristol-Myers Squibb's drug candidate, BMS-708163 (avagacestat; Figure 12) and its conformationally rigid analog **16** (Figure 12), are focused on developing GSIs that minimally interfere in Notch processing.[37b]

BMS-708163 exhibited 193 times relatively weaker potency for Notch processing as compared to its inhibitory potency for the GS, as monitored by the reduction in the $A\beta_{1-40}$ levels (IC_{50} = 58 nM (Notch) and 0.3 nM (Aβ), in mouse models).[37b] Cell assays using wild-type human HeLa cells showed similar IC_{50} values (10 to 38 nM) for the Notch processing.[36b] Further, due to its excellent oral bioavailability and due to its plasma- and brain-$A\beta_{1-40}$ lowering effects, it has entered into clinical trials for the treatment of AD at Bristol-Myers Squibb. However, these clinical studies are now being discontinued by Bristol-Myers Squibb, after phase II trials were completed, unrelated to the safety issues of this drug candidate.

3.1 Synthesis of BMS-708163

The Bristol-Myers Squibb synthesis of the BMS-708163 was achieved in eight steps starting from the 4,4,4-trifluorobutanal, in high yields, and with high enantiomeric purity.[37b] Diastereoselective Strecker synthesis is the key step in their synthesis. Diastereoselective Strecker reaction of 4,4,4-trifluorobutanal **17** with (R)-α-methylbenzylamine, followed by selective hydrolysis of the

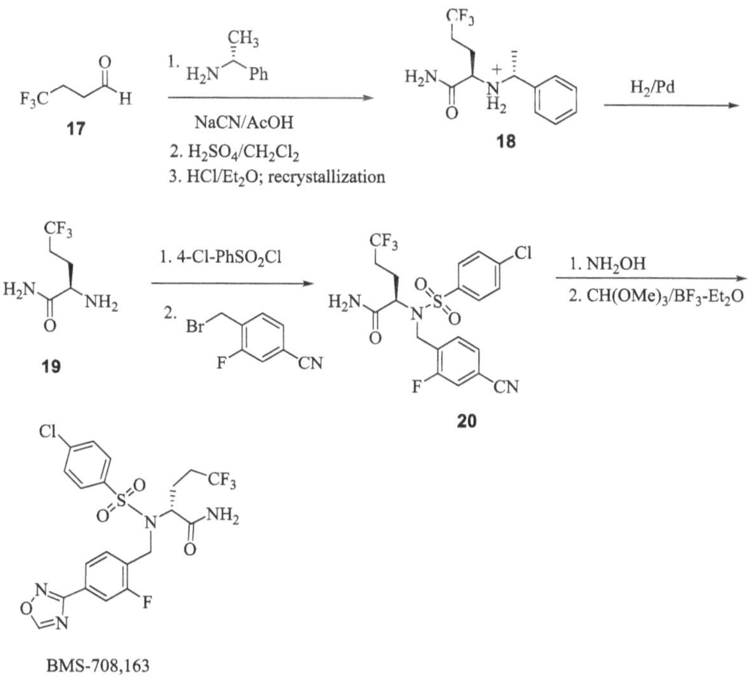

FIGURE 13 Synthesis of BMS-708163, a γ-secretase inhibitor (GSI).

nitrile moiety (to give the corresponding carboxamide), and recrystallization of the resulting hydrochloride salt afforded compound **18**, which after deprotection using catalytic hydrogenation afforded (3,3,3-trifluoropropyl)glycinamide **19**. Reaction of the latter compound with arylsulfonyl chloride followed by base-catalyzed alkylation (using 2-fluoro-4-cyanobenzyl bromide) gave compound **20**, which upon reaction with hydroxylamine followed by acid-catalyzed cycloalkylation (using trimethyl orthoformate ($CH(OMe)_3$)) afforded BMS-708163 (Figure 13).

An analog of BMS-708163, in which the fluorophenyl ring is replaced by the bicyclo[1.1.1]pentane bioisostere moiety (Compound **16**), showed improved GS inhibition, aqueous solubility, permeability, and in vitro metabolic stability, showing that the fluoroaryl ring is not an essential component for enzyme binding or for imparting lipophilicity to BMS-709163.[45] It has fourfold increased oral absorption characteristics in a mouse model over the BMS-708163. Thus, the bicyclo[1.1.1]pentane moiety in compound **16** has similar spatial features as the fluoroaryl ring, and not only it serves as a spacer group for connecting the oxadiazole ring with sulfonamide nitrogen, but also it exerts improved lipophilicity and other druglike characteristics to this compound.

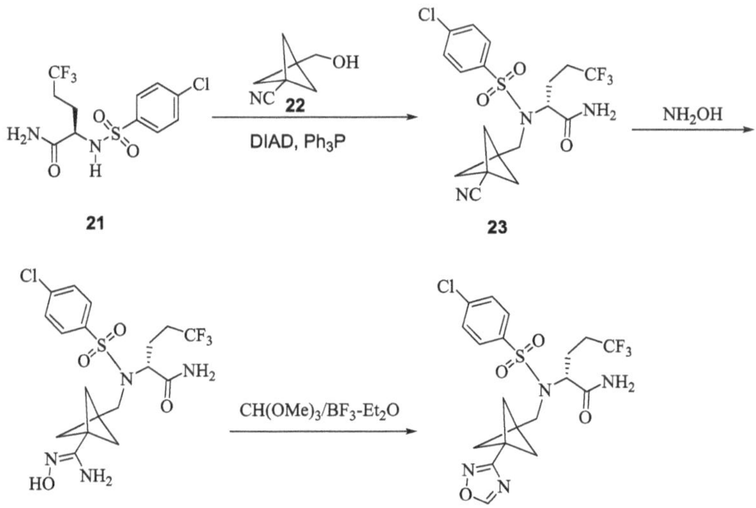

FIGURE 14 Synthesis of the BMS-709163 analog **16** (DIAD = diisopropyl azodicarboxylate).

The bioisostere bicyclo[1.1.1]pentane moiety is ideally suited for the coplanar arrangement of the oxadiazole and N-arylsulfonamide moieties that is required for the optimal binding of the substrate at the GS active site. The altered physicochemical characteristics of this drug analog (**16**), such as its significantly improved aqueous solubility and relatively increased intrinsic membrane permeability as compared to that of the BMS-708,163, may be attributable to the increased three dimensionality and thereby the disruption of the intermolecular π-stacking of the two aryl rings.[45]

3.2 Synthesis of the GS Inhibitor 16, a BMS-708163 Analog

The Pfizer group has reported the elegant synthetic procedure for compound **16**, which involves a Mitsunobu coupling of sulfonamide **21** to 4-cyanobicyclo[1.1.1]pentan-1-methanol (**22**), followed by transformation of the nitrile group to the oxadiazole moiety through a two-step sequence: reaction of compound **23** with hydroxylamine forms the intermediate amide oxime **24**, which undergoes Lewis-acid-catalyzed cycloalkylation in the presence of trimethyl orthoformate to give compound **16** (Figure 14).[45]

3.3 ELND006 and ELND007

Elan Pharmaceuticals developed two metabolically stable GSIs ELND006 and ELND007, which exhibit high selectivity for the inhibition of the GS over

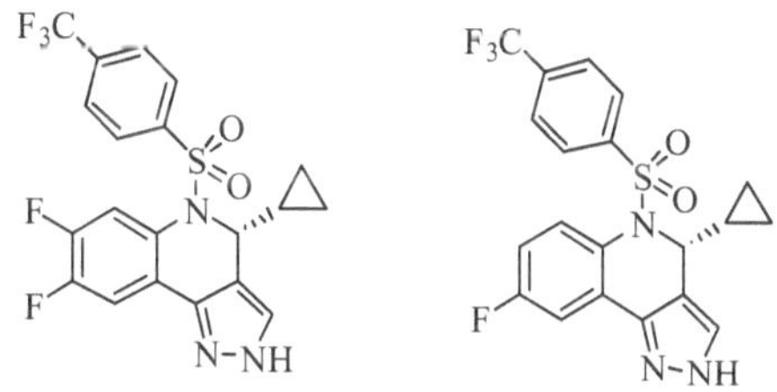

ELND006 ELND007

FIGURE 15 Structures of GSIs ELND006 and ELND007.

Notch signaling (Figure 15).[46] ELND006, at a single dose of 30 mg in a gelatin capsule lowered Aβ in the cerebrospinal fluid of healthy human volunteers by 31–46% 5 h after administration, due to its GS inhibitory activity.

Elan Pharmaceuticals initiated clinical trials of the ELND006 and ELND007. However, due to the adverse effects on the livers of patients, unrelated to their mechanisms of action, they have terminated their clinical trials.[46] This case, along with that of the BMS-avagacestat (BMS-708163), illustrates the hurdles involved in attempts of bringing well-designed Aβ-lowering drugs from clinical trials to therapeutics level.

3.3.1 Synthesis of ELND007

Elan Pharmaceuticals have developed elegant synthetic methods for their GSIs ELND006, ELND007, and other fluorinated analogs.[46] The synthesis of ELND007 is outlined in Figure 16. The key steps in the synthesis of ElanD007 involve condensation of the pyrazole aldehyde **25** with (S)-2-methylpropane-2-sulfinamide, followed by diastereoselective cyclopropanation, and then deprotection of the sulfinyl moiety to give compound **26**. Arylsulfonylation of amine **26** to give compound **27**, followed by Cu(II)-catalyzed intramolecular aryl amination of **27** afforded a 9:1 enantiomeric mixture of the ELND007 and its enantiomer (separated using chiral high-performance liquid chromatography) (Figure 16).[46]

4. CELL CYCLE INHIBITORS

It is now generally recognized that abnormal cell cycle reentry of neuronal cells leads to neuronal cell death, and it is hypothesized that such neuronal cell death precedes the formation of NFT and senile plaques.[47] This abnormal cell cycle reentry is initiated by the postmitotic CDK5 (a cyclin-dependent serine/threonine kinase) in complex with p25, a proteolytic segment containing the C-terminal region of the p35. Dysregulation of CDK5 is involved in the onset of neurological disorders, including PD, AD, and cancer. An X-ray structural

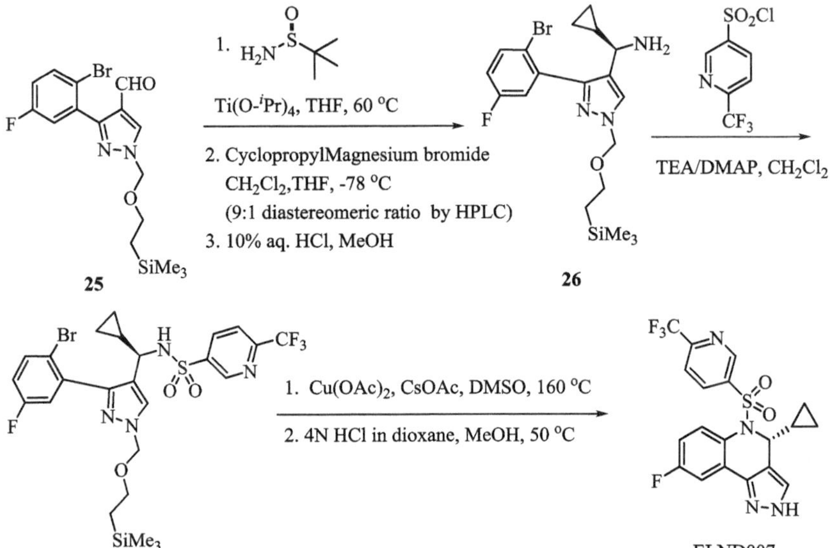

FIGURE 16 Synthesis of ELND007 (TEA = triethylamine; DMAP = 4-dimethylaminopyridine).

data on the CDK5/p25 complex with R-roscovitine (Figure 17A) paved the way for the investigation of the novel drugs that inhibit this kinase and thereby attenuate cell cycle reentry and cell-cycle-mediated neuronal death.[48,49] In AD patients, JNK (c-Jun N-terminal kinases), c-Jun, and CDK5 are upregulated. Although both c-Jun and CDK5 cause neurodegeneration in vivo, CDK5 has been shown to be an upstream regulator in the c-Jun NH_2 terminal kinase cascade, and therefore CDK5 is a preferable therapeutic target over JNK and c-Jun.[50,51]

Roscovitine (**28**), a potent, but non-selective, inhibitor of CDK5/p25 complex is currently undergoing clinical trials for treating cancers. It also exhibits antiangiogenic effects through its CDK5 inhibitory effect.[52] Because of its potential anticancer and anti-AD effects, search for similarly active, but more selective CDK5/p25 inhibitors is an active area of research in many academic and industrial laboratories.[53–59]

Purine-based fluorinated triazole analogs of Roscovitine inhibit the neuronal cell death in vitro in the cultured neuronal cells, and the cell cycle inhibition attained by these compounds (e.g., **29**) is comparable to that of Roscovitine.[49] The in vitro cell cycle inhibition of the fluorinated triazole compounds depends on the number of fluorines on the aryl ring. Thus, the triazole derivative with perfluoroaryl moiety (**29**-C_6F_5) has virtually no effect on cell cycle inhibition, whereas compound **29** shows optimal neuroprotective effects.

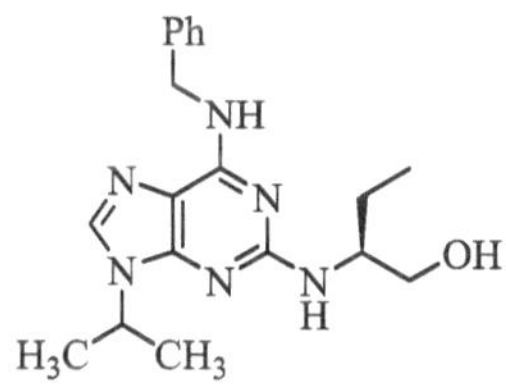

Roscovitine (**28**)

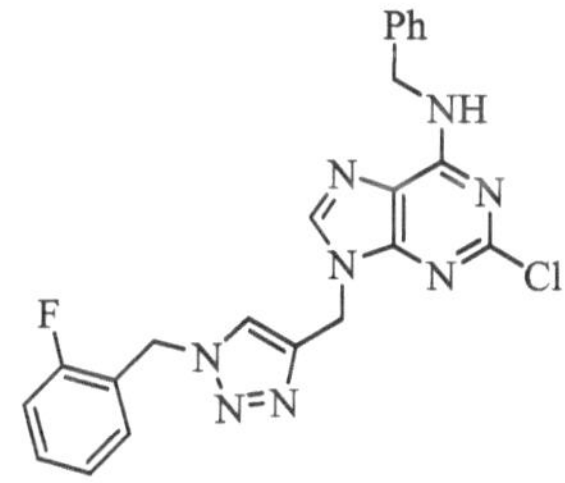

29

Docking studies show that binding affinities of these triazoles at the adenosine triphosphate binding site of the CDK5/p25 complex are comparable with that of Roscovitine.[49] The triazole analog **29** binds at the same binding site on CDK5/p25 as that on Roscovitine, but interacts strongly with other sets of amino acid residues, except for the Gln130 residue that interacts with both Roscovitine and compound **29** (Figure 17B).[49] Due to the relatively greater lipophilicity of the fluoroaryl triazole compounds, they would have potentially favorable BBB properties, and effective therapeutics may be developed in the near future using such a strategy.

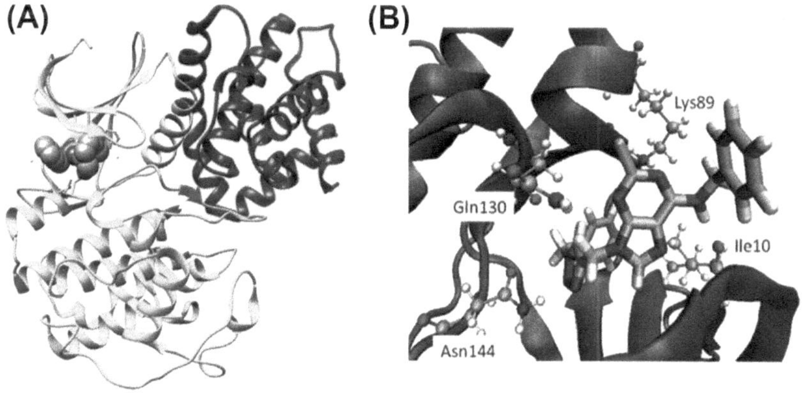

FIGURE 17 (A) X-ray structure of the CDK5 (in cyan)/p25 (in blue) complex bound to (R)-Roscovitine (spheres). Roscovitine is bound at the adenosine triphosphate binding pocket between the N- and C-terminal lobes of the kinase (the structure was created using UCSF Chimera software; PDB 1UNL); (B) expanded view of the simulated structure of the fluorotriazole (stick model) bound to CDK5; the binding site of the triazole is the same as that for Roscovitine, but has interactions with different sets of amino acid residues, except that for Gln130 residue. (For interpretation of the references to color in this figure legend, the reader is referred to the online version of this book.) *Reproduced with permission from Ref. 49. Copyright 2011, Elsevier.*

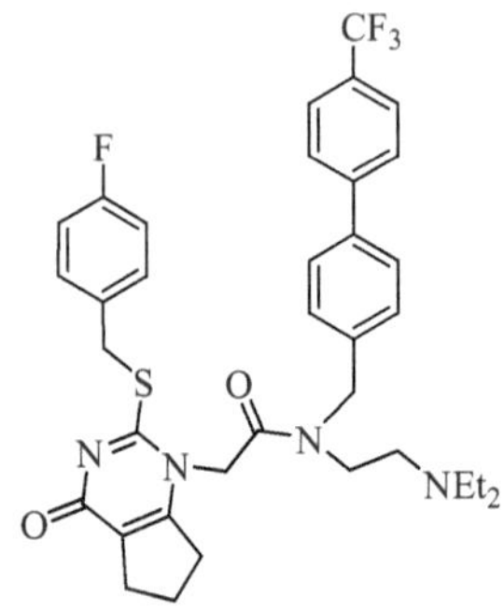

Darapladib

FIGURE 18 Structure of darapladib, an antiinflammatory compound, which lowers BBB permeability.

5. ANTIINFLAMMATORY COMPOUNDS AS THERAPEUTICS OF AD

Current evidence suggests that diabetes mellitus and hypercholesterolemia are the major risk factors for the AD, and that they may increase BBB permeability. This increased BBB permeability is evidenced by the leaking of IgG into the brain interstitium, and binding to the pyramidal neurons in cerebral cortex; the deposition of $A\beta_{1-42}$ peptide aggregates in the pyramidal neurons is also accelerated by the increased BBB permeabilization.[60] Increased lipoprotein-associated phospholipase A2 activity is correlated with the increased cardiovascular risk, including atherosclerosis. Darapladib (Figure 18), an antiinflammatory compound, inhibited lipoprotein-associated phospholipase A2 in the experimental models of diabetes- and hypercholesterolemia-induced pigs, and thereby significantly reduced the atherosclerotic lesions.[60,61] Darapladib, through its antiinflammatory effect, substantially reduced the expression of 24 genes that are responsible for the functioning of the macrophages and T lymphocytes. These data showing that the selective inhibition of Lipoprotein-associated phospholipase A2 (Lp-PLA2) inhibits progression to advanced coronary atherosclerotic lesions prompted the clinical trials of darapladib. GlaxoSmithKline's successive two phase III clinical trials of this compound against strokes and major coronary events, however, were not successful. In pig models of hypercholesterolemia and diabetes mellitus, darapladib reduced the amount of IgG-immunoprecipitation and the amount of $A\beta_{1-42}$-containing neurons, showing that it has the capability to reduce the BBB permeability. Thus, this drug candidate, and its analogs, may have potential applications in the treatment of AD[60]

6. SUMMARY AND OUTLOOK

Neurological disorders such as AD, PD, ALS, and Pick's disease are among the least understood in terms of their mechanisms of onset, although significant progress has been made in this direction. Currently available pharmaceuticals for treating AD, based on acetylcholinesterase inhibitors, NMDA receptor antagonists, and antiinflammatory drugs are based on symptomatic aspects of the disease rather than in a lasting cure. Current focus on developing AD therapeutics is targeted on β-secretase (BACE-1) inhibitors, GSMs and GSIs, and cell cycle inhibitors. Clinical trials based on some of the GSIs, such as BMS-708163, have been unsuccessful due to their adverse effects on patients, although in vivo studies show their excellent oral bioavailability and plasma- and brain-Aβ-lowering effects. BACE-1 inhibitors, based on the piperazinone moiety, are currently being evaluated for their therapeutic potential. Interestingly, a single fluorine substitution in the aryl ring in compound **6** improves its BACE-1 inhibition (K_i = 40 nM) by 135-fold over the nonfluorinated compound **5**. Thus, fluorinated compounds have significance in developing therapeutics for treating neurological disorders, when combined with other therapeutics.

REFERENCES

1. Tejada-Vera, B. Mortality from Alzheimer's Disease in the United States: Data for 2000 and 2010. *NCHS Data Brief.* **2013,** 1–8.
2. James, B. D.; Leurgans, S. E.; Hebert, L. E.; Scherr, P. A.; Yaffe, K.; Bennett, D. A. Contribution of Alzheimer Disease to Mortality in the United States. *Neurology* **2014,** *82,* 1045–1050.
3. Lopez, L. C.; Dos-Reis, S.; Espargaro, A.; Carrodeguas, J. A.; Maddelein, M.-L.; Ventura, S.; Sancho, J. Discovery of Novel Inhibitors of Amyloid β-Peptide 1-42 Aggregation. *J. Med. Chem.* **2012,** *55,* 9521–9530.
4. Oehlrich, D.; Berthelot, D. J. C.; Gijsen, H. J. M. γ-Secretase Modulators as Potential Disease Modifying Anti-Alzheimer's Drugs. *J. Med. Chem.* **2011,** *54,* 669–698.
5. Brookmeyer, R.; Johnson, E.; Ziegler-Graham, K.; Arrighi, H. M. Forecasting the Global Burden of Alzheimer's Disease. *Alzheimer's Dementia* **2007,** *3,* 186–191.
6. Aliev, G.; Obrenovich, M. E.; Reddy, V. P.; Shenk, J. C.; Moreira, P. I.; Nunomura, A.; Zhu, X.; Smith, M. A.; Perry, G. Antioxidant Therapy in Alzheimer's Disease: Theory and Practice. *Mini-Rev. Med. Chem.* **2008,** *8,* 1395–1406.
7. Blennow, K.; de Leon, M. J.; Zetterberg, H. Alzheimer's Disease. *Lancet* **2006,** *368,* 387–403.
8. Reddy, V. P.; Zhu, X.; Perry, G.; Smith, M. A. Oxidative Stress in Diabetes and Alzheimer's Disease. *J. Alzheimer's Dis.* **2009,** *16,* 763–774.
9. Sancataldo, G.; Vetri, V.; Fodera, V.; Di Cara, G.; Militello, V.; Leone, M. Oxidation Enhances Human Serum Albumin Thermal Stability and Changes the Routes of Amyloid Fibril Formation. *PLoS One* **2014,** *9.* e84552/84551–e84552/84511.
10. Guerrero-Munoz, M. J.; Castillo-Carranza, D. L.; Kayed, R. Therapeutic Approaches against Common Structural Features of Toxic Oligomers Shared by Multiple Amyloidogenic Proteins. *Biochem. Pharmacol.* **2014,** *88,* 468–478.
11. Gheysarzadeh, A.; Yazdanparast, R. Inhibition of H_2O_2-Induced Cell Death through FOXO1 Modulation by EUK-172 in SK-n-mc Cells. *Eur. J. Pharmacol.* **2012,** *697,* 47–52.
12. Ghosh, A. K.; Kumaragurubaran, N.; Hong, L.; Koelsh, G.; Tang, J. Memapsin 2 (Beta-Secretase) Inhibitors: Drug Development. *Curr. Alzheimer Res.* **2008,** *5,* 121–131.
13. Aguzzi, A.; O'Connor, T. Protein Aggregation Diseases: Pathogenicity and Therapeutic Perspectives. *Nat. Rev. Drug. Discov.* **2010,** *9,* 237–248.

14. Bolognesi, M. L.; Matera, R.; Minarini, A.; Rosini, M.; Melchiorre, C. Alzheimer's Disease: New Approaches to Drug Discovery. *Curr. Opin. Chem. Biol.* **2009,** *13,* 303–308.
15. Frisardi, V.; Solfrizzi, V.; Imbimbo, B. P.; Capurso, C.; D'Introno, A.; Colacicco, A. M.; Vendemiale, G.; Seripa, D.; Pilotto, A.; Capurso, A.; Panza, F. Towards Disease-Modifying Treatment of Alzheimer's Disease: Drugs Targeting β-Amyloid. *Curr. Alzheimer Res.* **2010,** *7,* 40–55.
16. Kreft, A. F.; Martone, R.; Porte, A. Recent Advances in the Identification of γ-Secretase Inhibitors to Clinically Test the Aβ Oligomer Hypothesis of Alzheimer's Disease. *J. Med. Chem.* **2009,** *52,* 6169–6188.
17 Mok, N. Y.; Chadwick, J.; Kellett, K. A. B.; Casas-Arce, E.; Hooper, N. M.; Johnson, A. P.; Fishwick, C. W. G. Discovery of Biphenylacetamide-Derived Inhibitors of BACE1 Using de Novo Structure-Based Molecular Design. *J. Med. Chem.* **2013,** *56,* 1843–1852.
18. Panza, F.; Solfrizzi, V.; Frisardi, V.; Capurso, C.; D'Introno, A.; Colacicco, A. M.; Vendemiale, G.; Capurso, A.; Imbimbo, B. P. Disease-Modifying Approach to the Treatment of Alzheimer's Disease: from α-Secretase Activators to γ-Secretase Inhibitors and Modulators. *Drugs Aging* **2009,** *26,* 537–555.
19. Vassar, R.; Kovacs, D. M.; Yan, R.; Wong, P. C. The β-secretase Enzyme BACE in Health and Alzheimer's Disease: Regulation, Cell Biology, Function, and Therapeutic Potential. *J. Neurosci.* **2009,** *29,* 12787–12794.
20. Boulton, M. E.; Cai, J.; Grant, M. B. γ-Secretase: a Multifaceted Regulator of Angiogenesis. *J. Cell. Mol. Med.* **2008,** *12,* 781–795.
21. Doraiswamy, P. M.; Xiong, G. L. Pharmacological Strategies for the Prevention of Alzheimer's Disease. *Expert Opin. Pharmacother.* **2006,** *7,* 1–10.
22. Stanton, M. G.; Hubbs, J.; Sloman, D.; Hamblett, C.; Andrade, P.; Angagaw, M.; Bi, G.; Black, R. M.; Crispino, J.; Cruz, J. C.; Fan, E.; Farris, G.; Hughes, B. L.; Kenific, C. M.; Middleton, R. E.; Nikov, G.; Sajonz, P.; Shah, S.; Shomer, N.; Szewczak, A. A.; Tanga, F.; Tudge, M. T.; Shearman, M.; Munoz, B. Fluorinated Piperidine Acetic Acids as γ-Secretase Modulators. *Bioorg. Med. Chem. Lett.* **2010,** *20,* 755–758.
23. Zlokovic, B. V. The Blood–Brain Barrier in Health and Chronic Neurodegenerative Disorders. *Neuron* **2008,** *57,* 178–201.
24. Walsh, D. M.; Selkoe, D. J. Aβ Oligomers – a Decade of Discovery. *J. Neurochem.* **2007,** *101,* 1172–1184.
25. Imbimbo, B. P.; Ottonello, S.; Frisardi, V.; Solfrizzi, V.; Greco, A.; Seripa, D.; Pilotto, A.; Panza, F. Solanezumab for the Treatment of Mild-to-Moderate Alzheimer's Disease. *Expert Rev. Clin. Immunol.* **2012,** *8,* 135–149.
26. Anonymous News in Brief. *Nat. Rev. Drug. Discov.* **2012,** *11,* 821.
27 Doody, R. S.; Thomas, R. G.; Farlow, M.; Iwatsubo, T.; Vellas, B.; Joffe, S.; Kieburtz, K.; Raman, R.; Sun, X.; Aisen, P. S.; Siemers, E.; Liu-Seifert, H.; Mohs, R. Phase 3 Trials of Solanezumab for Mild-to-Moderate Alzheimer's Disease. *N. Engl. J. Med.* **2014,** *370,* 311–321.
28. Klunk William, E.; Mathis Chester, A. The Future of Amyloid-Beta Imaging: a Tale of Radionuclides and Tracer Proliferation. *Curr. Opin. Neurol.* **2008,** *21,* 683–687.
29. Zhang, W.; Oya, S.; Kung, M.-P.; Hou, C.; Maier, D. L.; Kung, H. F. F-18 Stilbenes as PET Imaging Agents for Detecting β-Amyloid Plaques in the Brain. *J. Med. Chem.* **2005,** *48,* 5980–5988.
30. Wong, D. F.; Rosenberg, P. B.; Zhou, Y.; Kumar, A.; Raymont, V.; Ravert, H. T.; Dannals, R. F.; Nandi, A.; Brasic, J. R.; Ye, W.; Hilton, J.; Lyketsos, C.; Kung, H. F.; Joshi, A. D.; Skovronsky, D. M.; Pontecorvo, M. J. In Vivo Imaging of Amyloid Deposition in Alzheimer Disease Using the Radioligand 18F-av-45 (Flobetapir F-18). *J. Nucl. Med.* **2010,** *51,* 913–920.
31. Nordberg, A. PET Imaging of Amyloid in Alzheimer's Disease. *Lancet Neurol.* **2004,** *3,* 519–527.

32. Clark, C. M.; Schneider, J. A.; Bedell, B. J.; Beach, T. C.; Bilker, W. B.; Mintun, M. A.; Pontecorvo, M. J.; Hefti, F.; Carpenter, A. P.; Flitter, M. L.; Krautkramer, M. J.; Kung, H. F.; Coleman, R. E.; Doraiswamy, P. M.; Fleisher, A. S.; Sabbagh, M. N.; Sadowsky, C. H.; Reiman, P. E. M.; Zehntner, S. P.; Skovronsky, D. M. Use of Florbetapir-PET for Imaging β-Amyloid Pathology. *JAMA, J. Am. Med. Assoc.* **2011,** *305,* 275–283.
33. Choi, S. R.; Golding, G.; Zhuang, Z.; Zhang, W.; Lim, N.; Hefti, F.; Benedum, T. E.; Kilbourn, M. R.; Skovronsky, D.; Kung, H. F. Preclinical Properties of ^{18}F-av-45: a PET Agent for Aβ Plaques in the Brain. *J. Nucl. Med.* **2009,** *50,* 1887–1894.
34. McConlogue, L.; Buttini, M.; Anderson, J. P.; Brigham, E. F.; Chen, K. S.; Freedman, S. B.; Games, D.; Johnson-Wood, K.; Lee, M.; Zeller, M.; Liu, W.; Motter, R.; Sinha, S. Partial Reduction of BACE1 Has Dramatic Effects on Alzheimer Plaque and Synaptic Pathology in APP Transgenic Mice. *J. Biol. Chem.* **2007,** *282,* 26326–26334.
35. Hilpert, H.; Guba, W.; Woltering, T. J.; Wostl, W.; Pinard, E.; Mauser, H.; Mayweg, A. V.; Rogers-Evans, M.; Humm, R.; Krummenacher, D.; Muser, T.; Schnider, C.; Jacobsen, H.; Ozmen, L.; Bergadano, A.; Banner, D. W.; Hochstrasser, R.; Kuglstatter, A.; David-Pierson, P.; Fischer, H.; Polara, A.; Narquizian, R. β-Secretase (BACE1) Inhibitors with High in Vivo Efficacy Suitable for Clinical Evaluation in Alzheimer's Disease. *J. Med. Chem.* **2013,** *56,* 3980–3995.
36. (a) Yuan, J.; Venkatraman, S.; Zheng, Y.; McKeever, B. M.; Dillard, L. W.; Singh, S. B. Structure-based Design of β-Site APP Cleaving Enzyme 1 (BACE1) Inhibitors for the Treatment of Alzheimer's Disease. *J. Med. Chem.* **2013,** *56,* 4156–4180; (b) Albright, C. F.; Dockens, R. C.; Meredith, J. E., Jr.; Olson, R. E.; Slemmon, R.; Lentz, K. A.; Wang, J. -S.; Denton, R. R.; Pilcher, G.; Rhyne, P. W.; Raybon, J. J.; Barten, D. M.; Burton, C.; Toyn, J. H.; Sankaranarayanan, S.; Polson, C.; Guss, V.; White, R.; Simutis, F.; Sanderson, T.; Gillman, K. W.; Starrett, J. E., Jr.; Bronson, J.; Sverdlov, O.; Huang, S. -P.; Castaneda, L.; Feldman, H.; Coric, V.; Zaczek, R.; Macor, J. E.; Houston, J.; Berman, R. M.; Tong, G. Pharmacodynamics of selective inhibition of γ-secretase by avagacestat. *J. Pharmacol. Exp. Ther.* **2013,** *344,* 686–695.
37. (a) Toulokhonova, L.; Metzler, W. J.; Witmer, M. R.; Copeland, R. A.; Marcinkeviciene, J. Kinetic Studies on β-Site Amyloid Precursor Protein-cleaving Enzyme (BACE). *J. Biol. Chem.* **2003,** *278,* 4582–4589; (b) Gillman, K. W.; Starrett, J. E.; Parker, M. F; Xie, K.; Bronson, J. J; Marcin, L. R; McElhone, K. E; Bergstrom, C. P.; Mate, R. A; Williams, R.; Meredith, J. E; Nolan, S.; Burton, C. R; Barten, D. M; Toyn, J. H.; Roberts, S. B; Lentz, K. A.; Houston, J. G; Zaczek, R.; Albright, C. F; Decicco, C. P; Macor, J. E; Olson, R. E. Discovery and Evaluation of BMS-708163, a Potent, Selective and Orally Bioavailable γ-Secretase Inhibitor. *ACS Med. Chem. Lett.* **2010,** *1,* 120–124.
38. Ghosh, A. K.; Shin, D.; Downs, D.; Koelsch, G.; Lin, X.; Ermolieff, J.; Tang, J. Design of Potent Inhibitors for Human Brain Memapsin 2 (β-Secretase). *J. Am. Chem. Soc.* **2000,** *122,* 3522–3523.
39. Tresadern, G.; Delgado, F.; Delgado, O.; Gijsen, H.; Macdonald, G. J.; Moechars, D.; Rombouts, F.; Alexander, R.; Spurlino, J.; Van Gool, M.; Vega, J. A.; Trabanco, A. A. Rational Design and Synthesis of Aminopiperazinones as β-Secretase (BACE) Inhibitors. *Bioorg. Med. Chem. Lett.* **2011,** *21,* 7255–7260.
40. Ohki, Y.; Higo, T.; Uemura, K.; Shimada, N.; Osawa, S.; Berezovska, O.; Yokoshima, S.; Fukuyama, T.; Tomita, T.; Iwatsubo, T. Phenylpiperidine-type γ-secretase modulators target the transmembrane domain 1 of presenilin 1. *EMBO J* **2011,** *30,* 4815–4824.
41. Hall, A.; Elliott, R. L.; Giblin, G. M. P.; Hussain, I.; Musgrave, J.; Naylor, A.; Sasse, R.; Smith, B. Piperidine-derived γ-secretase modulators. *Bioorg. Med. Chem. Lett.* **2010,** *20,* 1306–1311.

42. Van Broeck, B.; Chen, J. M.; Treton, G.; Desmidt, M.; Hopf, C.; Ramsden, N.; Karran, E.; Mercken, M.; Rowley, A. Chronic treatment with a novel γ-secretase modulator, JNJ-40418677, inhibits amyloid plaque formation in a mouse model of Alzheimer's disease. *Br. J. Pharmacol* **2011,** *163,* 375–389.
43. Crump, C. J.; Johnson, D. S.; Li, Y.-M. Development and Mechanism of γ-Secretase Modulators for Alzheimer's Disease. *Biochemistry* **2013,** *52,* 3197–3216.
44. Rogers, K.; Felsenstein, K. M.; Hrdlicka, L.; Tu, Z.; Albayya, F.; Lee, W.; Hopp, S.; Miller, M.-J.; Spaulding, D.; Yang, Z.; Hodgdon, H.; Nolan, S.; Wen, M.; Costa, D.; Blain, J.-F.; Freeman, E.; De Strooper, B.; Vulsteke, V.; Scrocchi, L.; Zetterberg, H.; Portelius, E.; Hutter-Paier, B.; Havas, D.; Ahlijanian, M.; Flood, D.; Leventhal, L.; Shapiro, G.; Patzke, H.; Chesworth, R.; Koenig, G. Modulation of γ-secretase by EVP-0015962 reduces amyloid deposition and behavioral deficits in Tg2576 mice. *Mol. Neurodegener* **2012,** *7,* 61.
45. Stepan, A. F.; Subramanyam, C.; Efremov, I. V.; Dutra, J. K.; O'Sullivan, T. J.; DiRico, K. J.; McDonald, W. S.; Won, A.; Dorff, P. H.; Nolan, C. E.; Becker, S. L.; Pustilnik, L. R.; Riddell, D. R.; Kauffman, G. W.; Kormos, B. L.; Zhang, L.; Lu, Y.; Capetta, S. H.; Green, M. E.; Karki, K.; Sibley, E.; Atchison, K. P.; Hallgren, A. J.; Oborski, C. E.; Robshaw, A. E.; Sneed, B.; O'Donnell, C. J. Application of the Bicyclo[1.1.1]pentane Motif as a Nonclassical Phenyl Ring Bioisostere in the Design of a Potent and Orally Active γ-Secretase Inhibitor. *J. Med. Chem.* **2012,** *55,* 3414–3424.
46. Probst, G.; Aubele, D. L.; Bowers, S.; Dressen, D.; Garofalo, A. W.; Hom, R. K.; Konradi, A. W.; Marugg, J. L.; Mattson, M. N.; Neitzel, M. L.; Semko, C. M.; Sham, H. L.; Smith, J.; Sun, M.; Truong, A. P.; Ye, X. M.; Xu, Y.-z.; Dappen, M. S.; Jagodzinski, J. J.; Keim, P. S.; Peterson, B.; Latimer, L. H.; Quincy, D.; Wu, J.; Goldbach, E.; Ness, D. K.; Quinn, K. P.; Sauer, J.-M.; Wong, K.; Zhang, H.; Zmolek, W.; Brigham, E. F.; Kholodenko, D.; Hu, K.; Kwong, G. T.; Lee, M.; Liao, A.; Motter, R. N.; Sacayon, P.; Santiago, P.; Willits, C.; Bard, F.; Bova, M. P.; Hemphill, S. S.; Nguyen, L.; Ruslim, L.; Tanaka, K.; Tanaka, P.; Wallace, W.; Yednock, T. A.; Basi, G. S. Discovery of (R)-4-cyclopropyl-7,8-difluoro-5-(4-(trifluoromethyl)phenylsulfonyl)-4,5-dihydro-1H-pyrazolo[4,3-c]quinoline (ELND006) and (R)-4-Cyclopropyl-8-fluoro-5-(6-(trifluoromethyl)pyridin-3-ylsulfonyl)-4,5-dihydro-2H-pyrazolo[4,3-c]quinoline (ELND007): Metabolically Stable γ-Secretase Inhibitors that Selectively Inhibit the Production of Amyloid-β over Notch. *J. Med. Chem.* **2013,** *56,* 5261–5274.
47 Cash, A. D.; Perry, G.; Ogawa, O.; Raina, A. K.; Zhu, X.; Smith, M. A. Is Alzheimer's Disease a Mitochondrial Disorder? *Neuroscientist* **2002,** *8,* 489–496.
48. Mapelli, M.; Massimiliano, L.; Crovace, C.; Seeliger, M. A.; Tsai, L.-H.; Meijer, L.; Musacchio, A. Mechanism of CDK5/p25 Binding by CDK Inhibitors. *J. Med. Chem.* **2005,** *48,* 671–679.
49. Nair, N.; Kudo, W.; Smith, M. A.; Abrol, R.; Goddard, W. A., III; Reddy, V. P. Novel Purine-based Fluoroaryl-1,2,3-triazoles as Neuroprotecting Agents: Synthesis, Neuronal Cell Culture Investigations, and CDK5 Docking Studies. *Bioorg. Med. Chem. Lett.* **2011,** *21,* 3957–3961.
50. Sun, K.-H.; Lee, H.-g.; Smith, M. A.; Shah, K. Direct and Indirect Roles of Cyclin-Dependent Kinase 5 as an Upstream Regulator in the C-Jun NH2-terminal Kinase Cascade: Relevance to Neurotoxic Insults in Alzheimer's Disease. *Mol. Biol. Cell.* **2009,** *20,* 4611–4619.
51. Chang, K.-H.; Vincent, F.; Shah, K. Deregulated Cdk5 Triggers Aberrant Activation of Cell Cycle Kinases and Phosphatases Inducing Neuronal Death. *J. Cell. Sci.* **2012,** *125,* 5124–5137.
52. Liebl, J.; Krystof, V.; Vereb, G.; Takacs, L.; Strnad, M.; Pechan, P.; Havlicek, L.; Zatloukal, M.; Fuerst, R.; Vollmar, A. M.; Zahler, S. Anti-angiogenic Effects of Purine Inhibitors of Cyclin Dependent Kinases. *Angiogenesis* **2011,** *14,* 281–291.

53. Senderowicz, A. M. Novel Small Molecule Cyclin-dependent Kinases Modulators in Human Clinical Trials. *Cancer Biol. Ther.* **2003,** *2,* S84–S95.
54. Meijer, L.; Raymond, E. Roscovitine and Other Purines as Kinase Inhibitors. From Starfish Oocytes to Clinical Trials. *Acc. Chem. Res.* **2003,** *36,* 417–425.
55. Malumbres, M.; Pevarello, P.; Barbacid, M.; Bischoff, J. R. CDK Inhibitors in Cancer Therapy: What Is Next? *Trends Pharmacol. Sci.* **2008,** *29,* 16–21.
56. Ivanov, A.; Tyzio, R.; Zilberter, Y.; Ben-Ari, Y. (R)-Roscovitine, a Cyclin-Dependent Kinase Inhibitor, Enhances Tonic GABA Inhibition in Rat Hippocampus. *Neuroscience* **2008,** *156,* 277–288.
57 Ferreira, M. B. A.; Lima, J. P. S.N.; Cohen, E. E. W. Novel Targeted Therapies in Head and Neck Cancer. *Expert Opin. Invest. Drugs* **2012,** *21,* 281–295.
58. Blagden, S.; de Bono, J. Drugging Cell Cycle Kinases in Cancer Therapy. *Curr. Drug. Targets* **2005,** *6,* 325–335.
59. Benson, C.; White, J.; De Bono, J.; O'Donnell, A.; Raynaud, F.; Cruickshank, C.; McGrath, H.; Walton, M.; Workman, P.; Kaye, S.; Cassidy, J.; Gianella-Borradori, A.; Judson, I.; Twelves, C. A Phase I Trial of the Selective Oral Cyclin-dependent Kinase Inhibitor Seliciclib (CYC202; R-roscovitine), Administered Twice Daily for 7 Days Every 21 Days. *Br. J. Cancer* **2007,** *96,* 29–37.
60. Acharya, N. K.; Levin, E. C.; Clifford, P. M.; Han, M.; Tourtellotte, R.; Chamberlain, D.; Pollaro, M.; Coretti, N. J.; Kosciuk, M. C.; Nagele, E. P.; DeMarshall, C.; Freeman, T.; Shi, Y.; Guan, C.; MacPhee, C. H.; Wilensky, R. L.; Nagele, R. G. Diabetes and Hypercholesterolemia Increase Blood–Brain Barrier Permeability and Brain Amyloid Deposition: Beneficial Effects of the LpPLA2 Inhibitor Darapladib. *J. Alzheimer's Dis.* **2013,** *35,* 179–198.
61. Wilensky, R. L.; Shi, Y.; Mohler, E. R., III; Hamamdzic, D.; Burgert, M. E.; Li, J.; Postle, A.; Fenning, R. S.; Bollinger, J. G.; Hoffman, B. E.; Pelchovitz, D. J.; Yang, J.; Mirabile, R. C.; Webb, C. L.; Zhang, L.-F.; Zhang, P.; Gelb, M. H.; Walker, M. C.; Zalewski, A.; Macphee, C. H. Inhibition of Lipoprotein-Associated Phospholipase A2 Reduces Complex Coronary Atherosclerotic Plaque Development. *Nat. Med.* **2008,** *14,* 1059–1066.

Chapter 9

Organofluorine Compounds as Anticancer Agents

Chapter Outline

1. INTRODUCTION

During the past two decades, there has been a significant progress in the discovery of anticancer drugs. Majority of the current drug discovery efforts are focused on the most common types of cancers that include breast, lung, prostate, colorectal, and skin (melanoma) cancers. Modern drug design and discovery is usually based on the mechanism-based selective inhibition of kinases to inhibit the progress of cancer cell cycle. National Cancer Institute

Organofluorine Compounds in Biology and Medicine. http://dx.doi.org/10.1016/B978-0-444-53748-5.00009-5

statistics show that currently there are over 100 drugs or drug candidates that are being evaluated for treating various cancers.[1] Yet, despite the great progress made to date, long-term cancer survival rate is only modest in many cancers, especially those that metastasize. Addressing the severe cytotoxicity associated with chemotherapy is also another challenging goal in the area of cancer drug design. Fluorinated anticancer drugs such as gemcitabine and 5-fluorouracil (5-FU), for example, are highly cytotoxic and their use is accompanied by adverse side effects. Relatively nontoxic, hydrolytically stable derivatives of these compounds are often used as their prodrugs in order to minimize the cytotoxicity of these drugs (e.g., LY2334737 and tegafur are prodrugs for gemcitabine and 5-FU, respectively).[2–4] Intracellular enzymes convert these prodrugs to the active drugs at a slower rate so that at a given time the concentration of the active drugs is substantially diminished and thereby their toxicity is reduced (vide infra).

Due to the rapid metastasis of certain cancers such as small cell lung carcinoma, many of the commonly used anticancer drugs are ineffective at the late stage of the disease. Structurally related fluorinated drugs may have some advantage in these cases. Afatinib, an epidermal growth factor receptor (EGFR) tyrosine kinase inhibitor, for example, in single usage or when used as a combination drug, delays metastasis, and is effective for treating some of the epithelial cancers involving mutations. Similarly, vemurafenib,[5] a fluorinated molecule that inhibits B-Raf (BRAF) kinases, was designed for treating metastatic melanoma (vide infra). In general, the kinase inhibitory activity of the fluorinated drugs is relatively greater than that for the corresponding nonfluorinated analogs, and often due to the increased lipophilicity these compounds are relatively more cell permeable and, therefore, are relatively more potent.

2. NUCLEOSIDE-BASED THERAPEUTICS

2.1 Gemcitabine and Clofarabine

Originally discovered as antimetabolites, the nucleoside analogs are currently being used in the treatment of hematological malignancies, solid tumors, and other cancers. The mechanisms of the action of these drugs are multifaceted; they may either inhibit DNA polymerases or inhibit enzymes responsible for the synthesis of purine and pyrimidine nucleotides and ribonucleic acid synthesis, leading to cell death.[6] Gemcitabine (2′-deoxy-2′,2′-difluorocytidine; Gemzar; LY 188011; Gemcel) and clofarabine (2-chloro-9-(2-deoxy-2-fluoro-β-D-arabinofuranosyl)adenine; Clofarex, Clolar, Evoltra) are two of the important anticancer drugs in this class (Figure 1).

Gemcitabine is a cytidine analogue, and is used as its hydrochloride salt in the treatment of non–small cell lung cancer (NSCLC) and colon, breast, bladder, pancreatic, and ovarian cancers.[3,7–10] It is also used as an

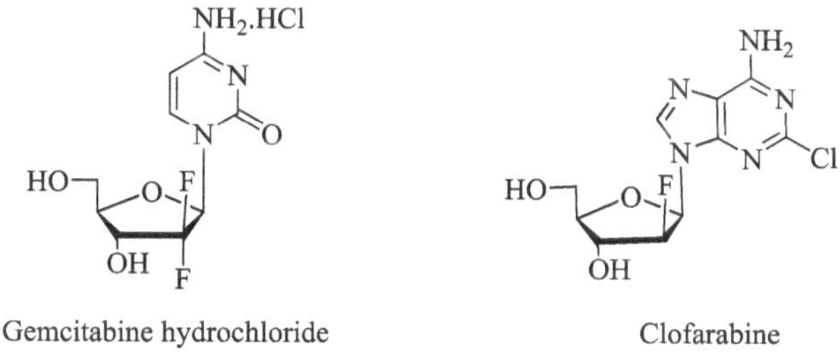

FIGURE 1 Structures of gemcitabine and clofarabine.

antiviral agent. Clofarabine (Figure 1), also an FDA-approved purine-based nucleoside antimetabolite, is in clinical trials for treating acute lymphoblastic leukemia, which is the most common malignant disease in children, and accounts for about 20% of acute leukemia cases in adults.[11] Clofarabine was approved by the FDA in 2004 for the treatment refractory patients (i.e., patients who are resistant to treatment using other anticancer drugs) with acute lymphoblastic leukemia and lymphoblastic lymphoma.[11]

Gemcitabine (administered as its hydrochloride salt) is phosphorylated in vivo to give the corresponding triphosphate, and through DNA-polymerase-catalyzed reactions gets incorporated into the growing DNA chain, and thereby suppresses chain elongation.[12] After incorporation of the gemcitabine into the growing DNA chain, efficiency of the incorporation of the next two correct nucleotides is decreased and the fidelity of the DNA polymerase at these sites is decreased by two orders of magnitude.[13] It also inhibits ribonucleotide reductase, thereby lowering the cellular deoxyribonucleotide triphosphate pools.[14] These combined effects result in cellular apoptosis.

Due to the high cytotoxicity of gemcitabine, it is preferable to have it administered in smaller dosage. However, it is rather unstable under physiological conditions, and is administered intravenously. Even in plasma, it undergoes extensive degradation by cytidine deaminase, which is an abundant constituent of plasma and liver, to form the 2′,2′-difluorodeoxyuridine (dFdU), thereby resulting in its decreased bioavailability. Attempts of derivatizing the gemcitabine into a suitable prodrug that would be relatively stable to oral administration and that would prolong drug action by slow release in the targeted cells are ongoing in many laboratories. Eli Lilly's group has introduced the prodrug LY2334737, a valpromide (2-propylpentanamide) derivative that is slowly hydrolyzed by the cellular esterases to give the active drug and the nontoxic byproduct valproic acid.[2,3] Preclinical studies on mice, rats, and dogs showed that upon oral administration, the prodrug is absorbed by intestinal epithelium, largely intact. Through slow hydrolysis of this prodrug by a cellular esterase (carboxylesterase 2), it is able to maintain sustained release of

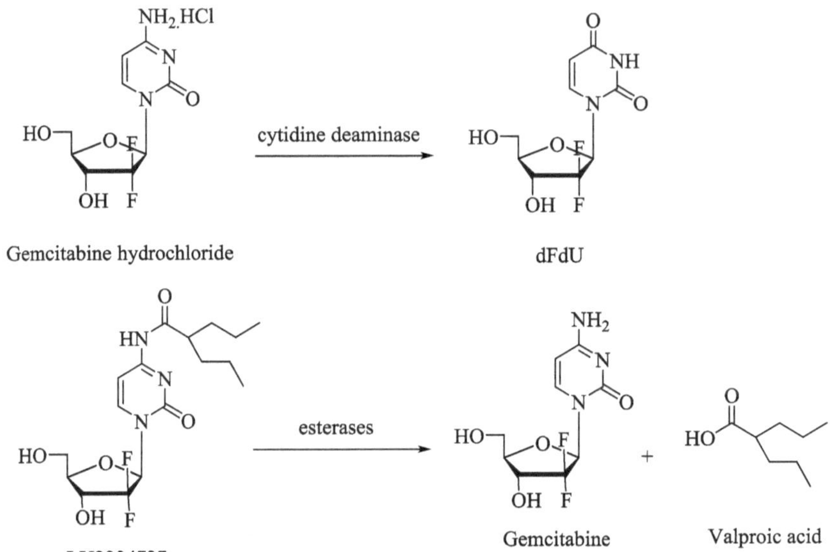

FIGURE 2 Enzymatic deamination of gemcitabine in its unprotected form (top), and the esterase-catalyzed hydrolysis of the prodrug LY2334737 (bottom); because of the slow release of gemcitabine from the prodrug, it does not undergo deamination reaction to give dFdU at an appreciable rate so that the bioavailability of the drug is significantly improved.

the active drug, which, because of its low concentrations at any given time, has relatively low clearance rate through cytidine-deaminase-catalyzed conversion into dFdU (Figure 2).

The use of hydrolytically stable prodrugs is a common practice in the development of active pharmaceutical candidates. For example, the prodrugs tegafur (a precursor of 5-FU), valacyclovir (a precursor of acyclovir), valganciclovir (a precursor of ganciclovir), and capecitabine (a precursor of Xeloda) are some of the widely used pharmaceuticals (Figure 3).

2.1.1 Synthesis of Gemcitabine

The original Eli Lilly's synthesis of gemcitabine and its analogs involves the Reformatsky reaction of 2,3-di-*O*-isopropylideneglyceraldehyde **1** with bromodifluoroacetate, as a key step (Figure 4).[15] Coupling of the intermediate product *gem*-difluororibofuranosyl mesylate **3** with trimethylsilyl-protected pyrimidine derivative **4**, followed by deprotection gives gemcitabine.

A stereoselective synthesis of gemcitabine and clofarabine has been reported starting from the 3,5-bis-*O*-[tris(1-methylethyl)silyl]-2-deoxyribonolactone (**5**) (Figure 5).[16] The deoxyribonolactone **5**, upon electrophilic fluorination using 1 and 2 mol equivalents of *N*-fluorobenzenesulfonimide gave the corresponding fluororibonolactones, **6** and **11**, respectively. Reduction of the lactone moiety of

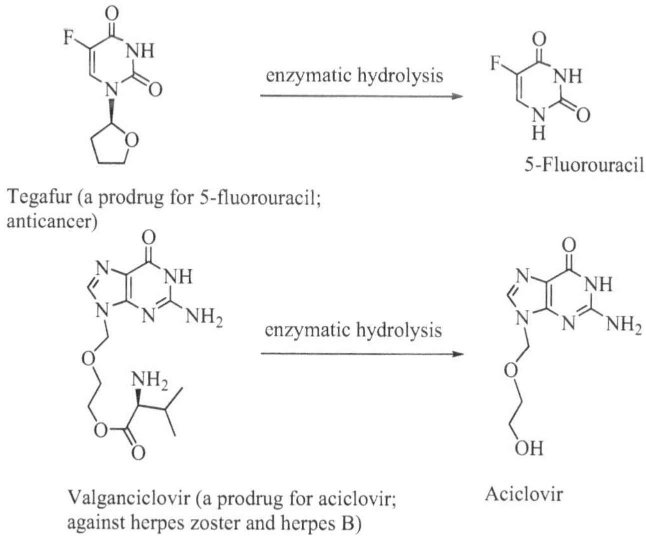

FIGURE 3 Prodrugs tegafur and valganciclovir.

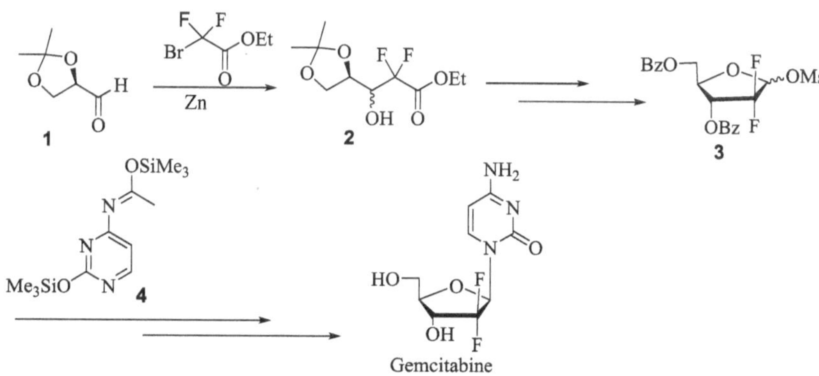

FIGURE 4 Eli Lilly's synthesis of gemcitabine.

6 using diisobutylaluminum hydride (DIBALH; DIBAL; DIBAL-H), followed by reaction with methanesulfonyl chloride gave **7**, which is formed through the spontaneous S_N2 reaction of the initially formed mesylate with the chloride ion. Reduction of compound **11**, followed by mesylation under the same reaction conditions, on the other hand, afforded the expected mesylate **12**. Coupling of **7** with 2,6-dichloropurine (**8**) gave **9**, which upon amination, followed by deprotection of the triisopropylsilyl group afforded an anomeric mixture of clofarabine. Similarly, coupling of **12** with O,*N*-bis(trimethylsilyl)cytosine (**13**), followed by deprotection gave gemcitabine and its α-anomer, which were further purified by high-performance liquid chromatography (HPLC).

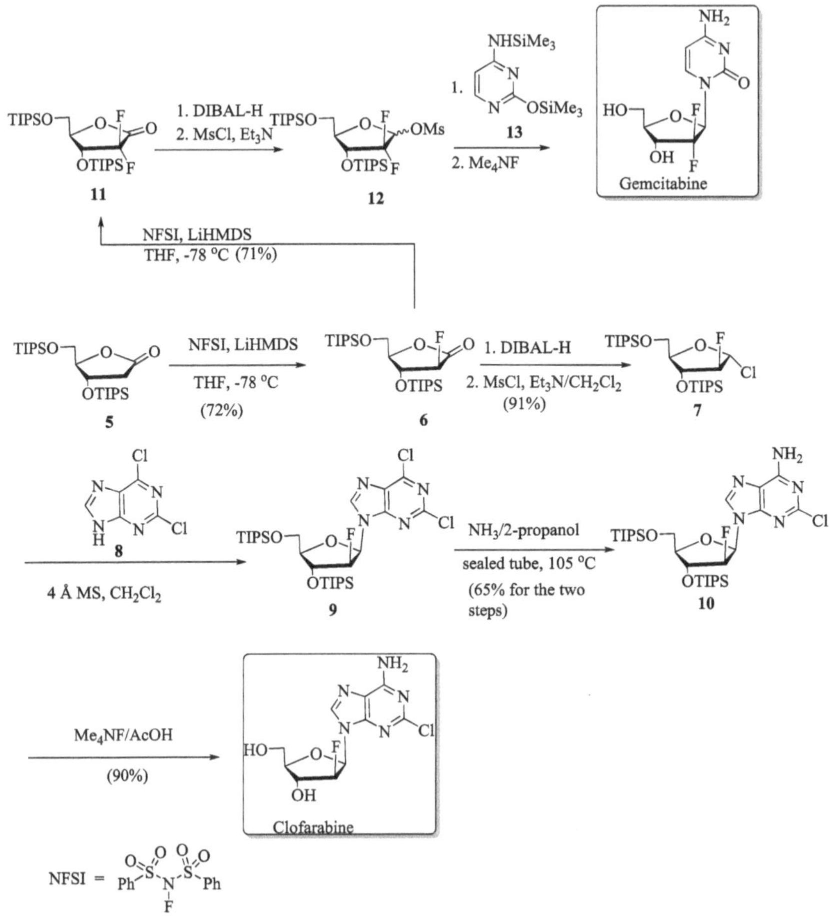

FIGURE 5 Stereoselective synthesis of gemcitabine and clofarabine.

3. RECEPTOR TYROSINE KINASE INHIBITORS

Majority of the lung cancer cases, especially cases of small cell lung carcinoma, are associated with tobacco smoking. Chemotherapy for this cancer is challenging as metastasis of this cancer is a common occurrence, and many of the currently available drugs are ineffective for treatment at this stage. Small cell as well as non–small cell lung carcinomas are due to the uncontrolled growth of the epithelial cells. Chemotherapy involves treatment with platinum-based cytotoxic drugs—such as cisplatin and carboplatin—which are usually combined with antimitotic drugs, paclitaxel, docetaxel, and gemcitabine. The drug candidates for many cancers, including lung and breast cancers, are targeted at the transmembrane EGFR tyrosine kinases, ErbB-1, ErbB-2, ErbB-3, and ErbB-4. Some of the drugs used for treating these cancers are discussed below.

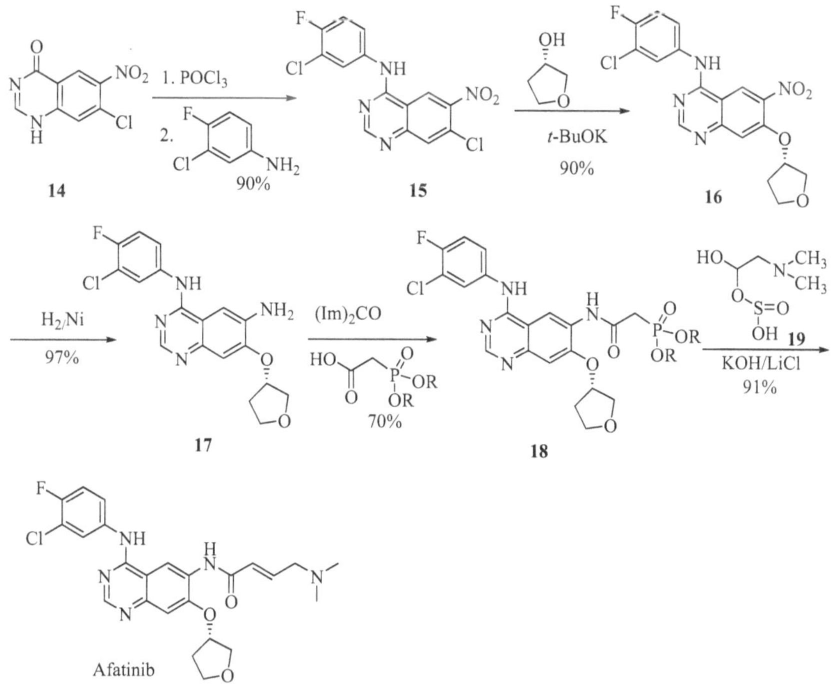

FIGURE 6 Synthesis of afatinib.

3.1 Afatinib

Afatinib (*N*-[4-[(3-chloro-4-fluorophenyl)amino]-7-[[(3*S*)-tetrahydro-3-furanyl] oxy]-6-quinazolinyl]-4-(dimethylamino)-(2*E*)-2-butenamide; Tovok; Gilotrif; Boehringer Ingelheim Pharmaceuticals, Inc; Figure 6) through single usage, or in combination therapy along with other kinase inhibitors, may delay or suppress the emergence of drug resistance caused by certain drug-induced ErbB (proto-oncogene B of the avian erythroblastosis virus) mutations. Structures of most of the transmembrane, EGFR tyrosine kinase inhibitors, including afatinib (approved by FDA in 2013 for non–small cell lung carcinoma), have the 4-(fluorophenylamino)quinazoline moiety, which exhibits strong hydrogen bonding interactions with the active site amino acid residues.[17] Afatinib is also being evaluated for other types of cancers such as head and neck and breast cancers.[17]

Afatinib is a potent inhibitor of ErbB family EGFR tyrosine kinases. In humans, there are four structurally similar EGFRs, which are HER-1 (ErbB-1), HER-2 (ErbB-2), HER-3 (ErbB-3), and HER-4 (ErbB-4). Aberrant signaling through these receptor kinases is responsible for various epithelial cancers, and thus targeting the ErbB family of growth factor receptors plays a key role in the development of therapeutics for epithelial cancers. Afatinib is an irreversible inhibitor of EGFR tyrosine kinases, and forms a covalent adduct with the active site sulfhydryl group through Michael addition reaction (vide infra).[17]

Afatinib exhibits nanomolar inhibitory concentrations on the EGFR tyrosine kinases: wild-type EGFR (EC_{50}=0.5 nM), HER-2 (EC_{50}=14 nM); EGFR-L858R/T790M (EC_{50}=10 nM), and EGFR-L858R (EC_{50}=0.4 nM). It is a potent inhibitor even for the mutated, drug-resistant versions of EGFR, and thereby is suitable for the treatment of certain metastatic epithelial cancers involving these mutations.[17] In 2013, the FDA has approved afatinib for the first-line treatment of metastatic NSCLC, which includes EGFR-L858R (on exon 21) mutations and EGFR exon 19 deletions. The approval of this drug was based on the relatively improved progression-free survival (PFS; 13.6 months as compared to 6.9 months in chemotherapy using pemetrexed/cisplatin) in a clinical study consisting of 345 patients with EGFR mutations.[19]

The conformational changes associated with the binding of afatinib at the active site brings the cysteine residues (Cys797, Cys805, and Cys803 in case of the ErbB-1/EGFR, ErbB-2/HER-2, and ErbB-4/HER-4, respectively) closer to the enamide functionality of afatinib, so that the sulfhydryl group of cysteine undergoes Michael addition to the enamide moiety to form a covalent adduct. The Boehringer Ingelheim group, Solca and coworkers, have characterized the formation of this covalent adduct of cysteine residue with afatinib by single-crystal X-ray crystallography (Figure 7).[17] Through this X-ray crystal structure determination and by tandem mass spectrometric studies, Solca and coworkers have shown that Cys797, Cys805, and Cys803 residues of ErbB-1/EGFR, ErbB-2/HER-2, and ErbB-4/HER-4, respectively, are involved in these Michael additions.[17]

3.1.1 Synthesis of Afatinib

Boehringer Ingelheim's synthesis of afatinib involves five steps starting from 7-chloro-6-nitro-3H-quinazoline-2-one (**14**).[18] Condensation of quinazolinone **14** with 3-chloro-4-fluoroaniline in the presence of $POCl_3$ gives compound **15**, which upon aromatic nucleophilic substitution by (S)-3-hydroxytetrahydrofuran, followed by reduction of the nitro group, gives compound **17** in good yields. N-acylation of compound **17**, using diethylphosphonoacetic acid and 1,1′-carbonyldiimidazole, gives compound **18** (70% yield), which upon Wittig–Horner–Emmons reaction with dimethylaminoacetaldehyde sodium bisulfite adduct (**19**) affords afatinib in 91% yield (Figure 6). Various analogs of afatinib could be synthesized through the above strategy.[18]

3.2 Afatinib Analogs

Compound BI 37781 and gefitinib (Figure 8), analogs of afatinib, are devoid of α,β-unsaturation at the amide carbonyl group, and thereby are incapable of irreversible inhibition of the receptor tyrosine kinases through Michael addition reactions. These compounds, therefore, serve only as reversible inhibitors, and are relatively less potent. Gefitinib (US brand name: Iressa) was approved to treat NSCLCs that are either metastasized or are at the locally advanced stage, and was recommended for cases where the disease was not curable with other chemotherapy.[1]

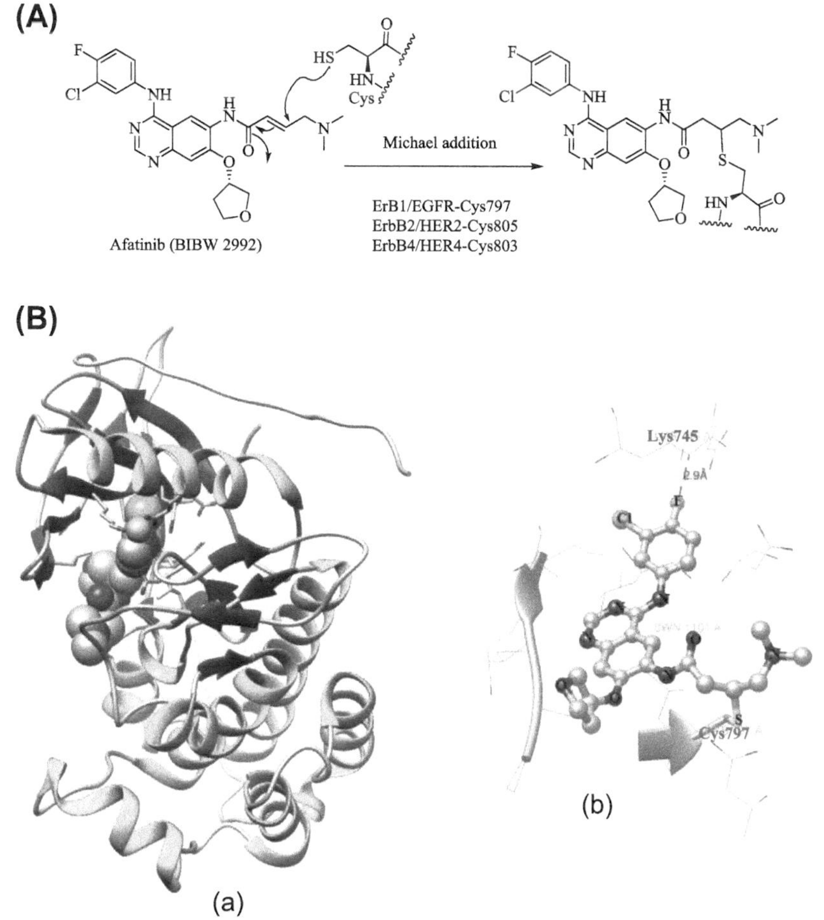

FIGURE 7 (A) Schematic representation of Michael addition of EGFR receptor kinases to afatinib; (B) X-ray structure of wild-type EGFR complexed to afatinib (spheres; (a)), and the expanded view of the active site region (b); afatinib (ball and stick model) binds covalently, through Michael addition, to the Cys797 sulfur (b); the structure was created using UCSF Chimera software; PDB 4G5P.

Although Boehringer Ingelheim's compound BI37781 has similar potency as afatinib on cells expressing wild-type EGFR or EGFR-L858R mutated forms, it has relatively lower or no activity against cells expressing the doubly mutated versions of EGFR, EGFR-L858R/T790M, and HER-2 (ErbB-1) or HER-4 (ErbB-4) (Figure 9).[17] These findings show the importance of the enamide moiety of afatinib and related drug candidates in their irreversible binding at the enzyme active sites, and thereby for their improved inhibitory action.

The mechanism of the enzyme inhibitory action of canertinib, which also has an enamide moiety, may be similar to that of afatinib, although canertinib is relatively less potent. Canertinib was not further developed as a therapeutic for

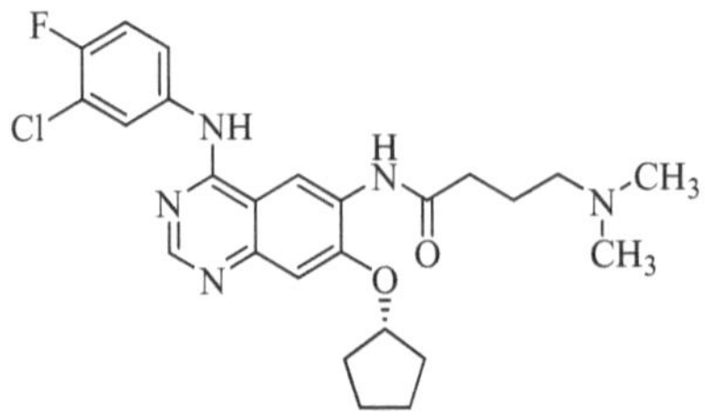

BI 37781

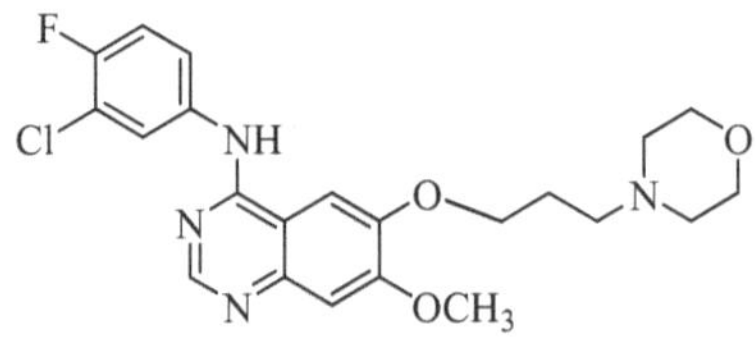

Gefitinib (ZD1839)

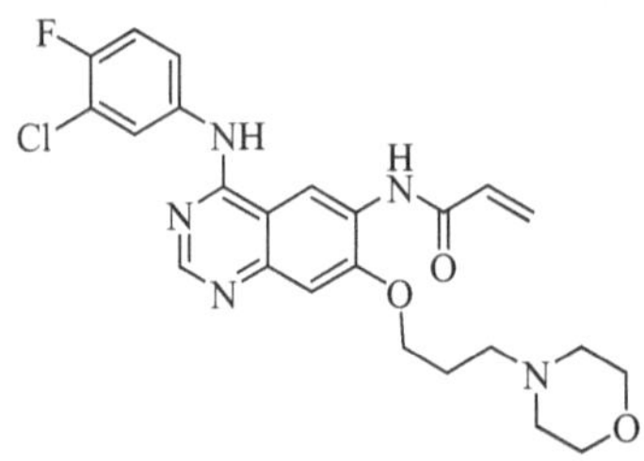

Canertinib (CI 1033; PD 183805)

FIGURE 8 Structures of selected afatinib analogs.

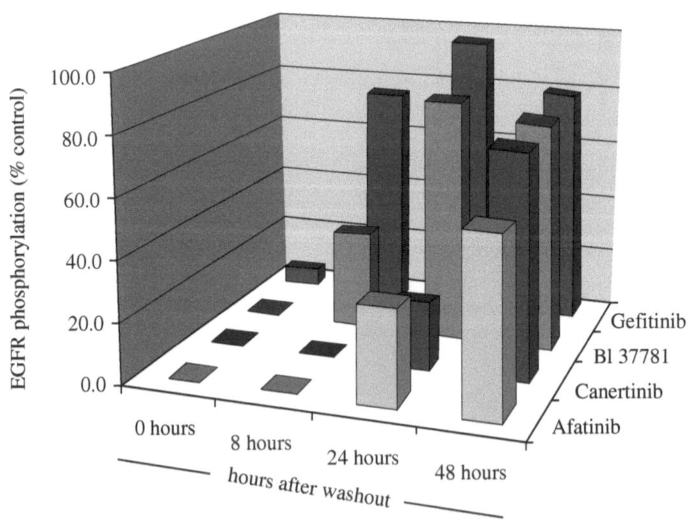

FIGURE 9 Cell studies comparing afatinib with other tyrosine kinase inhibitors; afatinib has the highest potency and prolonged duration of action in inhibiting tyrosine phosphorylation. *Adapted with permission from Ref. 17.*

the NSCLC due to its adverse side effects during phase II clinical trials, and due to its lack of efficacy.[20]

In these quinazoline-based anticancer pharmaceuticals (afatinib, BI37781, and gefitinib), the role of the aryl fluorine is not clear, although the strong C–F bond may contribute to their relatively higher metabolic and oxidative stabilities.

3.3 Sorafenib and Regorafenib

Sorafenib (Nexavar, Bayer; Figure 10) is used for the treatment of hepatocellular cancer (FDA approved in 2007) and renal cell carcinoma (FDA approved in 2005). The median PFS in patients treated with sorafenib is improved by about 3 months as compared to that of placebo cases. In 2013, the FDA approved sorafenib tosylate for the treatment of locally recurrent or metastatic differentiated thyroid carcinoma that is refractive to radioactive iodine treatment.[21]

Sorafenib is the first drug that targets BRAF mutant (V600E) melanoma, and in addition to its clinical use for kidney, liver, and thyroid cancers, is also in clinical trials for treating lung and brain cancers.[22] Sorafenib is a multikinase inhibitor (serine/threonine, Raf, and receptor tyrosine kinases), and targets receptor tyrosine kinase signaling. It inhibits vascular endothelial growth factor receptor (VEGFR), platelet-derived growth factor receptor (PDGFR), and receptors for glial-derived neurotrophic factor and stem cell factor, and thereby inhibits downstream Raf serine/threonine kinase activity, which leads to suppression of tumor growth by antiangiogenic, antiproliferative, and/or proapoptotic effects (Figure 11).[23]

Regorafenib (Stivarga, Bayer; Figure 10) is a closely related analog of sorafenib, differing from sorafenib by the presence of an aryl C–F bond *ortho* to the urea moiety. Regorafenib, as in the case of sorafenib, is an orally administered multikinase inhibitor, and was approved by the FDA, in 2012, for the treatment of metastatic colorectal cancer.

3.3.1 Synthesis of Sorafenib

Buchwald and coworkers have reported a convenient method of synthesis of sorafenib using palladium-catalyzed one-pot arylation-deprotection, followed by a second palladium-catalyzed arylation.[24] In the presence of a sterically crowded phosphine ligand, *t*-BuBrettPhos, these C–N cross-coupling reactions of the urea derivatives (**21** and **23**) proceeded in good yields (Figure 12).

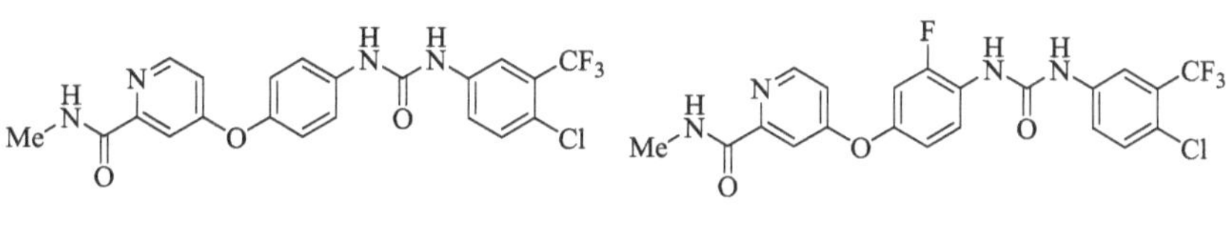

Sorafenib (Nexavar; Bayer)   Regorafenib (Stivarga; Bayer)

FIGURE 10 Structures of sorafenib and regorafenib.

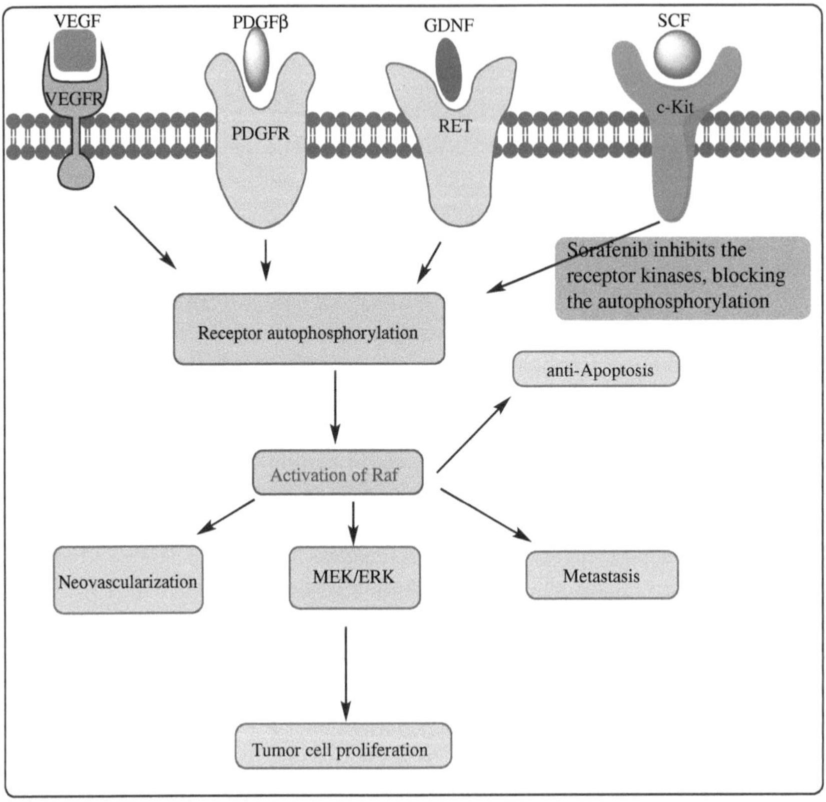

FIGURE 11 Schematic view of the anticancer effects of the multikinase inhibitor sorafenib; sorafenib blocks receptor autophosphorylation, and thereby inhibits downstream Raf serine/threonine kinase activity.[23] Thus, through targeting of receptor tyrosine kinases, sorafenib suppresses tumor growth by anti-angiogenic, antiproliferative, and/or proapoptotic effects.[23] VEGFR, vascular endothelial growth factor receptor; PDGFR, platelet-derived growth factor receptor; RET, rearranged during transfection proto-oncogene; GDNF, glial-derived neurotrophic factor receptor; c-Kit, (also referred as CD117) is a stem cell factor (SCF) receptor; MEK, mitogen-activated protein kinase; ERK, extracellular signal-regulated kinase.

3.4 Sunitinib

Sunitinib (Sutent; Pfizer; Figure 13) is a multiple receptor tyrosine kinase inhibitor, and is an orally used drug (FDA approval in 2006) to treat renal cell carcinoma and gastrointestinal stromal tumors that are resistant to imatinib. It inhibits several receptor tyrosine kinases, including VEGFR and PDGFR, KIT (c-Kit), RET, FLT3 (FMS-like tyrosine kinase 3), and CSF-1R (colony stimulating factor 1 receptor). The parallel inhibition of these tyrosine kinase receptors leads to attenuated tumor growth through reduced tumor vascularization and thereby apoptosis of cancer cells. The other small molecule VEGFR targeted drugs for treating renal cell carcinoma include axitinib (Pfizer) and sorafenib (Bayer; vide supra).[25] Gastrointestinal stromal cancers are caused by PDGFR and KIT activation. In cases of metastatic gastrointestinal stromal cancers, the

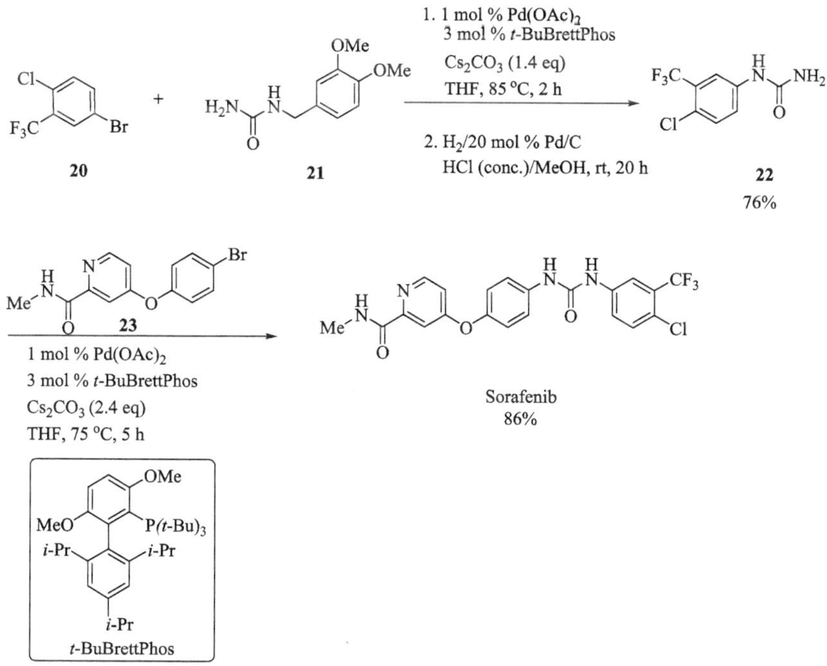

FIGURE 12 Synthesis of sorafenib.

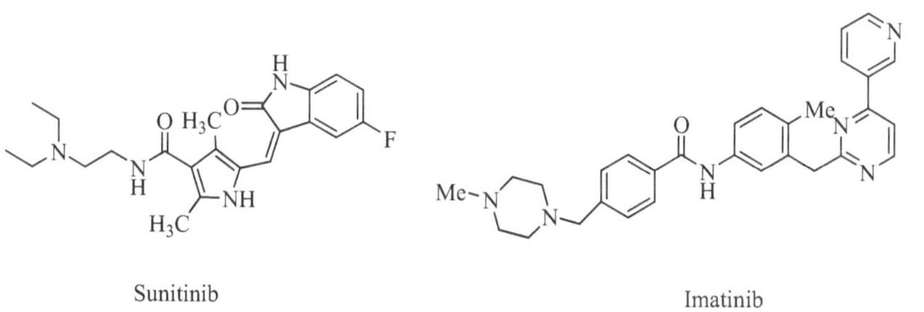

FIGURE 13 Structures of sunitinib and imatinib.

choice between sunitinib and imatinib is less clear-cut and therefore other factors such as tumor characteristics and drug pharmacokinetics are considered for individualized treatment of the affected patients.[26] For patients progressing on or for those intolerant to imatinib, continuing therapy using sunitinib and regorafenib is recommended.[27]

It is interesting to note that the cellular vascular endothelial growth factor and platelet-derived growth factor kinase activity of the sunitinib (0.005–0.05 μM, and 0.01 μM, respectively) is about 10 times greater than that for its nonfluorinated analog. Substitution of chlorine or bromine for the fluorine in sunitinib resulted in relatively increased cytotoxicity.[28]

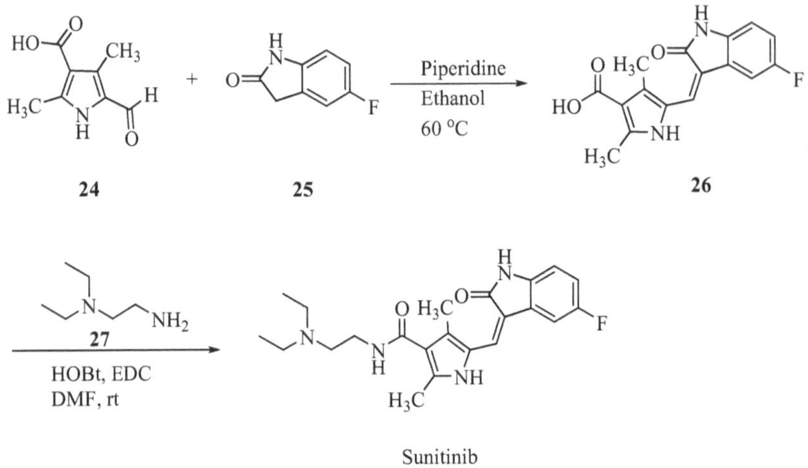

FIGURE 14 Synthesis of sunitinib (HOBt = 1-hydroxy-1H-benzotriazole; EDC = 1-ethyl-3-(dimethylaminopropyl)carbodiimide; DMF = N,N-dimethylformamide).

3.4.1 Synthesis of Sunitinib

The Knoevenagel condensation of the substituted pyrrole carbaldehyde **24** with 5-fluoro-indolin-2-one (**25**), followed by 1-ethyl-3-[3-(dimethylamino)propyl] carbodiimide (EDC)-mediated coupling of **26** with β-(diethylamino)ethylamine (**27**) afforded sunitinib (Figure 14).[28] This synthetic strategy is amenable for the synthesis of its various analogs.

3.5 Cabozantinib

Cabozantinib (Cometriq; Exelixis; Figure 15) was approved for treating medullary thyroid cancer in 2012. It inhibits tyrosine kinases of hepatocyte growth factor receptor (MET) and VEGFR-2 as well as a number of other receptor tyrosine kinases. In mouse models it decreases tumor and endothelial cell proliferation and exerts dose-dependent inhibition of tumor growth.[29] In phase III clinical trials of progressive medullary thyroid cancer, carbozantinib increased the PFS time to 11.2 months versus 4.0 months for placebo.[30] This drug also may be useful for treating various other cancers, including lung adenocarcinoma.[31] In phase II clinical trials of advanced prostate cancer, cabozantinib improved bone scans in 68% of patients (i.e., treatment with cabozantinib attenuated the bone response to the tumor, resulting in increased normal bone volume).[32]

3.5.1 Synthesis of Cabozantinib

Exelixis' group has reported a convenient synthesis of cabozantinib starting from 4-fluoroaniline. Thus, coupling of cyclopropane-1,1-dicarboxylic acid with 4-fluoroaniline using thionyl chloride, followed by conversion of the intermediate carboxylic acid to acid chloride **28** using oxalyl chloride and a second coupling with compound **29** gives cabozantinib (Figure 16).[31]

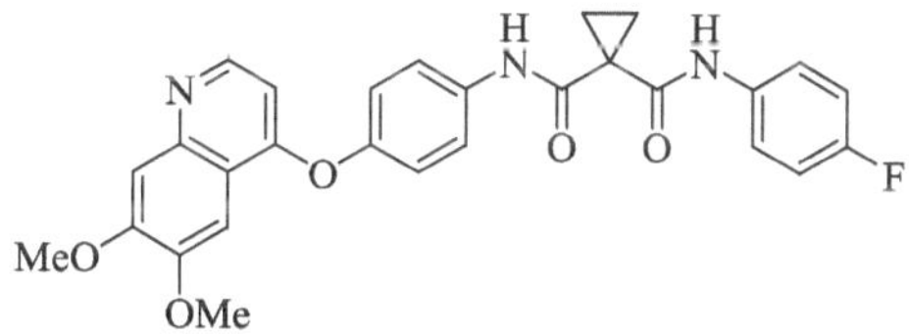

Cabozantinib

FIGURE 15 Structure of cabozantinib.

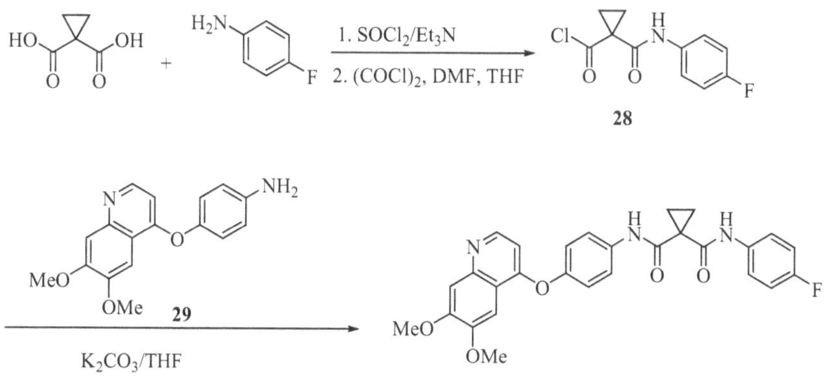

Cabozantinib

FIGURE 16 Synthesis of cabozantinib (DMF = N,N-dimethylformamide; THF = tetrahydrofuran).

4. CAPECITABINE AND LAPATINIB

Capecitabine (5′-deoxy-5-fluoro-*N*-((pentyloxy)carbonyl)cytidine; Xeloda; Roche Pharmaceuticals) is an orally administered prodrug for the otherwise cytotoxic drug, 5-FU, which acts as thymidylate synthase inhibitor and thereby inhibits DNA replication.[33] Capecitabine is hydrolyzed by carboxylesterase present in the liver to generate 5′-deoxy-5-fluorocytidine (5′-DFC), which upon a cytidine-deaminase-catalyzed reaction is transformed into 5′-deoxy-5-fluorouridine. The latter compound, upon entering the tumor cells, releases the cytotoxic drug 5-FU through thymidylate phosphorylase catalysis (Figure 17).[34] Capecitabine is thus a relatively nontoxic prodrug molecule and through controlled release of 5-FU interferes with the growth of the cancer cells, which are eventually removed by the normal cellular metabolism. It is used in the treatment of breast cancer and colorectal cancer.[35]

The final step for the conversion of capecitabine to 5-FU requires high concentrations of thymidylate phosphorylase, which is elevated in tumor cells as compared to that in the normal cells (vide supra). In this context, inhibition of EGFR by gefitinib results in an increase in the intracellular thymidylate phosphorylase so that the combination drug consisting of gefitinib and capecitabine has synergistic effect.

A potentially useful combination drug candidate molecule, ZRX1, (a conjugate of 2′,3′-di-*O*-acetyl capecitabine with the quinazoline EGFR inhibitor FD105), shows a relatively stronger growth inhibitory effect on tumor cells as compared to the use of binary mixtures of capecitabine and EGFR inhibitors

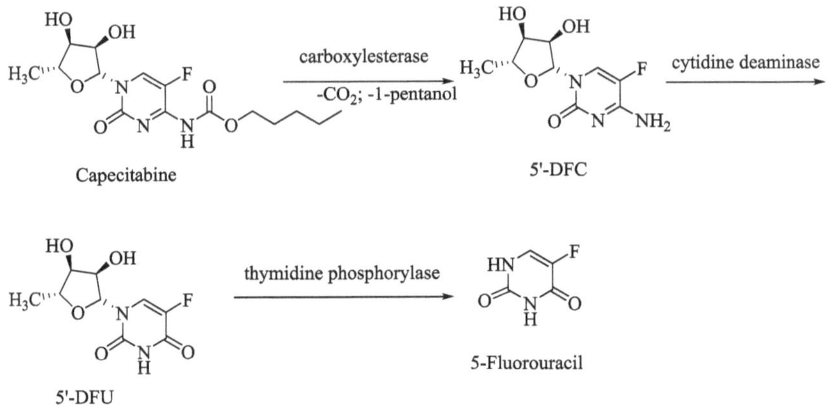

FIGURE 17 Enzyme-catalyzed release of 5-fluorouracil from capecitabine.

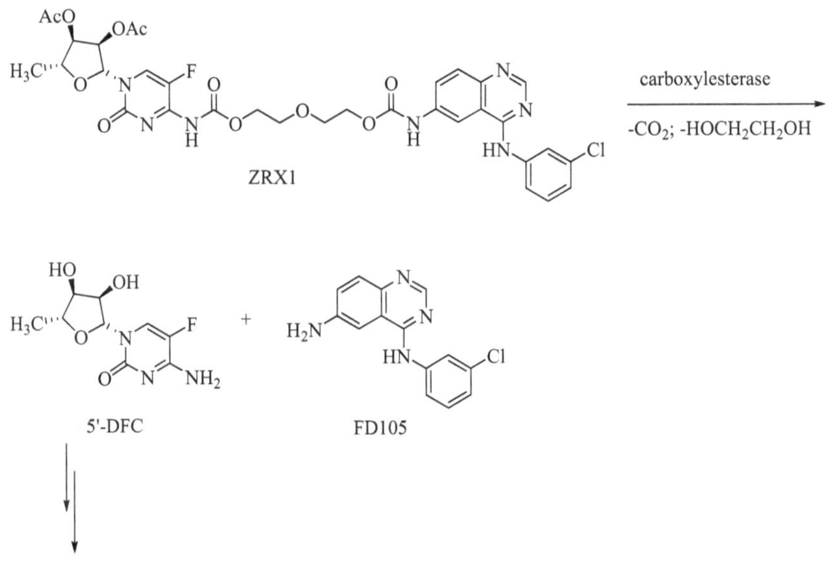

FIGURE 18 Esterase-catalyzed release of 5-fluorouracil from ZRX1.

(Figure 18).[34] The ZRX1 releases the FD105 and 5′-DFC through carboxylesterase catalysis. The 5′-DFC, eventually, is enzymatically (thymidylate phosphorylase) transformed to the cytotoxic drug 5-FU once inside the tumor cells.

Lapatinib (GW 2016, GSK) was approved by the FDA in 2007 for combination therapy, in conjunction with capecitabine (Xeloda, Roche), for breast cancer patients. Lapatinib is a reversible inhibitor of HER-4/ErbB-4 tyrosine kinase, and the protein–drug interactions at the protein active site are similar to those of other quinazoline derivatives.[37] All ErbB-4 residues contacted by

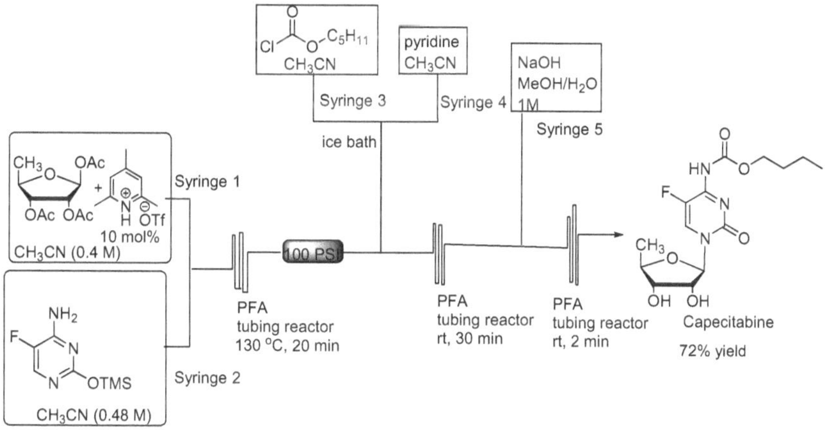

FIGURE 19 Schematic of the continuous flow reactor for the synthesis of capecitabine.[36]

lapatinib in the active site are also conserved in the ErbB-2 receptor active site, showing that the key elements of kinase activation and inhibition are similar in the ErbB members.[37] Through reversible, competitive binding of lapatinib at the active site of the EGFR, it blocks the normal binding of adenosine triphosphate that is required for its phosphorylation activity (Figure 20).

4.1 Synthesis of Capecitabine

Jamison and coworkers have developed a green and efficient synthesis of capecitabine in a continuous flow tube reactor, using Bronsted-acid-catalyzed glycosylation as the key step.[36] This synthetic strategy is amenable to the synthesis of several other 5′-deoxyribonucleoside pharmaceuticals in high yields. A freshly synthesized silylated 5-fluorocytosine amide derivative was reacted with 5-deoxyribose triacetate in the presence of the Bronsted acid 2,4,6-trimethylpyridinium triflate in a flow tube at 130 °C for 20 min, then a mixture of pentyl chloroformate and pyridine in acetonitrile was injected into the second flow reactor at 0 °C, and kept for 30 min. The final deprotection of the silyl moiety was effected in a third flow tube using methanolic sodium hydroxide at ambient temperature to afford capecitabine in an overall yield of 72% (Figure 19).

5. TRAMETINIB AND DABRAFENIB

In 2014, the FDA approved combination therapy using trametinib and dabrafenib (both from GSK; Figure 21) for the treatment of V600E mutated metastatic melanoma. Trametinib is a mitogen-activated kinase/extracellular signal-regulated kinase (ERK; MEK1 and MEK2) inhibitor, while dabrafenib is a BRAF (encoded by the gene BRAF) inhibitor. BRAF, a serine/threonine

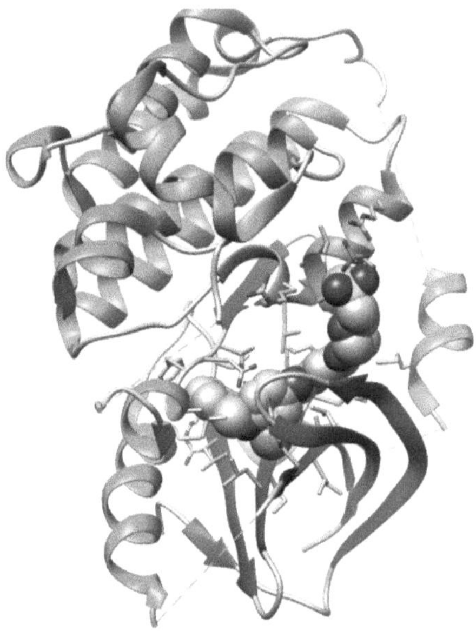

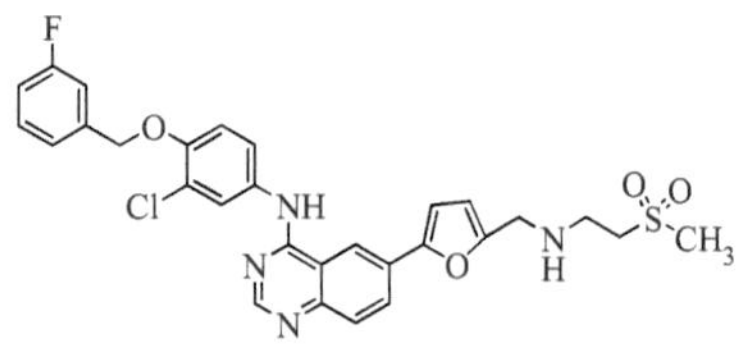

Lapatinib (GW 2016)

FIGURE 20 Lapatinib (spheres; F, green; N, blue; O, red) bound at the active site of EGFR; through competitive, reversible binding at the active site, it blocks the normal binding of adenosine triphosphate and thereby affects the kinase activity; the structure was created using UCSF Chimera software; PDB: 1XKK. *(For interpretation of the references to color in this figure legend, the reader is referred to the online version of this book.)*

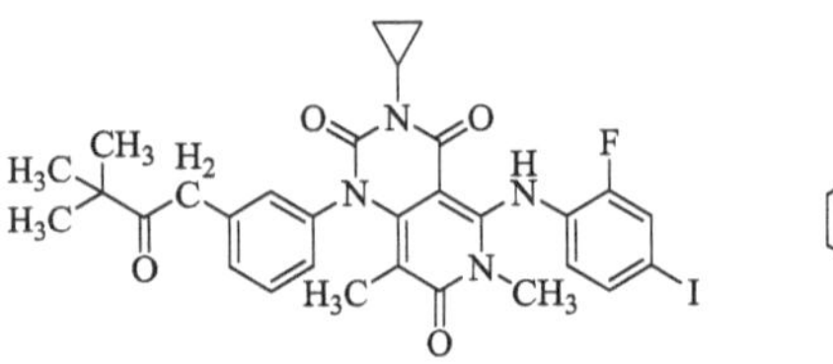

Trametinib (Mekinist; GSK)
MEK1 and MEK2 inhibitor; Melanoma drug

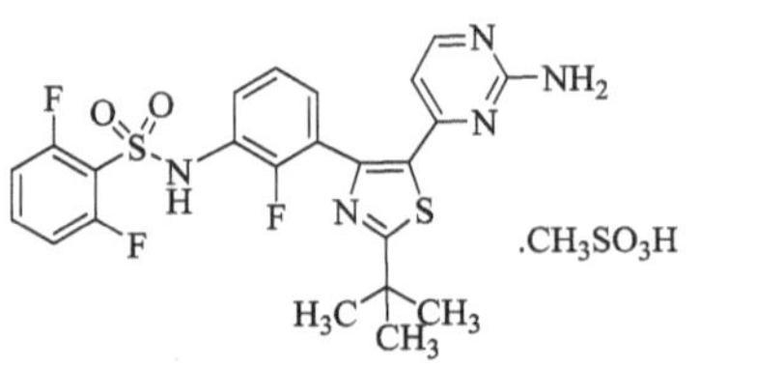

Dabrafenib (Tafinlar; GSK)
BRAF-inhibitor; used as methanesulfonate salt

FIGURE 21 Structures of trametinib and dabrafenib.

protein kinase, is involved in the regulation of cell growth, and is mutated in some human cancers.[38] The single use of trametinib (oral administration) for the treatment of metastatic melanoma involving BRAF mutations V600E and V600K (valine 600 residue substituted by glutamic acid and lysine, respectively), and the single use of dabrafenib (oral administration) for melanoma involving BRAF mutation V600E was approved by the FDA in 2013. Mutations such as BRAF V600E and V600K result in overactivation of the BRAF kinases, and thereby in the acceleration of the tumor growth. Inhibition of MEK

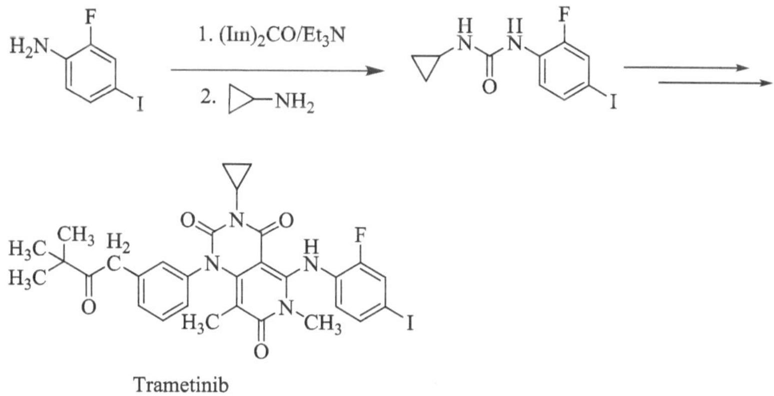

FIGURE 22 Synthesis of trametinib.

activation by trametinib causes activation of cell apoptosis, but not cell cycle arrest.[39] On the other hand, inhibition of BRAF kinase by dabrafenib causes cell cycle inhibition. Thus the combination therapy has mutually exclusive synergistic effect on the treatment of metastatic melanoma.

Dabrafenib has inhibitory effect on cell lines exhibiting BRAF V600K, BRAF V600D, and BRAF V600R mutations, with IC_{50} values in the subnanomolar to nanomolar range. Dabrafenib exhibits >500-fold selectivity for BRAF V600E over most other kinases.[40] The inhibition of the BRAF V600E kinase results in decreased ERK phosphorylation and thereby inhibition of cell proliferation at the G1 phase of the cell cycle. The coadministration of dabrafenib with trametinib in metastatic melanoma cases results in increased expression of melanoma antigens and a decrease in immunosuppressive cytokines, interleukin 6, and interleukin 8. This combination therapy is relatively more effective than treatment with either one alone in cases of metastatic melanoma, because the acquired resistance for either one is offset by the other.[40]

5.1 Synthesis of Trametinib and Dabrafenib

The fluoroaryl ring is usually introduced early in the synthesis of trametinib. A nine step synthesis of trametinib was reported starting from 2-fluoro-4-iodoaniline, with an overall yield of 2.9% for the combined nine steps (Figure 22).[41]

Dabrafenib was synthesized in four steps starting from the sulfonamide **32**, which was prepared by reaction of 2,6-difluorobenzenesulfonyl chloride (**30**) with methyl 3-amino-2-fluorobenzoate (**31**) (Figure 23).[40]

6. PACLITAXEL AND RELATED TAXOIDS

Taxol (paclitaxel, Bristol-Myers Squibb) and taxotere (docetaxel; Sanofi-Aventis) are FDA-approved drugs for the treatment of metastatic breast and ovarian

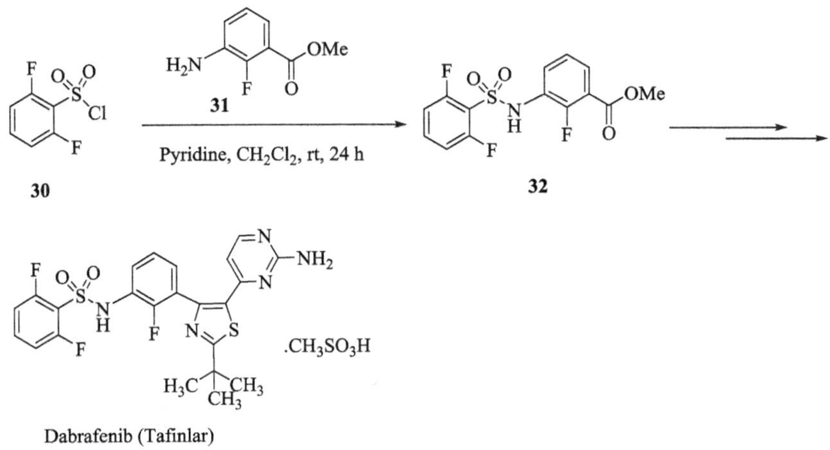

FIGURE 23 Synthesis of dabrafenib.

cancers. In 2006, docetaxel was approved as a combination therapy with cisplatin and fluorouracil for the treatment of patients with advanced squamous cell carcinoma of the head and neck; median overall survival for this combination therapy was about 18.6 months as compared to the 14.2 months for the cisplatin/fluorouracil therapy. Docetaxel, used as a combination therapy with other drugs, was also approved for treatment of gastric cancer, prostate cancer, and NSCLC.[42] Due to the enormous success of these Taxol drugs as antineoplastic agents, several of their structural analogs, including their fluorinated versions, were investigated for their improved anticancer activity. Ojima and coworkers have synthesized and evaluated a series of fluorine-containing taxoids, such as CF_3-Taxol, CF_3-taxotere, F-Ph-taxotere, mainly through the modification of the α-hydroxy-β-amino acid side chain moiety (Figure 24).[43–47] The aryl-fluoro derivatives of paclitaxel (SB-T-31031) and docetaxel (SB-T-3001 and F_2-10-Ac-docetaxel) were prepared and used as ^{19}F-nuclear magnetic resonance (NMR) probes to determine the solution conformations of these molecules and to determine their binding conformation in microtubules.[48] Solid-state rotational echo double resonance NMR studies of microtubule-bound fluorotaxoids provide the relevant ^{19}F–^{19}F distances (or ^{19}F–^{2}H distances in case of selectively deuterated compounds) so that, in combination with molecular modeling analysis, the bound state conformation can be determined. This information, in the absence of the X-ray crystal structures of the enzyme-bound taxoids, is valuable in the design of potent taxoids for cancer therapy.

The synthesis of a series of 3′-difluorovinyltaxoids **35** was achieved by Ojima-Holton coupling of 4-difluorovinyl-β-lactam **33** with the corresponding baccatin derivatives **34** (Figure 25). In vitro studies showed that these 3′-difluorovinyltaxoids exhibit superior activities in the in vitro apoptosis of cancer cells, mainly through activation of the caspases 2, 8, and 9.[47] These compounds show up to three orders more potency as compared to paclitaxel against drug-resistant

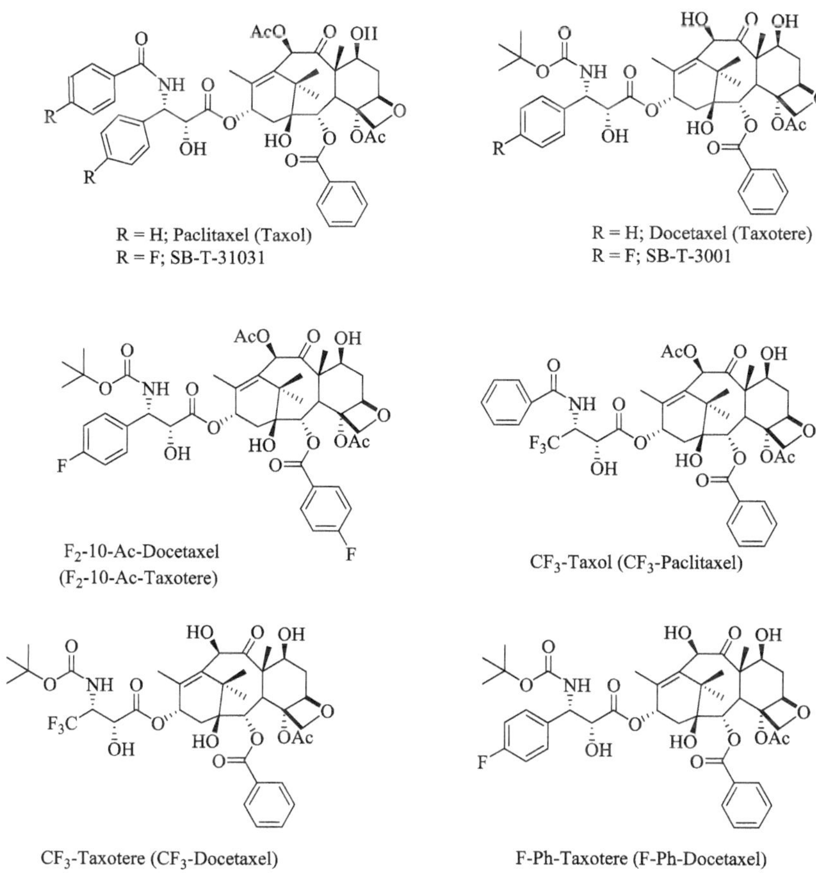

FIGURE 24 Structures of the paclitaxel, docetaxel, and their fluorinated derivatives.

cancer cell lines. Using a human colon cancer cell line, Ojima and coworkers have shown that the IC_{50} values of the difluorovinyltaxoids range from 0.14 to 0.28 nM, whereas paclitaxel has IC_{50} of 451 nM for these cell lines.[47]

Ojima and coworkers have developed "Trojan Horse" approaches for the drug delivery of various fluorinated taxoids, using biotin conjugates and single-walled carbon nanotube conjugates, consisting of self-immolative disulfide linkers.[47,48] Tumor cells express characteristic antigens on the cell surface, which are distinct from those of the normal cells. Making use of these biomarkers and using immunoglobulin G monoclonal antibodies (mAbs) that are specific to these tumor cell surface antigens, Ojima and coworkers have developed cancer-cell-specific drug delivery agents, consisting of the fluorotaxoid–mAb conjugates (Figure 26). Key to the success of this strategy is the use of a disulfide linker to connect the taxoid molecules to the mAb. Once inside the cancer cells, this disulfide linker rapidly reacts with cellularly abundant glutathione to release the taxoid drug molecules.[47,48]

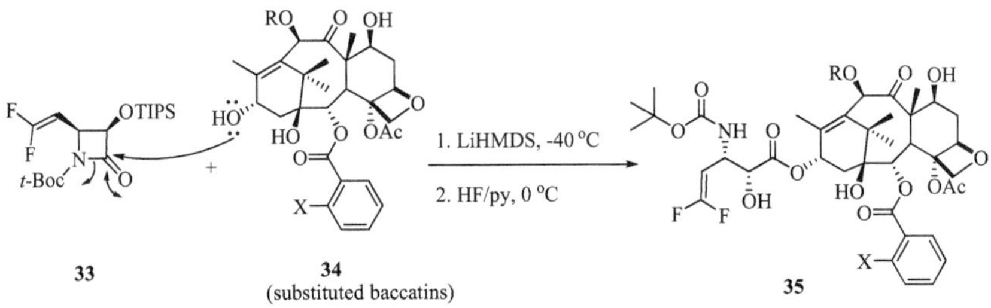

R = e.g., Ac, Me_2NCO, EtCO
X = e.g., OMe, F, Cl

FIGURE 25 Synthesis of the fluorotaxoid **35** (Ac = acetyl; *t*-Boc = *tert*-butoxycarbonyl; LiHMDS = lithium hexamethyldisilazide; py = pyridine; TIPS = triisopropylsilyl).

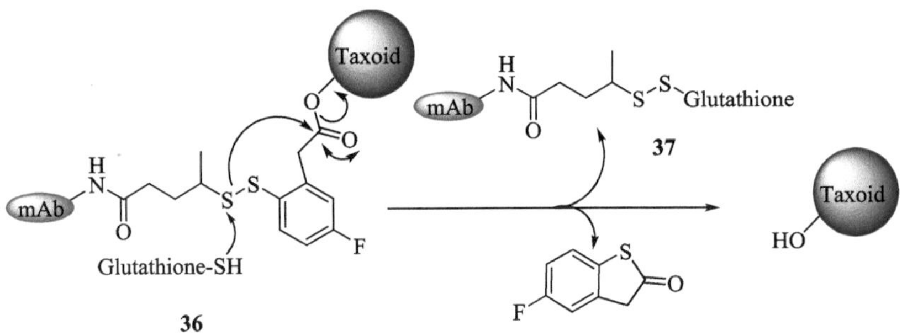

mAb = monoclinal antibody

FIGURE 26 Schematic representation of the mechanism of the taxoid drug delivery.

Ojima and coworkers also introduced the second-generation, vitamin–taxoid conjugates, consisting of self-immolative disulfide linkers for cancer-cell-specific drug delivery of the taxoids and fluorotaxoids (Figure 27). Tumor cells require abundant supply of vitamins such as biotin, folic acid, and vitamin B_{12}, and therefore these cells have excessive receptors for these vitamins on their cell surfaces, as compared with the normal cells. These drug conjugates upon binding to the cell surface receptors are internalized through endocytosis to release the "war head" taxoid molecules, through glutathione-mediated cleavage of the disulfide bond.[47] The advantage of using fluorine in the linker moiety is that it helps in the regioselective cleavage of the S–S bond to release the fluorotaxoids (or taxoids). The thionobenzolactone (**38**) and vitamin–glutathione conjugate (**41**) are formed as byproducts during this transformation.

The stability of the drug conjugates could also be monitored using ^{19}F NMR spectroscopy as a convenient alternative to the conventional HPLC or liquid chromatography–mass spectrometry. Due to the high sensitivity of the ^{19}F nucleus and due to the lack of any background ^{19}F signals in the medium, the progress of this transformation (disappearance of the drug conjugate and the simultaneous appearance of the fluorothionobenzolactone byproduct) could

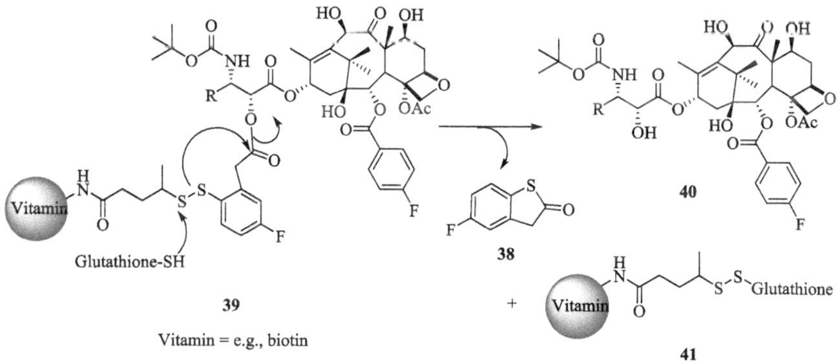

FIGURE 27 Schematic representation of the mechanism of the fluorotaxoid release from its vitamin conjugate.

be monitored by ^{19}F NMR. Biotin–taxoid conjugates consisting of ^{18}F-labeled alkyl side chain were also synthesized by Ojima and coworkers. The latter compounds serve as positron emission tomographic (PET) tracers for following the biodistribution of the drug conjugates.[47]

7. BICALUTAMIDE

The anticancer drug bicalutamide (Cazodex; AstraZeneca; FDA-approved in 1995) is used as a racemic mixture; the R-enantiomer is the active ingredient of the drug, whereas the S-enantiomer is inactive, but without any adverse effects. It is generally used as a palliative treatment in prostate cancers, in which case it acts as androgen antagonist, by binding to the androgen receptors and thereby blocking the synthesis of the endogenous testosterone. Prostate carcinoma is one of the most diagnosed cancers in men, and is, in most cases, treated using bicalutamide as the preferred drug. Mutation of tryptophan-746 by leucine (W746L), on the other hand, confers agonist activity to the R-bicalutamide, and apparently results in bicalutamide withdrawal syndrome.[49]

A 1.8-Å resolution X-ray crystal structure of the this mutant with R-bicalutamide shows that it binds at the dihydrotestosterone binding site for the wild-type androgen receptor (Figure 28).[49] The CF_3 moiety is buried in the hydrophobic cavity at the ligand binding site and has weak polar interactions with the neighboring side-chain C–H bonds, while the cyanide nitrogen is within the hydrogen bonding distance (3.0 Å) to the Arg752-NH_2 moiety.

Advanced prostate cancer treatment requires drugs that have superior binding affinity (more than that for bicalutamide) to the androgen receptor. Phase I/II clinical trials of two fluorinated drugs, RD162 and MDV3100 (Figure 29), show some promise in this regard as, in competition studies, they show relatively high 16β-[^{18}F]fluoro-5α-dihydrotestosterone equilibrium binding to androgen receptors as compared to bicalutamide (IC_{50} values are 30.9, 21.4, and 160 nM, respectively for RD162, MDV3100, and bicalutamide).[50]

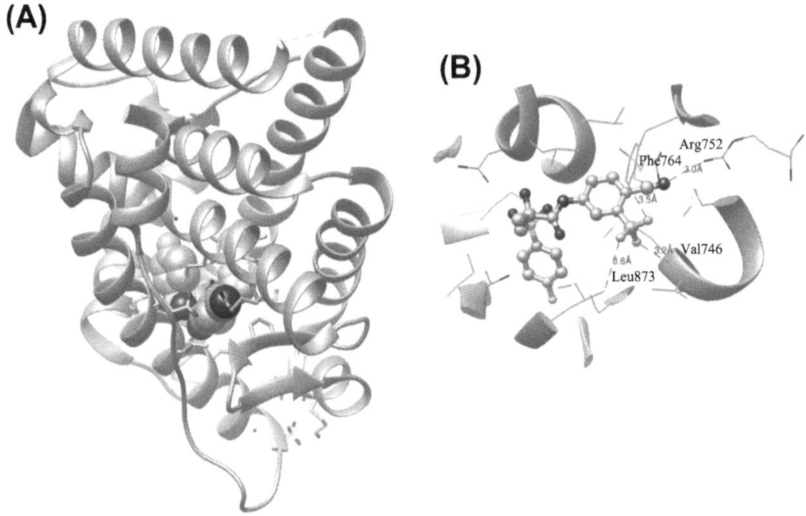

FIGURE 28 (R)-bicalutamide (shown as spheres in A and as ball and stick in B) bound to the ligand binding site of androgen receptor (A), and expanded view at the binding site (B); the cyanide nitrogen has hydrogen bonding interaction with Arg752, while the CF_3 moiety is in the hydrophobic pocket and has weak polar interactions with the neighboring side chain C–H bonds (F, green; N, blue; S, yellow) the structures were created using UCSF Chimera software; PDB 1Z95. *(For interpretation of the references to color in this figure legend, the reader is referred to the online version of this book.)*

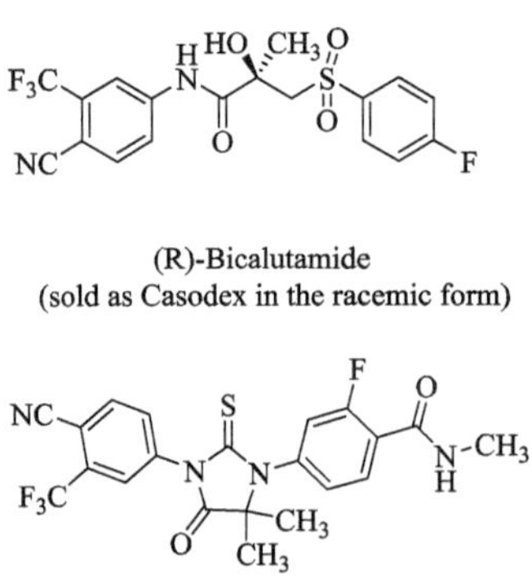

(R)-Bicalutamide
(sold as Casodex in the racemic form)

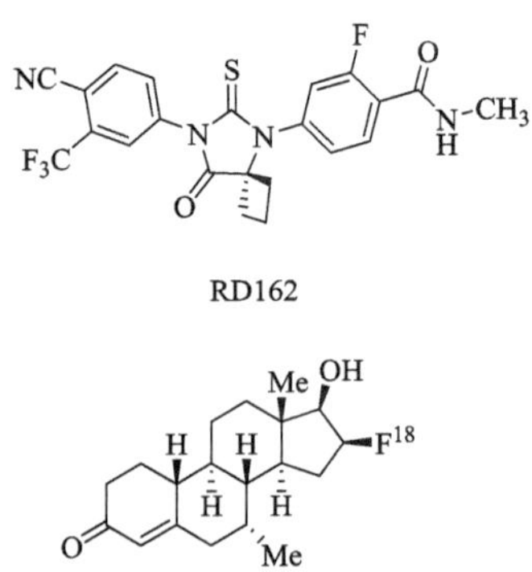

RD162

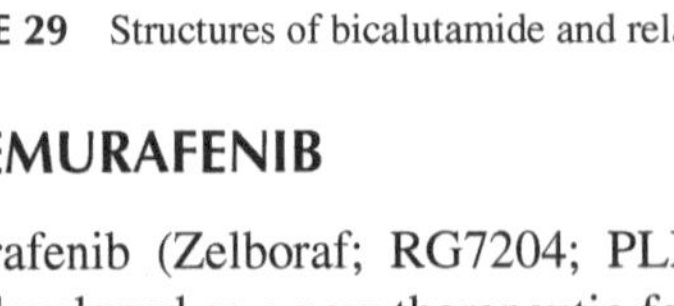

MDV3100

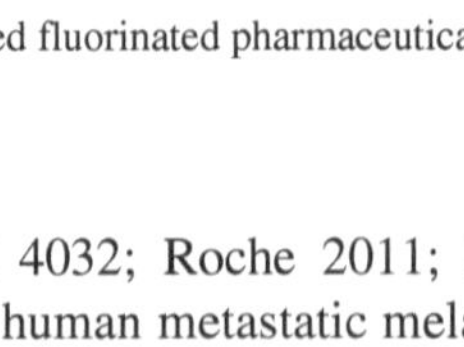

16β-[^{18}F]Fluoro-5α-dihydrotestosterone
(^{18}F-FDHT)

FIGURE 29 Structures of bicalutamide and related fluorinated pharmaceuticals.

8. VEMURAFENIB

Vemurafenib (Zelboraf; RG7204; PLX 4032; Roche 2011; Figure 30) has been developed as a new therapeutic for human metastatic melanoma by Hoffmann-La Roche.[51] It is the first small molecule inhibitor of serine/threonine BRAF kinases, and, in 2011, was approved by the FDA for treating metastatic

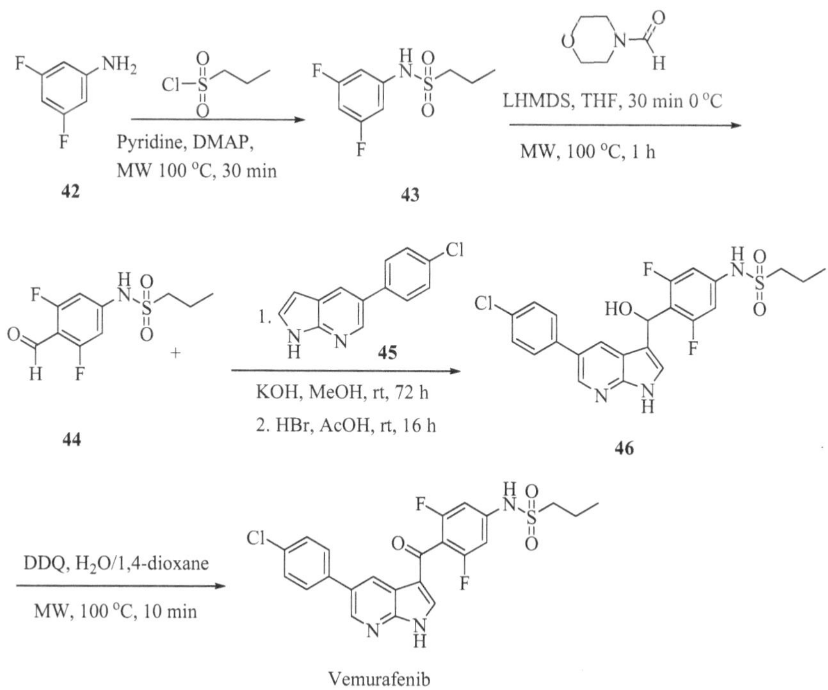

FIGURE 30 Synthesis of vemurafenib (DMAP = 4-dimethylaminopyridine; LHMDS = lithium hexamethyldisilazane; DDQ = 2,3-dichloro-5,6-dicyano-1,4-benzoquinone; MW = microwave irradiation).

melanomas. It complements other protein-based drugs, such as ipilimumab (Bristol-Myers Squibb), a human mAb specific for human cytotoxic T-lymphocyte-associated antigen 4, and interleukin 2.[5,52,53]

Vemurafenib is active against several human tumor serine/threonine protein kinase BRAF melanoma xenograft models.[54] BRAF V600E mutation is observed in melanoma and other human cancers, and vemurafenib is a potent, selective inhibitor of this mutated kinase. Vemurafenib has favorable pharmacokinetics as compared to the other structurally similar BRAF V600 mutated kinases. The regulation of mitogen-activated protein kinase (MAPK; also called as extracellular signal-regulated kinase (ERK)) has direct effect on the signaling pathways involved in the tumor cell proliferation. The Roche authors showed that vemurafenib, by inhibiting the mutated BRAF kinases (e.g., BRAF V600E or other mutant BRAF proteins), suppresses RAF–MEK–ERK signaling pathways. Thus, it suppresses gene expression corresponding to the cell cycle proteins, and thereby prevents cell proliferation. This drug shows no significant toxicity in the in vivo models tested. Thus, it has a great potential in the treatment of metastatic melanoma cancers resulting from BRAF V600E mutant gene.

Analysis of the phosphorylated ERK, cyclin D1, and Ki-67 proteins in tumor biopsy, and fluorodeoxyglucose (FDG)-PET are convenient methods for the follow-up of the progression or attenuation of the disease. All these three proteins virtually disappeared after day 15 of treatment with vemurafenib in cases of the metastatic melanoma. FDG-PET imaging further confirmed the dramatic decrease of the tumor populations after day 15.

It is significant that vemurafenib is selective inhibitor of the mutated BRAF in which Val600 is substituted by glutamic acid (i.e., BRAF V600E). However, vemurafenib is contraindicated for patients with BRAF mutations other than the V600E: in clinical studies, it was found that patients who did not have BRAF V600E mutation actually had tumor progression within the first 2 months of treatment using this drug. Thus, a thorough analysis of the BRAF mutations is required prior to the administration of this drug in metastatic melanoma cases.

8.1 Synthesis of Vemurafenib

Efficient synthesis of vemurafenib was achieved using microwave (MW) irradiation.[55] N-sulfonylation of 3,5-difluoroaniline (**42**) using propanesulfonyl chloride, followed by formylation of the aryl ring using LDA (or LHMDS) and morpholino carbaldehyde under MW conditions gives compound **44**. Compound **44** upon reaction with the azaindole **45**, followed by oxidation of the resulting alcohol **46** with DDQ gives vemurafenib in high yields at relatively short reaction times (Figure 30).

9. SONIDEGIB

Sonidegib (LDE225; erismodegib; Novartis; Figure 31) is an experimental drug for treating a variety of cancers, including basal cell carcinoma, melanoma, pancreatic cancer, and small cell lung cancer. Tobacco smoking is among the major risk factors for the onset of the latter two cancers. Sonidegib targets SONIC HEDGEHOG (an intercellular signaling peptide encoded by the SHH gene) and mitogen-activated protein kinase (MAPK) signaling pathway, which regulate many tumors including melanoma.[56] Sonidegib is active against human melanoma cell lines, mutated as well as nonmutated versions (i.e., with both BRAF V600E and BRAF (wild type)).[56] Phase II clinical trials of this compound by Novartis, in 2014, revealed marked decrease in tumor size in patients with advanced basal cell carcinoma.

9.1 Synthesis of Sonidegib

Novartis' synthesis of sonidegib involves Suzuki coupling of the arylboronic acid **48** with 3-bromo-2-methylbenzoic acid (**47**), followed by coupling of **49** with aminopyridine **50** using standard peptide coupling agents (e.g., HATU)

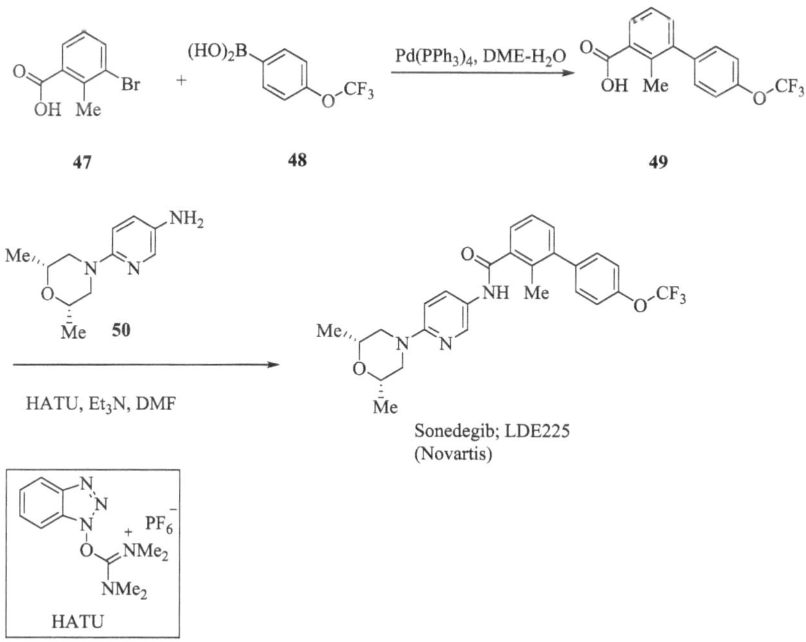

FIGURE 31 Synthesis of sonidegib.

(Figure 31).[57] This versatile synthetic methodology allows synthesis of many other analogs of this compound.

10. PONATINIB

Ponatinib (Iclusig; Ariad Pharmaceuticals; Figure 32) is approved by the FDA (2012) for the treatment of chronic myeloid leukemia (CML) and Philadelphia chromosome-positive acute lymphoblastic leukemia (Ph-pos ALL). Mutations in the BCR-ABL kinase is the frequent cause of the CML and Ph-pos ALL. Ponatinib inhibits the native as well as the mutant BCR-ABL, including the gatekeeper mutant BCR-ABL (T315I), which is resistant to other tyrosine kinase inhibitors such as imatinib.[58]

10.1 Synthesis of Ponatinib

Huang and coworkers have developed versatile synthetic methodology for the preparation of the ponatinib and its analogs, which involves Pd(0)-catalyzed Sonogashira reaction as the key step. Thus, S_N2 reaction of 1-methylpiperazine with 4-nitro-2-(trifluoromethyl)benzyl bromide **51** gives **52**, which upon reduction of the nitro group, followed by reaction with 3-iodo-4-methylbenzoyl

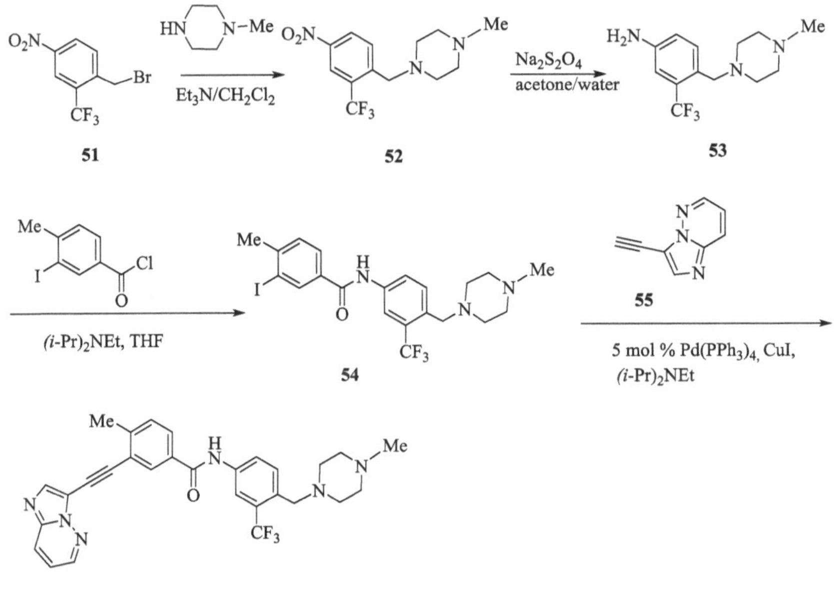

FIGURE 32 Synthesis of ponatinib.

chloride gives compound **54**. Sonogashira coupling of **54** with the alkyne **55**, at ambient temperature, gave ponatinib.[59]

11. FASLODEX

Tamoxifen has been successfully used as a therapeutic for the treatment of hormone-dependent breast cancer. Unfortunately, its side effects include increased risks for endometrial cancer. It was recognized that it is an estrogen antagonist in breast tissue, but retains its estrogen agonist characteristics in bones and endometrium. Faslodex (fulvestrant; AstraZeneca; Figure 33), on the other hand is an efficient estrogen antagonist and has no agonist properties, and therefore serves as a better alternative to tamoxifen.[60] Faslodex, by competitively binding to estrogen receptors (ERs), that is, with greater affinity than estradiol, attenuates tumor cell proliferation. With some exceptional serious side effects, faslodex is well tolerated and is recommended as a useful treatment alternative for patients with advanced breast cancer and for following progress (reoccurrence) in prior endocrine treatment.[61]

Due to the emergence of the multiple drug resistance endocrine receptors in ER-negative tumors, standard chemotherapy using taxoids (e.g., paclitaxel or docetaxel) is usually not long lasting in patients.[62] It has been shown that Faslodex sensitizes the ER-negative multidrug-resistant breast cancer cells, and when used in combination with other chemotherapeutic agents such as paclitaxel reverses the multidrug resistance in ER-negative breast cancer cells.[63] Fulvestrant also increases gefitinib

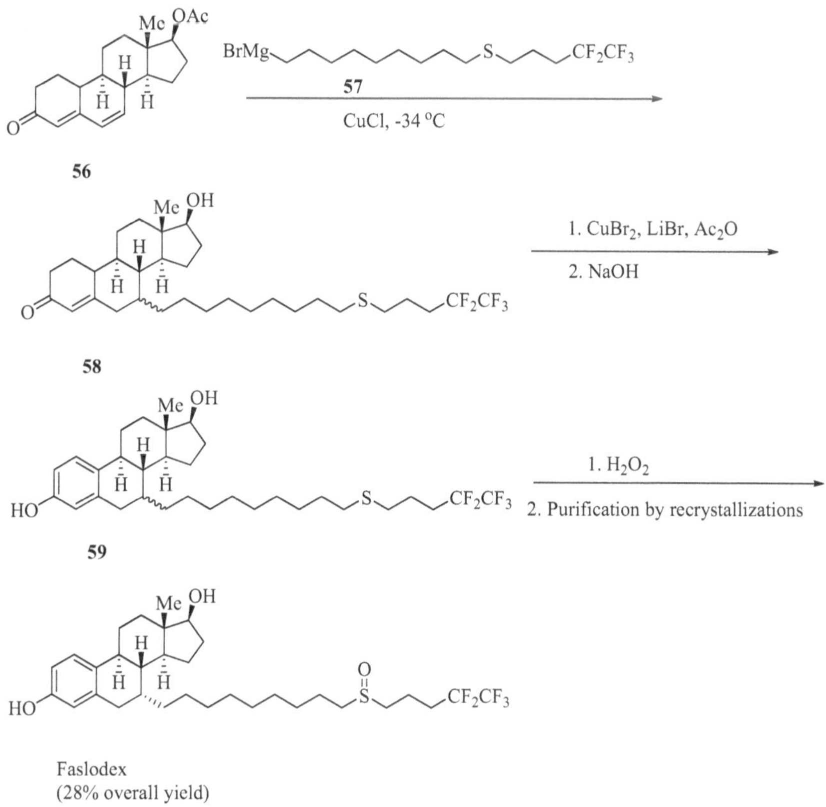

FIGURE 33 Synthesis of Faslodex.

sensitivity in NSCLC cells that have acquired resistance to gefitinib as a result of mutation in the EGFR genes (e.g., those corresponding to T790M mutations).[64]

11.1 Synthesis of Faslodex

The key steps of the AstraZeneca's synthesis of faslodex involve stereoselective 1,6-addition of the organocuprate **57** to the 19-nortestosterone-17-acetate (**56**), followed by Cu(II)-mediated A-ring aromatization of **58** to give **59**. A final step involving the oxidation of the side chain thioether moiety of **59** to the corresponding sulfoxide using hydrogen peroxide, and purification of the diastereomeric product mixtures by repeated recrystallizations gives faslodex in 28% overall yield starting from **56** (Figure 33).[65]

12. VINFLUNINE

Vinflunine (Figure 34), a *gem*-difluoromethylenated derivative of vinca alkaloid, is currently used as a second-line chemotherapy for advanced transitional cell

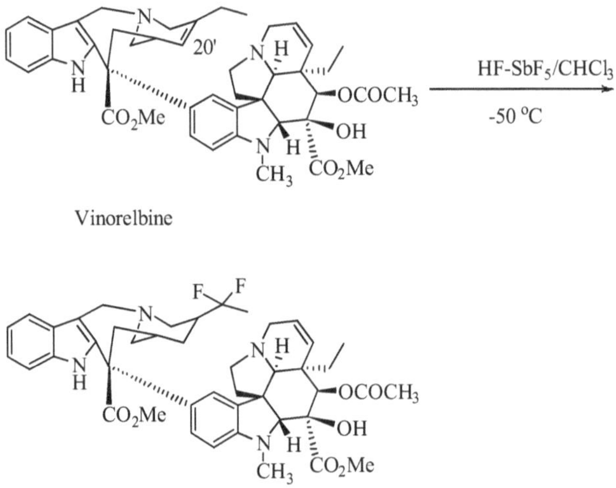

Vinflunine (~35% yield)

FIGURE 34 Superacid-catalyzed synthesis of vinflunine.

carcinoma.[66] Vinflunine is also under clinical trials for the treatment of breast cancer.[67] The most successful compounds in the treatment of breast cancer are the taxanes and the vinca alkaloids. Vinorelbine, a vinca alkaloid derivative, is currently used in advanced breast cancer patients, and has a remarkable clinical activity, increased response, and manageable toxicity. However, it is susceptible to drug resistance by cancer cells. Of the many analogs of vinorelbine, vinflunine, a *gem*-difluoro analog, is currently showing promise in this regard, as its drug resistance is developed at relatively slower rates. Phase II trials of this compound show favorable safety profile. Further, the in vitro neurotoxicity of vinflunine is lower than that of vincristine and of vinorelbine, warranting further clinical investigations of this compound in combination with other active agents in the treatment of metastatic breast cancer.[67] Vinflunine has moderate activity against upper urinary tract urothelial cell carcinoma, which is a rare aggressive urological cancer with a propensity for metastasis.[68]

12.1 Synthesis of Vinflunine

Vinflunine is the major product of the superacid-catalyzed *gem*-fluorination of vinorelbine in chloroform solutions (Figure 34). The observed 35% yield is remarkable in light of the other competing protonations of olefin, hydroxyl, and acetoxy moieties by the superacid that would be expected to give numerous rearrangement products.[69,70] Synthesis of vinflunine exemplifies the use of superacids in the synthesis of complex natural product derivatives.

The rearrangements involved in the conversion of vinorelbine into vinflunine can be rationalized as going through a series of the intermediate carbocation intermediates (Figure 35).

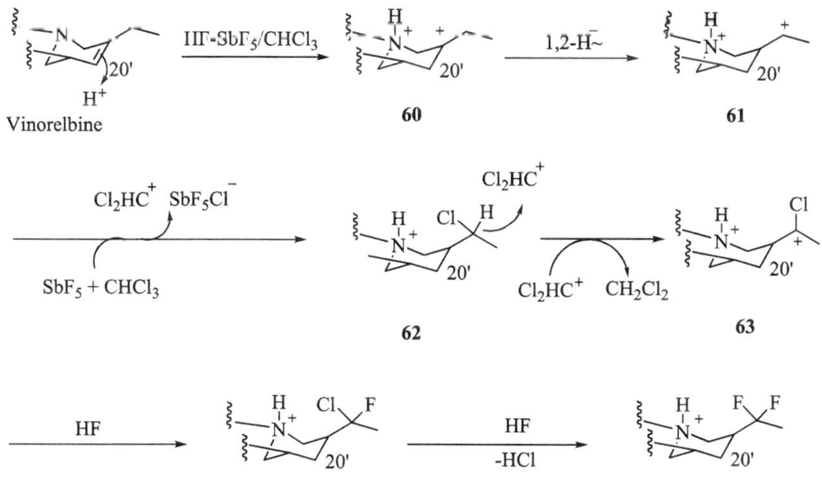

FIGURE 35 Mechanistic rationalization of the superacid-catalyzed conversion of vinorelbine to vinflunine.

Protonation of the vinorelbine at C20′ gives the corresponding tertiary carbocation **60**, which upon 1,2-hydrogen migration followed by chloride ion capture from SbF_5Cl^- gives the chloro derivative **62**. The hydride ion abstraction of **62** by the relatively mild Lewis acid Cl_2HC^+ (formed through the reaction of chloroform with SbF_5) gives the α-chlorocarbenium ion **63**, which is captured by the fluoride anion to give **64**. Anion exchange of chloride by fluoride then gives the *gem*-difluorinated vinflunine. Thus, in chloroform solution the strong Lewis acid SbF_5 is transformed into the relatively milder Lewis acid Cl_2HC^+, so that the formation of the otherwise complex product mixture is avoided.[70]

13. SUMMARY AND OUTLOOK

Modern therapeutics for cancer are generally based on mechanism-based selective inhibition of kinases, or inactivation of DNA polymerases to inhibit cancer cell proliferation. *gem*-Difluorinated nucleoside analogs, such as gemcitabine, are mimetics of the corresponding nonfluorinated nucleosides and thus competitively bind to the DNA polymerase, thereby suppressing the DNA-chain elongation. Afatinib, an FDA-approved (2013) pharmaceutical for the first-line treatment of metastatic NSCLC, is a mechanism-based inhibitor of EGFR tyrosine kinases, and is also a potent inhibitor of the drug-resistant versions of EGFR tyrosine kinases. Fluorinated taxoid derivatives conjugated to vitamins, or mAbs, are designed for cancer-cell-specific drug delivery. Biotin conjugates of taxoids with a ^{18}F-labeled alkyl side chain serve as PET tracers for monitoring the biodistribution of these drug conjugates. Despite the significant progress made in the discovery of anticancer agents, the currently available pharmaceuticals have limited scope in improving the PFS time, although fluorinated pharmaceuticals such as afatinib have relatively improved performance.

REFERENCES

1. http://www.cancer.gov/.
2. Pratt, S. E.; Durland-Busbice, S.; Shepard, R. L.; Donoho, G. P.; Starling, J. J.; Wickremsinhe, E. R.; Perkins, E. J.; Dantzig, A. H. Efficacy of Low-Dose Oral Metronomic Dosing of the Prodrug of Gemcitabine, LY2334737, in Human Tumor Xenografts. *Mol. Cancer Ther.* **2013,** *12,* 481–490.
3. Wickremsinhe, E.; Bao, J.; Smith, R.; Burton, R.; Dow, S.; Perkins, E. Preclinical Absorption, Distribution, Metabolism, and Excretion of an Oral Amide Prodrug of Gemcitabine Designed to Deliver Prolonged Systemic Exposure. *Pharmaceutics* **2013,** *5,* 261–276.
4. Sayed, Y. M. E.; Sadee, W. Metabolic Activation of R,S-1-(Tetrahydro-2-furanyl)-5-fluorouracil (Ftorafur) to 5-Fluorouracil by Soluble Enzymes. *Cancer Res.* **1983,** *43,* 4039–4044.
5. Flaherty, K. T.; Yasothan, U.; Kirkpatrick, P. Vemurafenib. *Nat. Rev. Drug Discov.* **2011,** *10,* 811–812.
6. Galmarini, C. M.; Mackey, J. R.; Dumontet, C. Nucleoside Analogues: Mechanisms of Drug Resistance and Reversal Strategies. *Leukemia* **2001,** *15,* 875–890.
7. Kelland, L. R. Emerging Drugs for Ovarian cancer. *Expert Opin. Emerg. Drugs* **2005,** *10,* 413–424.
8. Gesto, D. S.; Cerqueira, N. M.; Fernandes, P. A.; Ramos, M. J. Gemcitabine: A Critical Nucleoside for Cancer Therapy. *Curr. Med. Chem.* **2012,** *19,* 1076–1087.
9. Cohen, M. H.; Williams, G. A.; Sridhara, R.; Chen, G.; Pazdur, R. FDA Drug Approval Summary: Gefitinib (ZD1839) (Iressa) Tablets. *Oncologist* **2003,** *8,* 303–306.
10. Karampelas, T.; Argyros, O.; Sayyad, N.; Spyridaki, K.; Pappas, C.; Morgan, K.; Kolios, G.; Millar, R. P.; Liapakis, G.; Tzakos, A. G.; Fokas, D.; Tamvakopoulos, C. GnRH-Gemcitabine Conjugates for the Treatment of Androgen-Independent Prostate Cancer: Pharmacokinetic Enhancements Combined with Targeted Drug Delivery. *Bioconjug. Chem.* **2014,** *25,* 813–823.
11. Robak, T. New Purine Nucleoside Analogs for Acute Lymphoblastic Leukemia. *Clin. Cancer Drugs* **2014,** *1,* 2–10.
12. Hui, Y. F.; Reitz, J. Gemcitabine: A Cytidine Analogue Active against Solid Tumors. *Am. J. Health Syst. Pharm.* **1997,** *54,* 162–170.
13. Fowler, J. D.; Brown, J. A.; Johnson, K. A.; Suo, Z. Kinetic Investigation of the Inhibitory Effect of Gemcitabine on DNA Polymerization Catalyzed by Human Mitochondrial DNA Polymerase. *J. Biol. Chem.* **2008,** *283,* 15339–15348.
14. Huang, P.; Plunkett, W. Induction of Apoptosis by Gemcitabine. *Semin. Oncol.* **1995,** *22,* 19–25.
15. Sauve, A. A.; Cen, Y. *Preparation and SirT Activity of Halogenated 2-deoxy-lactones, 2′-deoxy-d-nucleosides*, WO2011003018A2; Cornell University, USA, 2011.
16. Cen, Y.; Sauve, A. A. Efficient Syntheses of Clofarabine and Gemcitabine from 2-Deoxyribonolactone. *Nucleosides Nucleotides Nucleic Acids* **2010,** *29,* 113–122.
17. Solca, F.; Dahl, G.; Zoephel, A.; Bader, G.; Sanderson, M.; Klein, C.; Kraemer, O.; Himmelsbach, F.; Haaksma, E.; Adolf, G. R. Target Binding Properties and Cellular Activity of Afatinib (BIBW 2992), an Irreversible ErbB Family Blocker. *J. Pharmacol. Exp. Ther.* **2012,** *343,* 342–350.
18. Schroeder, J.; Dziewas, G.; Fachinger, T.; Jaeger, B.; Reichel, C.; Renner, S. *A Process for Producing Aminocrotonylamino-substituted Quinazoline Derivatives and Their Intermediates*, WO2007085638A1; Boehringer Ingelheim International G.m.b.H., Germany; Boehringer Ingelheim Pharma G.m.b.H. & Co. K.-G., 2007.

19. http://www.fda.gov/Drugs/DrugSafety/ucm365050.htm.
20. Janne, P. A.; von Pawel, J.; Cohen, R. B.; Crino, L.; Butts, C. A.; Olson, S. S.; Eiseman, I. A.; Chiappori, A. A.; Yeap, B. Y.; Lenehan, P. F.; Dasse, K.; Sheeran, M.; Bonomi, P. D. Multicenter, Randomized, Phase II Trial of CI-1033, an Irreversible Pan-ERBB Inhibitor, for Previously Treated Advanced Non-Small-Cell Lung Cancer. *J. Clin. Oncol.* **2007,** *25,* 3936–3944.
21. http://www.fda.gov/drugs/informationondrugs/approveddrugs/ucm376547.htm.
22. Al-Marrawi, M. Y.; Saroya, B. S.; Brennan, M. C.; Yang, Z.; Dykes, T. M.; El-Deiry, W. S. Off-Label Use of Cetuximab Plus Sorafenib and Panitumumab Plus Regorafenib to Personalize Therapy for a Patient with V600E BRAF-Mutant Metastatic Colon Cancer. *Cancer Biol. Ther.* **2013,** *14,* 703–710.
23. Wilhelm, S.; Carter, C.; Lynch, M.; Lowinger, T.; Dumas, J.; Smith, R. A.; Schwartz, B.; Simantov, R.; Kelley, S. Discovery and Development of Sorafenib: A Multikinase Inhibitor for Treating Cancer. *Nat. Rev. Drug Discov.* **2006,** *5,* 835–844.
24. Breitler, S.; Oldenhuis, N. J.; Fors, B. P.; Buchwald, S. L. Synthesis of Unsymmetrical Diarylureas via Pd-Catalyzed C–N Cross-Coupling Reactions. *Org. Lett.* **2011,** *13,* 3262–3265.
25. Vano, Y.-A.; Tartour, E.; Fournier, L. S.; Beuselinck, B.; Mejean, A.; Oudard, S. Prognostic Factors in Patients with Advanced Renal Cell Carcinoma Treated with VEGF-Targeted Agents. *Expert Rev. Anticancer Ther.* **2014,** *14,* 523–542.
26. Reichardt, P. Optimal Use of Targeted Agents for Advanced Gastrointestinal Stromal Tumours. *Oncology* **2010,** *78,* 130–140.
27. Le Cesne, A.; Blay, J.-Y.; Reichardt, P.; Joensuu, H. Optimizing Tyrosine Kinase Inhibitor Therapy in Gastrointestinal Stromal Tumors: Exploring the Benefits of Continuous Kinase Suppression. *Oncologist* **2013,** *18,* 1192–1199.
28. Sun, L.; Liang, C.; Shirazian, S.; Zhou, Y.; Miller, T.; Cui, J.; Fukuda, J. Y.; Chu, J.-Y.; Nematalla, A.; Wang, X.; Chen, H.; Sistla, A.; Luu, T. C.; Tang, F.; Wei, J.; Tang, C. Discovery of 5-[5-Fluoro-2-oxo-1,2- dihydroindol-(3*Z*)-ylidenemethyl]-2,4- dimethyl-1*H*-pyrrole-3-carboxylic Acid (2-Diethylaminoethyl)amide, a Novel Tyrosine Kinase Inhibitor Targeting Vascular Endothelial and Platelet-Derived Growth Factor Receptor Tyrosine Kinase. *J. Med. Chem.* **2003,** *46,* 1116–1119.
29. Yakes, F. M.; Chen, J.; Tan, J.; Yamaguchi, K.; Shi, Y.; Yu, P.; Qian, F.; Chu, F.; Bentzien, F.; Cancilla, B.; Orf, J.; You, A.; Laird, A. D.; Engst, S.; Lee, L.; Lesch, J.; Chou, Y.-C.; Joly, A. H. Cabozantinib (XL184), a Novel MET and VEGFR2 Inhibitor, Simultaneously Suppresses Metastasis, Angiogenesis, and Tumor Growth. *Mol. Cancer Ther.* **2011,** *10,* 2298–2308.
30. Elisei, R.; Schlumberger, M. J.; Mueller, S. P.; Schoffski, P.; Brose, M. S.; Shah, M. H.; Licitra, L.; Jarzab, B.; Medvedev, V.; Kreissl, M. C.; Niederle, B.; Cohen, E. E. W.; Wirth, L. J.; Ali, H.; Hessel, C.; Yaron, Y.; Ball, D.; Nelkin, B.; Sherman, S. I. Cabozantinib in Progressive Medullary Thyroid Cancer. *J. Clin. Oncol.* **2013,** *31,* 3639–3646.
31. Aftab, D. T. *Inhibitors of Met, VEGFR and Ret for Use in the Treatment of Lung Adenocarcinoma,* WO2014039971A1; Exelixis, Inc., USA, 2014.
32. Nguyen, H. M.; Ruppender, N.; Zhang, X.; Brown, L. G.; Gross, T. S.; Morrissey, C.; Gulati, R.; Vessella, R. L.; Schimmoller, F.; Aftab, D. T.; Corey, E. Cabozantinib Inhibits Growth of Androgen-Sensitive and Castration-Resistant Prostate Cancer and Affects Bone Remodeling. *PLoS One* **2013,** *8,* e78881.
33. Geyer, C. E.; Forster, J.; Lindquist, D.; Chan, S.; Romieu, C. G.; Pienkowski, T.; Jagiello-Gruszfeld, A.; Crown, J.; Chan, A.; Kaufman, B.; Skarlos, D.; Campone, M.; Davidson, N.; Berger, M.; Oliva, C.; Rubin, S. D.; Stein, S.; Cameron, D. Lapatinib Plus Capecitabine for HER2-Positive Advanced Breast Cancer. *N. Engl. J. Med.* **2006,** *355,* 2733–2743.

34. Ait-Tihyaty, M.; Rachid, Z.; Larroque-Lombard, A.-L.; Jean-Claude, B. J. ZRX1, the First EGFR Inhibitor-Capecitabine Based Combi-Molecule, Requires Carboxylesterase-Mediated Hydrolysis for Optimal Activity. *Invest. New Drugs* **2013,** *31,* 1409–1423.
35. www.drugs.com.
36. Shen, B.; Jamison, T. F. Rapid Continuous Synthesis of 5′-Deoxyribonucleosides in Flow via Bronsted Acid Catalyzed Glycosylation. *Org. Lett.* **2012,** *14,* 3348–3351.
37. Qiu, C.; Tarrant, M. K.; Choi, S. H.; Sathyamurthy, A.; Bose, R.; Banjade, S.; Pal, A.; Bornmann, W. G.; Lemmon, M. A.; Cole, P. A.; Leahy, D. J. Mechanism of Activation and Inhibition of the HER4/ErbB4 Kinase. *Structure* **2008,** *16,* 460–467.
38. Combination Therapy Approved for Melanoma. *Cancer Discovery* **2014,** *4,* 262.
39. Kwong, L. N.; Costello, J. C.; Liu, H.; Jiang, S.; Helms, T. L.; Langsdorf, A. E.; Jakubosky, D.; Genovese, G.; Muller, F. L.; Jeong, J. H.; Bender, R. P.; Chu, G. C.; Flaherty, K. T.; Wargo, J. A.; Collins, J. J.; Chin, L. Oncogenic NRAS Signaling Differentially Regulates Survival and Proliferation in Melanoma. *Nat. Med.* **2012,** *18,* 1503–1510.
40. Rheault, T. R.; Stellwagen, J. C.; Adjabeng, G. M.; Hornberger, K. R.; Petrov, K. G.; Waterson, A. G.; Dickerson, S. H.; Mook, R. A.; Laquerre, S. G.; King, A. J.; Rossanese, O. W.; Arnone, M. R.; Smitheman, K. N.; Kane-Carson, L. S.; Han, C.; Moorthy, G. S.; Moss, K. G.; Uehling, D. E. Discovery of Dabrafenib: A Selective Inhibitor of Raf Kinases with Antitumor Activity against B-Raf-Driven Tumors. *ACS Med. Chem. Lett.* **2013,** *4,* 358–362.
41. Sakai, T.; Kawasaki, H.; Abe, H.; Hayakawa, K.; Iida, T.; Kikuchi, S.; Yamaguchi, T.; Nanayama, T.; Kurachi, H.; Tamaru, M.; Hori, Y.; Takahashi, M.; Yoshida, T. *Preparation of 5-amino-2,4,7-trioxo-3,4,7,8-tetrahydro-2H-pyrido[2,3-d]pyrimidines and Related Compounds as MEK Inhibitors and P15 and P27 Protein Inducers for the Treatment of Cancer and Rheumatic Diseases*, WO2005121142A1; Japan Tobacco Inc., Japan, 2005.
42. http://www.cancer.gov/cancertopics/druginfo/fda-docetaxel.
43. Voborilova, J.; Nemcova-Fuerstova, V.; Neubauerova, J.; Ojima, I.; Zanardi, I.; Gut, I.; Kovar, J. Cell Death Induced by Novel Fluorinated Taxanes in Drug-Sensitive and Drug-Resistant Cancer Cells. *Invest. New Drugs* **2011,** *29,* 411–423.
44. Pepe, A.; Kuznetsova, L.; Sun, L.; Ojima, I. Fluoro-Taxoid Anticancer Agents. *Fluorine Med. Chem. Chem. Biol.* **2009,** 117–139.
45. Kuznetsova, L.; Sun, L.; Chen, J.; Zhao, X.; Seitz, J.; Das, M.; Li, Y.; Veith, J. M.; Pera, P.; Bernacki, R. J.; Xia, S.; Horwitz, S. B.; Ojima, I. Synthesis and Biological Evaluation of Novel 3′-Difluorovinyl Taxoids. *J. Fluorine Chem.* **2012,** *143,* 177–188.
46. Kamath, A.; Ojima, I. Advances in the Chemistry of β-Lactam and Its Medicinal Applications. *Tetrahedron* **2012,** *68,* 10640–10664.
47. Seitz, J.; Vineberg, J. G.; Zuniga, E. S.; Ojima, I. Fluorine-Containing Taxoid Anticancer Agents and Their Tumor-Targeted Drug Delivery. *J. Fluorine Chem.* **2013,** *152,* 157–165.
48. Ojima, I. Exploration of Fluorine Chemistry at the Multidisciplinary Interface of Chemistry and Biology. *J. Org. Chem.* **2013,** *78,* 6358–6383.
49. Bohl, C. E.; Gao, W.; Miller, D. D.; Bell, C. E.; Dalton, J. T. Structural Basis for Antagonism and Resistance of Bicalutamide in Prostate Cancer. *Proc. Natl. Acad. Sci. U.S.A.* **2005,** *102,* 6201–6206.
50. Tran, C.; Ouk, S.; Clegg Nicola, J.; Chen, Y.; Watson Philip, A.; Arora, V.; Wongvipat, J.; Smith-Jones Peter, M.; Yoo, D.; Kwon, A.; Wasielewska, T.; Welsbie, D.; Chen Charlie, D.; Higano Celestia, S.; Beer Tomasz, M.; Hung David, T.; Scher Howard, I.; Jung Michael, E.; Sawyers Charles, L. Development of a second-generation antiandrogen for treatment of advanced prostate cancer. *Science* **2009,** *324,* 787–790.

51. Yang, H.; Higgins, B.; Kolinsky, K.; Packman, K.; Go, Z.; Iyer, R.; Kolis, S.; Zhao, S.; Lee, R.; Grippo, J. F.; Schostack, K.; Simcox, M. E.; Heimbrook, D., Bollag, G., Su, F. RG7204 (PLX4032), A Selective BRAF V600E Inhibitor, Displays Potent Antitumor Activity in Preclinical Melanoma Models. *Cancer Res.* **2010,** *70,* 5518–5527.
52. Ribas, A.; Flaherty, K. T. BRAF Targeted Therapy Changes the Treatment Paradigm in Melanoma. *Nat. Rev. Clin. Oncol.* **2011,** *8,* 426–433.
53. Sondak Vernon, K.; Flaherty Lawrence, E. Targeted Therapies: Improved Outcomes for Patients with Metastatic Melanoma. *Nat. Rev. Clin. Oncol.* **2011,** *8,* 513–515.
54. Zambon, A.; Niculescu-Duvaz, I.; Niculescu-Duvaz, D.; Marais, R.; Springer, C. J. Small Molecule Inhibitors of BRAF in Clinical Trials. *Bioorg. Med. Chem. Lett.* **2012,** *22,* 789–792.
55. Buck, J. R.; Saleh, S.; Imam Uddin, M.; Manning, H. C. Rapid, Microwave-Assisted Organic Synthesis of Selective V600E BRAF Inhibitors for Preclinical Cancer Research. *Tetrahedron Lett.* **2012,** *53,* 4161–4165.
56. Jalili, A.; Mertz, K. D.; Romanov, J.; Wagner, C.; Kalthoff, F.; Stuetz, A.; Pathria, G.; Gschaider, M.; Stingl, G.; Wagner, S. N. NVP-LDE225, a Potent and Selective SMOOTHENED Antagonist Reduces Melanoma Growth In vitro and In vivo. *PLoS One* **2013,** *8,* e69064.
57. Pan, S.; Wu, X.; Jiang, J.; Gao, W.; Wan, Y.; Cheng, D.; Han, D.; Liu, J.; Englund, N. P.; Wang, Y.; Peukert, S.; Miller-Moslin, K.; Yuan, J.; Guo, R.; Matsumoto, M.; Vattay, A.; Jiang, Y.; Tsao, J.; Sun, F.; Pferdekamper, A. C.; Dodd, S.; Tuntland, T.; Maniara, W.; Kelleher, J. F., III; Yao, Y.-m.; Warmuth, M.; Williams, J.; Dorsch, M. Discovery of NVP-lde225, a Potent and Selective Smoothened Antagonist. *ACS Med. Chem. Lett.* **2010,** *1,* 130–134.
58. Cortes, J. E.; Kantarjian, H.; Shah, N. P.; Bixby, D.; Mauro, M. J.; Flinn, I.; O'Hare, T.; Hu, S.; Narasimhan, N. I.; Rivera, V. M.; Clackson, T.; Turner, C. D.; Haluska, F. G.; Druker, B. J.; Deininger, M. W. N.; Talpaz, M. Ponatinib in Refractory Philadelphia Chromosome-Positive Leukemias. *N. Engl. J. Med.* **2012,** *367,* 2075–2088.
59. Huang, W.-S.; Metcalf, C. A.; Sundaramoorthi, R.; Wang, Y.; Zou, D.; Thomas, R. M.; Zhu, X.; Cai, L.; Wen, D.; Liu, S.; Romero, J.; Qi, J.; Chen, I.; Banda, G.; Lentini, S. P.; Das, S.; Xu, Q.; Keats, J.; Wang, F.; Wardwell, S.; Ning, Y.; Snodgrass, J. T.; Broudy, M. I.; Russian, K.; Zhou, T.; Commodore, L.; Narasimhan, N. I.; Mohemmad, Q. K.; Iuliucci, J.; Rivera, V. M.; Dalgarno, D. C.; Sawyer, T. K.; Clackson, T.; Shakespeare, W. C. Discovery of 3-[2-(Imidazo[1,2-b]pyridazin-3-yl) ethynyl]-4-methyl-*N*-[4-((4-methylpiperazin-1-yl)methyl)-3-(trifluoromethyl)phenyl]benzamide (AP24534), a Potent, Orally Active Pan-Inhibitor of Breakpoint Cluster Region-Abelson (BCR-ABL) Kinase Including the T315I Gatekeeper Mutant. *J. Med. Chem.* **2010,** *53,* 4701–4719.
60. Robertson, J. F. R. Selective Oestrogen Receptor Modulators/New Antioestrogens: A Clinical Perspective. *Cancer Treat. Rev.* **2004,** *30,* 695–706.
61. Wang, J.; Jain, S.; Coombes, C. R.; Palmieri, C. Fulvestrant in Advanced Breast Cancer Following Tamoxifen and Aromatase Inhibition: A Single Center Experience. *Breast J.* **2009,** *15,* 247–253.
62. Hirano, A.; Shimizu, T.; Watanabe, O.; Kinoshita, J.; Kimura, K.; Kamimura, M.; Domoto, K.; Aiba, M.; Ogawa, K. Epirubicin and Cyclophosphamide Followed by Docetaxel as Primary Systemic Chemotherapy in Locally Advanced Breast Cancer. *Anticancer Res.* **2008,** *28,* 4137–4142.
63. Jiang, D.; Huang, Y.; Han, N.; Xu, M.; Xu, L.; Zhou, L.; Wang, S.; Fan, W. Fulvestrant, A Selective Estrogen Receptor Down-regulator, Sensitizes Estrogen Receptor Negative Breast Tumors to Chemotherapy. *Cancer Lett.* **2014,** *346,* 292–299.
64. Shen, H.; Liu, J.; Wang, R.; Qian, X.; Xu, R.; Xu, T.; Li, Q.; Wang, L.; Shi, Z.; Zheng, J.; Chen, Q.; Shu, Y. Fulvestrant Increases Gefitinib Sensitivity in Non-Small Cell Lung Cancer Cells by Upregulating Let-7c Expression. *Biomed. Pharmacother.* **2014,** *68,* 307–313.

65. Brazier, E. J.; Hogan, P. J.; Leung, C. W.; O'Kearney-McMullan, A.; Norton, A. K.; Powell, L.; Robinson, G. E.; Williams, E. G. Fulvestrant: From the Laboratory to Commercial-Scale Manufacture. *Org. Process Res. Dev.* **2010,** *14,* 544–552.
66. Rozzi, A.; Santini, D.; Salerno, M.; Bordin, F.; Mancuso, A.; Minniti, G.; Nardoni, C.; Corona, M.; Falbo, P. T.; Recine, F.; Lanzetta, G. Pegylated Liposomal Doxorubicin as Third-Line Chemotherapy in Patients with Metastatic Transitional Cell Carcinoma of Urothelial Tract: Results of a Phase II Study. *Med. Oncol.* **2013,** *30,* 1–4.
67. Fumoleau, P.; Guiu, S. New Vinca Alkaloids in Clinical Development. *Curr. Breast Cancer Rep.* **2013,** *5,* 69–72.
68. Audenet, F.; Yates, D. R.; Cussenot, O.; Roupret, M. The Role of Chemotherapy in the Treatment of Urothelial Cell Carcinoma of the Upper Urinary Tract (UUT-UCC). *Urol. Oncol. Semin. Orig. Invest.* **2013,** *31,* 407–413.
69. Jacquesy, J. C.; Jouannetaud, M. P. Vinflunine: A New Anti-Cancer Fluorinated Agent Derived from Vinca-Alkaloids. *Ann. Pharm. Fr.* **2005,** *63,* 28–34.
70. Fahy, J.; Duflos, A.; Ribet, J.-P.; Jacquesy, J.-C.; Berrier, C.; Jouannetaud, M.-P.; Zunino, F. Vinca Alkaloids in Superacidic Media: A Method for Creating a New Family of Antitumor Derivatives. *J. Am. Chem. Soc.* **1997,** *119,* 8576–8577.

Index

Note: Page numbers followed by "f" and "t" indicate, figures and tables respectively.

B

C

D

E

G

O

P

Q

R

S

T

U

V

X

Y

Z

FSC
www.fsc.org
MIX
Papier | Fördert gute Waldnutzung
FSC® C083411

Kolibri 360 GmbH
Ferdinand-Jühlke-Straße 7
99095 Erfurt, Deutschland
produktsicherheit@kolibri360.de